FOOD HANDBOOK

ELLIS HORWOOD SERIES IN FOOD SCIENCE AND TECHNOLOGY

Editor-in-Chief: I. D. MORTON, Professor and formerly Head of Department of Food and Nutritional Science, King's College, London.
Series Editors: D. H. WATSON, Ministry of Agriculture, Fisheries and Food; and
M. J. LEWIS, Department of Food Science and Technology, University of Reading

Fats for the Future   R.C. Cambie
Food Handbook   C.M.E. Catsberg & G.J.M. Kempen-van Dommelen
Principles and Applications of Gas Chromatography in Food Analysis   M.H. Gordon

Forthcoming titles
Food Biochemistry   C. Alais & G. Linden
Traditional Fermented Foods   M.Z. Ali & R.K. Robinson
Food Microbiology, Volumes 1 & 2   C.M. Bourgeois, J.F. Mescle & J. Zucca
Determination of Veterinary Residues in Food   N.T. Crosby & C.M. Clark
Food Container Corrosion   D.R. Davis & A.V. Johnston
Technology of Meat and Meat Products   J. P. Girard
Dairy Technology   A. Grandison, M.J. Lewis, R.A. Wilbey & R.K. Robinson
Separation Processes: Principles and Applications   A. Grandison & M.J. Lewis
Microbiology of Chilled and Frozen Foods   W.F. Harrigan
Nitrate and Nitrites in Food and Water   M.J. Hill
Modern Food Processing   J. Lamb
Food Technology Data   M.J. Lewis
Technology of Biscuits, Crackers and Cookies, 2nd Edition   D.J.R. Manley
Second European Conference on Food Science and Technology   I.D. Morton
Modified Atmosphere Packaging of Food   B. Ooraikul
Food, Volumes 1 & 2   P. Patel
Feta and Related Cheeses   R.K. Robinson
Handbook of Edible Gums   K.R. Stauffer
Vitamins and Minerals   M. Tolonon
Applied Human Nutrition: For Food Scientists and Home Economists   A.F. Walker
Natural Toxicants in Food   D.H. Watson

FOOD HANDBOOK

C. M. E. CATSBERG
G. J. M. KEMPEN-VAN DOMMELEN
both of the College Hogeschool Nijmegen
The Netherlands

Translator
F. M. OUTER-LEAFTINK, Reading, Berks

Translation Editor
M. J. LEWIS, University of Reading

ELLIS HORWOOD
NEW YORK   LONDON   TORONTO   SYDNEY   TOKYO   SINGAPORE

This English edition first published in 1990 by
ELLIS HORWOOD LIMITED
Market Cross House, Cooper Street,
Chichester, West Sussex, PO19 1EB, England
A division of
Simon & Schuster International Group

This English edition is translated from the original Dutch edition
Levensmiddelenleer Produktinformatie over voedings en genotmiddelen,
published in 1987 by Nijgh & Van Ditmar Educatief,
© the copyright holders

© English Edition, Ellis Horwood 1990

All rights reserved. No part of this publication may be reproduced, stored
in a retrieval system, or transmitted, in any form, or by any means, electronic,
mechanical, photocopying, recording or otherwise, without the prior permission,
in writing, of the publisher

Typeset in Times by Ellis Horwood Limited
Printed and bound in Great Britain
by Bookcraft (Bath) Limited, Midsomer Norton

Exclusive distribution by Van Nostrand Reinhold/AVI London:
Australia and New Zealand:
CHAPMAN AND HALL AUSTRALIA
Box 4725, 480 La Trobe Street, Melbourne 3001, Victoria, Australia
Canada:
NELSON CANADA
1120 Birchmount Road, Scarborough, Ontario, Canada, M1K 5G4
Europe, Middle East and Africa:
VAN NOSTRAND REINHOLD/AVI LONDON
11 New Fetter Lane, London EC4P 4EE, England
North America:
VAN NOSTRAND REINHOLD/AVI NEW YORK
115 Fifth Avenue, 4th Floor, New York, New York 10003, USA
Rest of the world:
THOMSON INTERNATIONAL PUBLISHING
10 Davis Drive, Belmont, California 94002, USA

British Library Cataloguing in Publication Data

Catsberg, C.M.E.
Food handbook
1. Food
I. Title. II. Kempen-Van Dommelen, G.J.M.
641.3
ISBN 0-7476-0054-6

Library of Congress Cataloging-in-Publication Data

Catsberg, C.M.E., 1954–
Food handbook / C.M.E. Catsberg, G.J.M. Kempen-Van Dommelen.
p. cm. — (Ellis Horwood series in food science and technology)
ISBN 0-7476-0054-6
1. Food industry and trade. I. Kempen-Van Dommelen, G.J.M., 1943– . II. Title.
III. Series.
TP370.C37   1989
664—dc20   89-19911
                                                                              CIP

Table of contents

Preface 13

Acknowledgements of illustrations 15

1 Production, distribution and legislation 17
    1.1   Introduction 17
    1.2   Food production and food distribution. 18
    1.3   Purchasing 22
    1.4   Packaging 24
    1.5   Food legislation 26
    1.6   Consumerism. 30

2 Quality deterioration and spoilage 31
    2.1   Introduction 31
    2.2   Physical spoilage. 32
    2.3   Chemical, biochemical or enzymatic spoilage 32
    2.4   Microbial spoilage. 34
    2.5   Spoilage by parasites 39
    2.6   Spoilage by vermin 40

3 Methods of preservation 42
    3.1   Introduction 42
    3.2   Stimulation of the activity of enzymes and microorganisms. 44
    3.3   Removal of pathogenic and spoilage microorganisms 44
    3.4   Retarding the activity of enzymes and microorganisms 45
    3.5   Killing microorganisms. 54

4 Additives and contaminants 58
    4.1   Introduction 58
    4.2   Additives 58
    4.3   Legislation in connection with additives 62

Table of contents

- 4.4 Contaminants 63
- 4.5 Legislation concerning contaminants 66

5 Meat and meat products 67
- 5.1 Introduction 67
- 5.2 The slaughter process 70
- 5.3 Kinds of meat 73
- 5.4 Organs 80
- 5.5 Hanging and maturation 81
- 5.6 Quality deterioration, spoilage and storage 83
- 5.7 Colour formation and colour preservation in meat and meat products 85
- 5.8 Minced or chopped meat 86
- 5.9 Meat products 86
- 5.10 Cooked products and gelatine 94
- 5.11 Flavours, sauces and soups 95
- 5.12 Standards by law 96

6 Game and fowl 97
- 6.1 Introduction — game 97
- 6.2 Kinds of game 98
- 6.3 Introduction — poultry or fowl 101
- 6.4 Kinds of poultry 103
- 6.5 Quality deterioration, spoilage and storage 105
- 6.6 Products of game and fowl 105
- 6.7 Legislation 107

7 Fish 108
- 7.1 Introduction 108
- 7.2 Production and distribution 111
- 7.3 Kinds of fish 112
- 7.4 Fish products 117
- 7.5 Quality deterioration, spoilage and storage 122
- 7.6 Legislation 123

8 Crustaceans and shellfish 124
- 8.1 Introduction 124
- 8.2 Kinds of crustaceans 125
- 8.3 Kinds of shellfish 128
- 8.4 Products of crustaceans and shellfish 133
- 8.5 Quality deterioration, spoilage and storage 133

9 Milk and some milk products 134
- 9.1 Introduction 134
- 9.2 Production and processing 136
- 9.3 Milk and milk products 138
- 9.4 Quality deterioration, spoilage and storage 143
- 9.5 Legislation 144
- 9.6 Control 144

10 Butter . . . 145
- 10.1 Introduction . . . 145
- 10.2 Production and distribution . . . 146
- 10.3 Kinds of butter . . . 148
- 10.4 Quality deterioration, spoilage and storage . . . 150
- 10.5 Legislation . . . 150

11 Cheese . . . 151
- 11.1 Introduction . . . 151
- 11.2 The production of Dutch cheese (cows' milk) . . . 153
- 11.3 Classification of cheese . . . 155
- 11.4 Kinds of cheese . . . 158
- 11.5 Fresh cheese . . . 162
- 11.6 Processed cheese . . . 164
- 11.7 Legislation . . . 166

12 Eggs . . . 168
- 12.1 Introduction . . . 168
- 12.2 Structure of the egg . . . 169
- 12.3 Production and distribution . . . 170
- 12.4 Eggs and egg products . . . 172
- 12.5 Quality deterioration, spoilage and storage . . . 173
- 12.6 Legislation . . . 174

13 Oils and fats . . . 176
- 13.1 Introduction . . . 176
- 13.2 Production of oils and fats . . . 177
- 13.3 Fat and oil types . . . 179
- 13.4 Products which contain oils or fats . . . 180
- 13.5 Quality deterioration, spoilage and storage . . . 181
- 13.6 Legislation . . . 181

14 Margarine . . . 183
- 14.1 Introduction . . . 183
- 14.2 Manufacture . . . 184
- 14.3 Kinds of margarine . . . 184
- 14.4 Quality deterioration, spoilage and storage . . . 185
- 14.5 Legislation . . . 185

15 Cereals . . . 186
- 15.1 Introduction . . . 186
- 15.2 Structure of cereal fruit . . . 188
- 15.3 Production and distribution . . . 189
- 15.4 Kinds of cereals and their products . . . 191
- 15.5 Quality deterioration, spoilage and storage . . . 203
- 15.6 Legislation . . . 204

16 Thickeners 205
- 16.1 Introduction 205
- 16.2 Products based on carbohydrates 205
- 16.3 Products based on fats 209
- 16.4 Products based on proteins 209
- 16.5 Pudding powders 210
- 16.6 Legislation 210

17 Bread 211
- 17.1 Introduction 211
- 17.2 Production and distribution 213
- 17.3 The bread production process based on yeast dough 217
- 17.4 Kinds of bread 218
- 17.5 Bakers' products 221
- 17.6 Quality deterioration, spoilage and storage 222
- 17.7 Legislation 224

18 Pulses 225
- 18.1 Introduction 225
- 18.2 Production and distribution 226
- 18.3 Kinds of pulses 226
- 18.4 Pulse products 228
- 18.5 Quality deterioration, spoilage and storage 230
- 18.6 Legislation 230

19 Potatoes 231
- 19.1 Introduction 231
- 19.2 Potato cultivation 232
- 19.3 Potatoes and potato products 234
- 19.4 Quality deterioration, spoilage and storage 236
- 19.5 Legislation 238

20 Vegetables 239
- 20.1 Introduction 239
- 20.2 Production and distribution 240
- 20.3 Kinds: division according to edible part 242
- 20.4 Vegetables and vegetable products 254
- 20.5 Quality deterioration, spoilage and storage 256
- 20.6 Legislation 258

21 Fruit 259
- 21.1 Introduction 259
- 21.2 Production and distribution 260
- 21.3 Kinds of fruit 261
- 21.4 Quality deterioration, spoilage and storage 274
- 21.5 Nuts and seeds 275
- 21.6 Fruit preserves 278

Table of contents

    21.7   Legislation 282

22   Fungi 283
    22.1   Introduction 283
    22.2   Kinds 284
    22.3   Preserved fungi 288
    22.4   Quality deterioration, spoilage and storage 288
    22.5   Legislation 289

23   Herbs and spices 290
    23.1   Introduction 290
    23.2   Flavour compounds in herbs and spices 292
    23.3   Types: division of herbs and spices 293
    23.4   Products 299
    23.5   Quality deterioration, spoilage and storage 300
    23.6   Legislation 301

24   Salt 302
    24.1   Introduction 302
    24.2   Production and distribution 302
    24.3   Kinds of salt 303
    24.4   Quality deterioration, spoilage and storage 304
    24.5   Legislation 304

25   Sugar, syrup, confectionery and sweeteners 305
    25.1   Introduction 305
    25.2   The production of beet sugar 306
    25.3   The production of cane sugar 308
    25.4   Kinds of sugar 309
    25.5   Kinds of syrup 309
    25.6   Confectionery, liquorice and wine gums 310
    25.7   Sweeteners 311
    25.8   Quality deterioration, spoilage and storage 314
    25.9   Legislation 314

26   Honey 315
    26.1   Introduction 315
    26.2   Production and distribution 316
    26.3   Kinds of honey 318
    26.4   Quality deterioration, spoilage and storage 319

27   Coffee 320
    27.1   Introduction 320
    27.2   Production and distribution 321
    27.3   Products 323
    27.4   Quality deterioration, spoilage and storage 324
    27.5   Legislation 324

Table of contents

28 Tea 325
- 28.1 Introduction 325
- 28.2 Production and distribution 326
- 28.3 Kinds of tea 327
- 28.4 Quality deterioration, spoilage and storage ... 329
- 28.5 Legislation 329

29 Cocoa and chocolate 330
- 29.1 Introduction 330
- 29.2 Production and distribution 331
- 29.3 Cocoa and cocoa products 333
- 29.4 Quality deterioration, spoilage and storage ... 334
- 29.5 Legislation 334

30 Mineral water and beverages 335
- 30.1 Introduction 335
- 30.2 Natural mineral water 336
- 30.3 Fruit juices 337
- 30.4 Soft drinks 338
- 30.5 Quality deterioration, spoilage and storage ... 339
- 30.6 Legislation 340

31 Alcoholic drinks 341
- 31.1 Introduction 341
- 31.2 Beer 342
- 31.3 Wine 345
- 31.4 Distilled liquors 354
- 31.5 Legislation 358

32 Vinegar 360
- 32.1 Introduction 360
- 32.2 Vinegar production 360
- 32.3 Kinds of vinegar 361
- 32.4 Quality deterioration, spoilage and storage ... 361
- 32.5 Legislation 362

Appendix: Literature review 363

Index 367

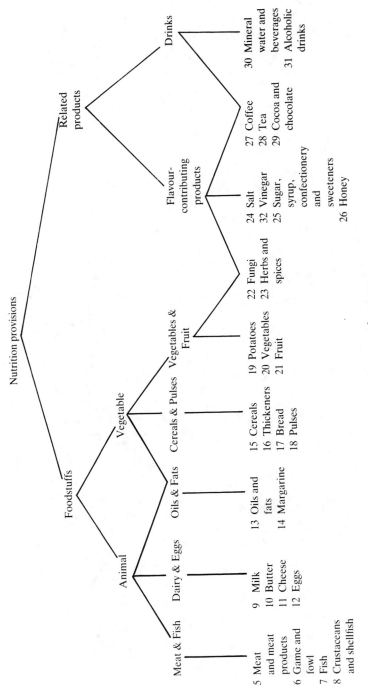

Summary of contents.

Preface

This textbook is intended for students of dietetics and applied home economics, for use in teacher training, higher hotel schools and for students of human nutrition in agricultural departments of universities. Students and others with a professional or personal interest who want to know more about foodstuffs and related products can also use it as a reference work.

The first four chapters give a general introduction and deal with, in this order, production, distribution and legislation (Chapter 1), potential forms of quality deterioration and spoilage (Chapter 2), methods of preservation (Chapter 3), and the presence of additives and contaminants in foodstuffs (Chapter 4). The main part of the book (see the scheme on page 11) describes the various product groups and, group by group, deals with technology, composition, potential use and storage advice. Where appropriate, a description of typical and special products follows, including those that are not essential dietary items.

We are indebted to the Board of Directors of the Hogeschool Nijmegen Akademie Dietetiek. We were allowed to use as the basis for this book a thesis written previously. We also thank our colleagues, especially those from the subject department of Food, and students from the Akademie Dietetiek for remarks and marginal notes, which they made in the aforementioned thesis.

Mrs S. H. van Oosten-van der Neut, lecturer at the Haagse Hogeschool voor Beroepsonderwijs, in the field of Nutrition and Dietetics, has commented on an earlier version of this text. Her accuracy was a stimulus for us. Sary, we thank you for this.

Finally we think it of importance to mention that we both have contributed equally to the completion of this book. Consequently the authors' names are given in alphabetical order.

Nijmegen, 1987
                                                                                 Carla Catsberg
                                                          Gerda Kempen-van Dommelen

Acknowledgements of illustrations

We thank the following persons, businesses and institutions for their kind cooperation with regard to the illustrations.

Bayer Nederland B.V., Arnhem: Fig. 2.6.
Bedrijfsschap Slagersbedrijf 'Het Slagershuis' Rijswijk: Fig. 5.12.
Bejo Zaden B.V., Noordscharwoude: Figs 20.1 to 20.9.
Bongrain Nederland B.V., Breda: Fig. 11.2.
Centraal Bureau van de Tuinbouwveilingen in Nederland, 's-Gravenhage: Figs 21.7 to 21.10.
Gedistilleerd en Wijngroep Nederland B.V., Zoetermeer: Figs 31.5 and 31.6.
Het kleine Loo, 's-Gravenhage: Fig. 15.10.
Het Nederlandse Zuivelbureau, Rijswijk: Figs 9.1, 9.2 and 11.1.
Ministerie van Landbouw en Visserij, 's-Gravenhage: Figs 7.3, 8.1, 8.2, 8.3 and 8.6.
Nederlands verpakkingscentrum, 's-Gravenhage: Fig. 1.7.
Nederlands Voorlichtingsinstitut voor Aardappelen (NIVAA), 's-Gravenhage: Fig. 19.1.
Produktschap voor pluimvee en eieren, 's-Gravenhage: Fig. 12.1.
Proefstation voor de Champignoncultuur, Horts: Figs 22.2 to 22.6.
Slachthuis Nijmegen, R.V.V.-kring 6, Nijmegen: Figs 5.1 to 5.6.
Sopexa Nederland, 's-Gravenhage: Figs: 31.2, 31.3 and 31.4.
Herman Stegeman, Amsterdam: Fig. 6.1.
Stichting Propaganda Groenten en Fruit, 's-Gravenhage: Figs 20.10 and 21.1 to 21.6.
Stichting Scharreleierencontrole, 's-Gravenhage: Figs 12.6 and 12.7.
Stichting Voorlichting Brood, 's-Gravenhage: Fig. 17.1.
A.G.J. van der Tang, 's-Gravenhage: Figs 17.1 and 18.1.
Vereniging Promotie Inormatie Traditioneel Bier, Amsterdam: Fig. 31.1.
Vereniging tot bevordering der bijenteelt in Nederland, Wageningen: Figs 26.1 to 26.3.
Vereniging van Nederlandse Reformhuizen, Ede: Fig. 1.6.
Voorlichtingsbureau voor de Voeding, 's-Gravenhage: Figs 10.1 and 23.1.
Centrum voor Champignonteeltonderwijs, Horst: Fig. 22.1.

1

Production, distribution and legislation

1.1 INTRODUCTION

Until recently, the assortment of available foodstuffs was reasonably limited. For products like coffee or margarine, only a few brands were known. Now, tens of brands of coffee or margarine are to be found on the shelves of any supermarket. What are the reasons for the enormous growth in the foodstuff market?

— The developments in food technology play a large part: new products, such as margarine or whipped cream, are made, while other existing products are treated or processed in many new ways, for instance coffee can be decaffeinated or freeze-dried.
— Increasing affluence is also a reason for the foodstuff explosion: people are able, for instance, to spend money on more expensive or luxury items such as pastries, snacks, drinks and prepared products.
— Increased travel is also a factor: many exotic products are now available, introduced, for example, by foreign workers, or because people have become acquainted with foreign products and eating habits while holidaying abroad.
— The extension and improvement of transport and cooling systems have contributed to the enlargement of the assortment of foodstuffs, particularly with regard to the import of fresh vegetables and fruit from tropical or subtropical countries.
— As a result of a reaction against excessive consumption and with the subsequent trend to more simple eating and the use of products in their original form, wholefood and natural products have become more popular.
— The fear of the "illnesses of affluence" such as cancer and heart and circulatory illnesses, and also the fear of additives, such as colour and preservatives are the reasons for the appearance of products rich in linoleic acid for instance, and products with reduced levels of, or free from, additives.

1.2 FOOD PRODUCTION AND FOOD DISTRIBUTION

Food production

All our food stems from plants. With the help of solar energy the basic food components are deposited in the plants. We eat plant parts, which often have to be treated, and animal products, which again in their turn have grown with the help of plants.

Formerly, man's dependence on agriculture for his food was secondary to his dependence on fishing and hunting. Because of increased knowledge and technical innovations, the importance of agriculture has increased greatly, dating from the beginning of the nineteenth century.

Currently, *general agriculture* is characterized by its use of fertilizers and remedial chemicals, fungicides, large-scale intensiveness of culture, mechanization and specialization. However, agriculture can also be practised in a different way. Some of these other methods, to which the term *alternative agriculture* is given, are based on a natural science concept of agriculture, which differs from the prevailing policy, and reflects another vision of society. Most alternative methods in agriculture have as their characteristic that their cultivation procedures, so far as is possible, maintain the existing biological processes in and above the soil; small-scale operations are also a feature.

In **organic agriculture** the use of fertilizers and chemical remedies is as much as possible limited or rejected. The soil is treated with animal manure and there is frequent crop rotation so that the soil does not get exhausted of nutrients. These starting points, among other concerns, are to be found also in ecological and biological-dynamic agriculture and horticulture.

In **ecological agriculture** the well-being of man, plant and animal and their interrelationship are the central factors. Ecological agriculture works with natural remedies and manuring. The crops are grown in time-honoured combinations, for example onions next to carrots, so as to combat the onion fly as well as the carrot fly. In addition, attention is also paid to:

— the stimulation of all aspects of people-friendly and environment-friendly work methods, for example democratic work structures in the connected industries;
— the development of an efficiently organized production, preparation and distribution chain;
— openess about the origin of products, the methods of production, the price-breakdown, and the reporting of a complete declaration of ingredients for composite products;
— stimulation of the use of environment-friendly packaging material;
— the here-and-now principle, which means there is a preference for products in season and locally grown;
— avoidance of importation from countries with a totalitarian regime.

In the Netherlands, for organizations which work according to ecological methods, the *Stichting Ekomerk Controle* (SEC) has developed special legally registered quality logos, the Ekologos.

Organizations which are changing over from general agriculture (see section 1.2) to ecological agriculture use the Eko 1- and 2-star logos (see Fig. 1.1).

If an agriculture concern complies completely with the ecological agriculture standards and code of behaviour, then the 3-star logo is used (see Fig. 1.2).

Products from the ecological agriculture also carry a product logo (see Fig. 1.3). Products carrying the Ekologo are manufactured from or composed of raw materials which come from recognized ecological concerns. It is prohibited to add compounds foreign to the product, such as colour or preservatives. The products also have to satisfy the eco-agricultural quality standards laid down.

The founder of the **biological-dynamic** (BD) agriculture method was Rudolf Steiner, founder of the present-day anthroposophy. The **biological** aspect of the BD cultural method concerns attempts to aim at optimum growing possibilities for plants and excludes the use of artificial fertilizer and chemical remedies. For fertilization, manure is used, often enriched with silica and plant preparations, for instance the horsetail plant. The **dynamic** element concerns the effect of forces from the ether or cosmos, (light–air–water–earth). Soil treatment fertilization, sowing and harvesting take place according to a sowing calender which is based on the position of the planets.

The Dutch Association for the Advancement of Biological-dynamic Agriculture has laid down guidelines which a BD-based business has to satisfy.

Products of BD agriculture and horticulture are labelled with an internationally copyrighted brand logo (see Fig. 1.4).

Farms, which do not yet quite comply to the requirements of the Association, but are changing over, are allowed to label their products with the Biodyn logo (see Fig. 1.5).

Other less well-known methods of biological agriculture are organic-biological agriculture, Lemaire–Boucher agriculture, Anog-agriculture, vegan agriculture and macrobiotic agriculture.

Food distribution

Food can come to the consumer in several different ways.

— From the land or the farm, products go to the factory where they are processed. After that they go to the wholesaler, who further distributes the products to the shops. Finally they reach the consumer.
— The distribution of vegetables, fruit and potatoes takes place through auctions, where the market gardeners sell their products to exporters, canneries or direct to the retailers. Vegetables, fruit and potatoes are also grown on a contract basis. This means that the farmer or horticulturist grows an agreed amount of produce of a specified quality for supply to the food industry, for a previously agreed price.
— In the meat trade, business is done on a contract basis between the cattle owner and the slaughter-house. From the slaughter-house it can be distributed direct to the meat-processing industry or to the butcher.
— Trade in eggs and fish is also conducted through auctions.
— Alternatively produced products are distributed through their own centres. At present the consumer may choose from 'reform', ecological, biological-dynamic and macrobiotic products.

Reform products have emerged in the Netherlands as a consequence of the

Fig. 1.1 — Eko agricultural quality 1- and 2-star logos.

Fig. 1.2 — Eko agricultural quality 3-star logo.

Fig. 1.3 — Eko product logo.

Fig. 1.4 — Demeter logo.

Fig. 1.5 — Biodyn logo.

Reform Movement. This movement developed in Germany during the last century under the name *Lebensreformbewegung*. The Reform Movement was a clear reaction to excess, not only in food (too much meat and alcohol) but also in dress and general lifestyle. In the first instance this reaction began among the rich upper class of society. In time, the idea appealed to a much larger group as well. This reaction led, in the area of nutrition, to the use of products in their natural state and to abstinence from meat and alcohol. Because the need arose for shops where one could buy foodstuffs prepared according to reform principles, several organizations were set up. In Germany, at the end of the last century, the so-called *Neuform* organization started which imported and regulated these reform products. In the Netherlands in 1935 the VNR (Association of Dutch Reformhouses) was started as a regulating organization and a production and distribution business for reform products, which are recognizable through the VNR logo (see Fig. 1.6)

Fig. 1.6 — VNR logo.

The principal aim of the VNR is the supply of vegetable products of good quality, which have been kept as far as possible with their natural qualities intact and which are free from chemical additives such as colours and preservatives. However, the reform banner flies over a very wide-ranging field. Reform products include not only biological-dynamic grown foodstuffs and products which contain unusual compounds (for instance algae or soya), but also everyday products from the food industry, such as wholemeal bread or brown rice.

Ecological products carrying the Eko logo are distributed by cooperative ecological distribution centres. From these centres, ecological shops, which are spread over the whole country, are supplied.

Biological-dynamic products have their own distribution system, for example Lima, a Belgian undertaking which buys in the harvest of BD-products and processes and distributes them on its own behalf. Dutch distribution centres include, among others, Akwarius and Proserpina. Besides biological-dynamic products the distribution centres also trade in biologically grown products.

Macrobiotic products are distributed through Manna, a chain of macrobiotic shops with their own distribution centres. The macrobiotic life-style is based on a Japanese nutrition- and life-style, by which one strives after a balance between two poles, Yin and Yang. The foodstuffs used are produced by biological-dynamic agricultural methods.

1.3 PURCHASING

With the increase in the quantity of foodstuffs, often attractively packaged and seductively promoted, responsible buying has become increasingly more difficult. Before discussing the factors which play a part in the buying of food, a summary is given of the selling places.

Retail businesses

The most recognizable feature of these shops is their good accessibility. The service is still largely personal and friendly. One can expect, in the Netherlands, knowledgeable service because there, to open a shop, a licence is obligatory, and this is only given to those who have followed a course in the subject. Competition with larger stores is often financially not possible. Not much in the way of advertising or sensational price-reduction campaigns is to be expected. The range of goods is often limited, depending on the area where the shop is situated. The prices are often high, following the recommended prices without gimmicks, prizes or loss-leaders. By bulk buying the selling price can be reduced to the consumer. A number of these corner shops have formed themselves into trading combines, by which means they are better able to compete with larger stores.

Specialist shops have a large assortment of one or more products, such as delicatessens or confectionery shops.

Supermarkets

In supermarkets, besides groceries, fresh meat, vegetables, fruit and potatoes are also sold and many goods other than food, such as clothes, books and stationery. Their great advantage over a smaller shop is the much larger range of brands of any specific product. Large supermarkets sell their 'own brand' products which are often cheaper than the more well-known branded products. A personal and knowledgeable service is sometimes found at the meat, vegetable or cheese counters, but the customers mainly serve themselves. This saves time and staff labour, which influences the prices of the products.

Cash-and-carry stores
These are mostly housed in simply furnished buildings with a self-service system. Articles are offered at greatly reduced prices. Often they are still packaged in big cardboard boxes. The shops are mostly situated away from the centre of a village or town because of lack of space and high land prices or rents. Also, branded goods, 'white' products, are often included in the range as loss-leaders. The assortment is often limited to products with a long shelf-life.

Normal retail businesses, supermarkets and cash-and-carry shops vary so much, as to overlap each other, which makes it difficult to distinguish the three types of selling places. For large-scale customers, for instance catering-, hospital- and factory-kitchens there are special stores where a pass is required for entry. These stores are not intended for sales to private citizens and are more wholesale than retail.

Department stores
These sell mainly goods other than food. Some large stores, however, have a separate food department, with the same characteristics as the supermarkets.

Street markets
Buying in markets can be much cheaper than in shops because of their low overheads, which decrease selling prices. The choice can vary greatly. Fresh products such as vegetables and fruit are strongly seasonally dependent. It is sometimes possible to buy very cheaply especially at the end of the day.

Mobile shops
Formerly, the greengrocer, the milkman and the baker delivered products to the door. These days in some areas there are 'mobile' shops for all daily shopping. The choice cannot be so large because of the limited space. However the accessibility is great, as is the service. These factors, however, will very likely increase the price.

There are a number of factors which play a part in purchasing and influence food choice.

— *The composition of the foodstuff* What materials are used in the manufacture? Are there colours in the product? Have preservatives been added?
— *The needs of the purchaser* Family size and age, taste, ability to cook, and customs or habits play a part.
— *The purpose that is intended* For example, is a festive meal being prepared, or a simple one?
— *The time one is willing to spend preparing the food* There are ready-made off-the-shelf products such as canned apple sauce, hamburgers or faggots, or oven ready products such as prepared vegetables, pre-peeled potatoes or frozen meals which may still need some preparation.

- *The available storage space* Has the buyer sufficient cupboard space or a cellar for a stock of tins? Is there a refrigerator or a deep-freeze in the home?
- *The equipment* For the preparation of deep-frozen meals a cooker is needed; pre-fried croquettes and similar products presuppose a deep fryer.
- *The amount of money available.*
- *The quality of the product* The term quality is difficult to define. Indicators of quality can be: appearance, taste, freshness, brand, packaging, reputation of the shop. For instance: fresh vegetables are sorted for quality into class I or class extra. This takes into account factors such as the shape, the colour, or the number of superficial blemishes (see Chapter 2.1).
- *The size and the weight of the product* In buying vegetables, one must also take into account the wastage factor or the shrinkage.
- *The price* Prices of vegetables and fruit are particularly subject to seasonal variation — the first strawberries or the first asparagus of the season are dearly paid for. The weather plays a part as well as supply and demand. Preparation such as cleaning vegetables or partly cooking cereals increases the price. The sales site, size and turnover of the firm, the size or amount packaged, competition and advertising offers are also factors which go to determine the price.

1.4 PACKAGING

At present, nearly all foodstuffs come pre-packaged.
Packaging has many advantages:

- The products are directly and hygienically packaged after processing and cannot be contaminated or soiled after that.
- Products that are sealed in air- and moisture-proof packing deteriorate less quickly, through reduced loss of aroma, so avoiding becoming limp or drying out.
- Packaging promotes a rapid turnover.
- Packaging provides a site for printed information such as that required by law, specific information on the product or advertising.

Packaging can also be disadvantageous:

- Packaged foodstuffs are dearer than unpackaged, especially if the packaging is luxurious.
- Examination of the goods when shopping is nearly impossible.
- The packaging can mislead, for instance by having a thick base or wall and a pretty colour.
- Sometimes there is a charge for bottles and jars, which is refunded when they are returned (refundable deposit).

Packaging materials
The most important materials which are used for packaging are: paper and cardboard, glass, tinned-steel, plastics, aluminium and wood (especially for transport packaging). They have specific characteristics. Often they have to complement each other to obtain a good result.

— **Glass** is transparent, can be sterilized, is impenetrable to gases but is breakable, heavy and translucent. Returnable and disposable glass each have their positive and negative aspects.
— **Paper** and **cardboard** are reasonably cheap, easy to process into sacks, wrappers and so forth and easy to print on, but allow gas and moisture to pass through if no extra protection is provided.
— **Metals** (aluminium and cans) are non-breakable and can be sterilized, and they do protect against air and moisture; but the contents are not visible, and they are expensive. Cans are sometimes produced so that they can be opened without a can opener.
— **Cellophane** or viscose film is as clear as glass, airtight and can resist oils and fats. To make cellophane suitable for deep-freeze packaging, it is covered on one or both sides with a thin layer of plastic.

Fig. 1.7 — Packaging machine for paperpacks.

— **Plastic material**, including plastic sheeting and polyethylene, polyvinylchloride and polystyrene containers. Plastic materials offer many advantages such as little or no permeability to light, air and water; are often resistant to heat; are pliable and have good insulation; but can also cause danger to health. A problem with plastic packaging is the movement of materials from the one site to the other; this is called **migration**. This happens, for instance, with tubs of margarine and with plastic bottles containing oil, where fat and plastic react. The problem also occurs with recycled paper where residues of toxic ink move from the paper to the product which is packaged in it.

Sometimes combinations of materials which are called **laminates** are chosen. For instance, cardboard with a thin layer of aluminium foil for packaging orange juice, cardboard with a thin layer of plastic for milk.

1.5 FOOD LEGISLATION

It has been common practice in most countries for unscrupulous traders to adulterate foods, examples being the mixing of sand with pepper and the watering down of milk. Also, food unfit for human consumption has been passed for sale. Such happenings have been going on for centuries and still occur, although to a much lesser extent. Over a long period of time, rules and regulations have been introduced to counteract these practices. These started at a local level and in some cases have spread to cover districts, countries and even parts of continents. The development of food legislation to its present state has been a very long and complex process.

The two principal aims of legislation are the protection of the health of the consumer and the prevention of fraud. The details of food legislation will differ from one country to another. In some parts of the world, for example North America, Northern Europe and Japan, legislation will be at a more advanced state than it would be in some of the developing countries.

The main aspects that should be covered by food legislation are listed below (source: Jukes, 1987):

Primary legislation
Regulations and their development
Compositional standards
Additives
Contaminants
Labelling
Hygiene and health
Weights and measures
Enforcement

This covers the basic framework for a legislative system. Obviously the differences in detail from one country to another necessitate that anyone intending to provide food for consumption should consult their national regulations to ensure that they are working within the law.

Primary legislation
In England and Wales the primary legislative powers are contained in the Food Act 1984, which sets the criteria that food should be safe and of the quality demanded by the purchaser. In some countries this may be termed a Goods Law Act, and cover some non-food items.

Regulations
The majority of the detailed technical requirements for food products are contained in the regulations issued by Ministers and laid before Parliament or the appropriate governing body, for approval.

Development of regulations
Before regulations are issued, a great deal of consultation takes place between Ministers, those parties who will be affected by the regulations and scientific experts. Usually, independent expert scientific or advisory committees are established to give advice to Ministers.

Compositional standards
Compositional standards exist in the UK for staple products such as flour, bread and sausages as well as more obscure ones such as mustard and curry powder. They help ensure uniformity of product and prevent fraud. Since joining the EC, the recent trend has been towards restricting the number of compositional standards and ensuring consumer protection by more stringent labelling requirements.

Additives
The use of additives in food products is controlled by regulations which adapt a 'positive list' system. All additives are thoroughly assessed before they are allowed onto the list.

Contaminants
Controls on specific contaminants in the UK are of a limited nature, although there are regulations for heavy metal residues in several foods. There is extensive monitoring of pesticides and other agricultural residues, radioactive particles and mycotoxins in foods. There is increasing consumer awareness of these matters, and the 'green' movement is becoming more of a serious pressure group.

Labelling
Labelling regulations can be very complex but the consumer is now given a much more comprehensive listing of the ingredients and additives used in processed foods. Nutritional labelling also provides a further important source of information.

Hygiene and health
The Food Hygiene Regulations ensure the safety of fresh and processed foods. In the UK the emphasis has been on monitoring and controlling rather than imposing statutory microbiological standards. The most important forms of control come through heat treatment regulations, designed to eliminate pathogenic organisms, such as *Salmonella* and *Lysteria* (pasteurization) and *Clostridium botulinum* (sterilization), combined with good hygienic practice. Recent reported serious outbreaks of food poisoning may pressurize the Government to introduce more statutory microbiological standards.

Weights and measures
The main aspects that concern food products are the control of specified weights which restrict pack sizes on a number of products to certain weights and the application of an average weight system for packaged goods.

Enforcement structure
Enforcement of the food legislation is very important and in the UK is the responsibility of the local authorities, who employ trained personnel to ensure that the Acts and regulations are enforced. In any country, enforcement involves inspection, sampling and analysis. There may also be specialized inspectorates, for example for meat inspection or pesticides and herbicides.

Many of these aspects will be discussed in greater detail for specific foods, within the text.

There have been many attempts at harmonizing food legislation over greater geographical areas. For example, the EC Communities and the Codex Alimentarius. This is important in promoting trade between countries as it becomes difficult to export products manufactured in one country if they do not comply with the regulations of the intended country of export. For example, currently it is forbidden to import irradiated foods into the United Kingdom. Therefore any country which irradiates foods as part of their processing will in theory not be allowed to send them to the United Kingdom. However it is likely that food irradiation will be permitted in the UK in the near future.

International law
Food and beverages are imported and exported. The Benelux (Belgium, the Netherlands and Luxembourg) is an economic union. The Group of Ministers from the Benelux can advise or lay down regulations. A few examples of such regulations are those for meat extract, liquid flavours, powdered flavours and soups.

Inside the European Community (EC) the member countries have a communal agricultural ruling. Because each country has its own rules for composition and labelling of products, there must be some compromise in order to reach communal rulings. Ministers of the EC court of agriculture work towards harmonizing the European rules in the area of foodstuffs and legislation. Some examples of agreements already reached are the additive laws, whereby, in all EC countries, the same anti-oxidants and colours are permitted; and the quality classification system for vegetables and eggs, for instance the grading of a cauliflower as Extra, Class 1 and Class 2.

On a world level, the **Codex Alimentarius** exists. This has been set up by the World Health Organization (WHO) and the Food and Agriculture Organization (FAO) of the United Nations. The aim is to set up norms for foodstuffs. The norms of the Codex are not themselves binding, but the work of the Codex Committee has a standardizing influence on the norms developed and enforced throughout the world. This is particularly valuable for smaller and poorer countries, which have not yet established their own laws for foodstuffs, as the Codex can be used as for guidance.

Finally, legislation is a dynamic process: it is changing all the time in order to keep pace with new knowledge and technologies and a rapidly changing world. For

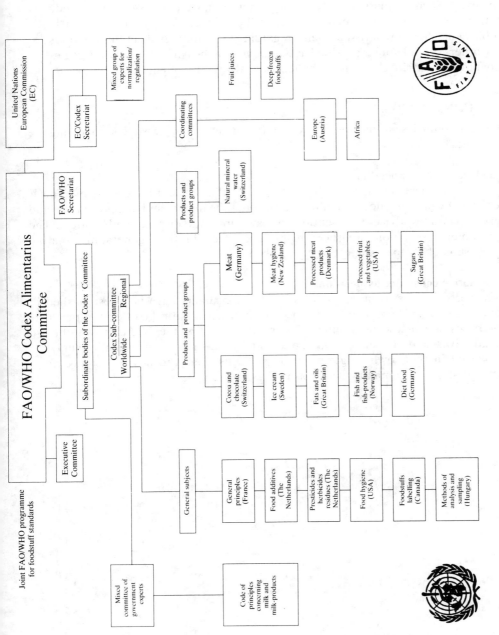

Fig. 1.8 — Joint FAO/WHO programme for foodstuff standards.

example, the laws on irradiated foods within the United Kingdom are currently being reviewed. Such a situation makes producing written statements of law exceedingly difficult and any such information could well be out-of-date by the time it is finally published. Therefore any statements regarding legislation made in this textbook, or any other, should not be regarded as definitive and the current legislation regarding any product, additive or contaminant should always be sought out.

1.6 CONSUMERISM

In many countries various forms of consumer organizations exist. The following are examples of such.

The Consumer Association

The aim of consumer organizations is to protect the rights of the consumer. This is achieved by investigation and information about the quality and prices of goods and services, and by advice, intervention and law support in disputes with traders. The Consumer Association publishes a magazine, *Which*, in which general information and results of comparison tests are published.

Alternative consumer associations

Other associations for alternative or 'green' consumers have been established in some countries. These associations occupy themselves with the effects of products and processing on the environment; for instance, the way the food is grown, the use of additives and the effects on the world's food-sharing. Information may be published in their own magazines. 'Green' consumers are now becoming more influential in policy-making decisions in Western Europe.

2

Quality deterioration and spoilage

2.1 INTRODUCTION

Our foodstuffs, from both plants and animals, are subject to changes in quality. This is the result of transformations which originate through, among other things, enzymes which are in the product or microorganisms which come into contact with the product via air, water or soil (see Fig. 2.1). The smell, colour, taste and

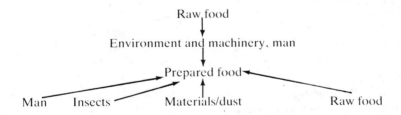

Fig. 2.1 — Cycle of contamination. (Source: Brochure 402, *Hygiene in de keuken*, deel 2 Voorlichtings bureau voor de voeding, 's-Gravenhage.)

consistency of the foodstuff can change through this. When it is a question of beneficial change, then the processes can be stimulated, as in the making of yoghurt or sauerkraut. If the foodstuff changes in the unfavourable sense or should the changes result in danger to health, then we speak of *spoilage*. Spoilage can be prevented by preservation.

The concept of quality is difficult to describe. Factors to be noted are: taste, smell, appearance, weight and firmness; but such aspects as the freshness, the price, the keeping qualities or the nutritional value also play a part. All these characteristics do not count equally for everyone. Some can be measured objectively, whilst some are much more subjective. For several aspects of quality, such as chemical composition, standards have been laid down by law, such as the Agriculture Qualities Law (Netherlands) or the Food Standards (United Kingdom).

Forms of spoilage are:

— physical spoilage (section 2.2);
— chemical, biochemical or enzymatic spoilage (section 2.3);
— microbial spoilage (section 2.4);
— spoilage by parasites (section 2.5);
— spoilage by vermin (section 2.6).

2.2 PHYSICAL SPOILAGE

This is mostly not harmful. The foodstuff deteriorates in quality, a change which is quickly taken in by the senses. Forms of physical spoilage are:

— **Dehydration** The foodstuff looks less attractive; for example, leaves wilt, cheese cracks.
— **Absorption of moisture** Powdery products start to form clumps, or biscuits and similar crisp products go soft.
— **Crystallization** This happens, for instance, with sugar in honey or on the outside of dried plums, or with salt for example on the surface of smoked meat or strongly brined ham.
— **Sweating** Fat can separate out with increased temperature. It is visible as small droplets on the surface, for instance on cheese. With chocolate which has been stored too warmly, the cocoa butter solidifies on the outside as a white layer (bloom).
— **Retrogradation** This is the releasing of moisture that was previously bound. This, for instance, takes place in bread a few days old. It becomes dry and crumbly ('stale').
— **Freezing** By lowering the temperature below 0°C the water freezes in the product. Large sharp ice crystals can be formed which damage the cell walls. On thawing, the cells drain, and so the product becomes wet and limp and can quickly spoil.

2.3 CHEMICAL, BIOCHEMICAL OR ENZYMATIC SPOILAGE

The changes which we regularly come across in foodstuffs may be chemical reactions, occurring for instance as a result of heating; but many are the result of enzyme activity. It is sometimes difficult to say if a certain reaction is purely chemical, or is being initiated by enzymes. The most important reactions which we see occurring are the following.

Auto-oxidation of fats
This is a reaction of oxygen from the product itself with the foodstuff. The reaction is speeded up by light, increase in temperature, and heavy metals.

Hydrolysis of fats
Through the presence of water in fatty foodstuffs the fats can wholly or partly hydrolyse. Triglycerides are separated into glycerol and free fatty acids. Some of these fatty acids have a disagreeable smell.

Oxidation of fats
These can be subdivided into ketone and aldehyde rancidity.

- **Ketone rancidity** If fat is partly hydrolysed, freed saturated fatty acids can oxidize. The compounds which result have a certain smell (perfume). Bacteria and moulds which secrete enzymes activate the reaction.
- **Aldehyde rancidity** The fatty acids in unsaturated fats oxidize. No hydrolysis has to take place first. The reaction is influenced by light, a high temperature and some metals, among which are copper, silver and iron.

Acrolein formation in fats
If fats are heated for a long time in the presence of water, **glycerol** can be formed by hydrolysis. At high temperatures glycerol is dehydrated to acrolein, which is detrimental to health. Acrolein formation can also occur in frying chips.

Polymerization of fats
With heating for a long time and in the presence of oxygen, unsaturated fats can polymerize in air, which means the fat molecules form long chains. After cooling, the fat or the oil is viscous. When next used, the fat will quickly smoke.

Non-enzymic browning as result of the Maillard reaction and caramelization
- The **Maillard reaction** occurs between reducing sugars and proteins, amines and amino acids. Brown pigments are formed, and smell and taste compounds are formed, which are sometimes desirable and sometimes undesirable. The reaction proceeds faster at higher temperature.
- **Caramelization** By dry-heating of sugars, brown colouring and taste changes occur.

Enzymatic browning
These reactions are found in vegetables and fruit. Several enzymes are active during this process.

Breakdown of pectins
The enzyme pectinase causes a softening of firm vegetables and fruit varieties.

Reactions of foodstuffs with their packaging
There is a distinction between:

- reactions with plastic and paper packaging (see section 1.4) and
- reactions with metal can packaging. A can consists of steelplate with a thin coating of tin. The foodstuff can react with the mild-steel as well as with the tin.

Reaction with tin

Protein-rich products especially can react. The inside of the can is attacked, and this is seen as an etched pattern. Also, traces of tin can go into solution and can give a change of taste. Sometimes the solubilized tin is visible as a white precipitate, for instance on frankfurters. To prevent these reactions the cans are mostly laquered on the inside.

Reactions with iron

The tin layer can be damaged by falling or bumping, so that the contents come into contact with the iron. Protein-rich products can react with iron and among other things FeS is formed. With acid products such as tomato paste H_2 gas is formed. Because the gas cannot escape, the can will bulge (blow). This is called **chemical blowing**. It is not advisable to open such a can, the contents will shoot out with some force. Also, taste and colour changes will have taken place, so that the contents are not usable any more (see Fig. 2.2).

Fig. 2.2 — Cans blown by gas formation through spoilage.

The can may start to rust externally, if it is kept in damp conditions. In the end, this can lead to leakage, and also microorganisms will be able to enter the can.

The bulging of a tin can have a different cause. Blowing can also occur through insufficient sterilization, especially of protein-rich products. The gas formation is not the result of a chemical reaction, but of microbial spoilage. We call this phenomenon **bacteriological blowing**. The rounding of the can is in this case not related to physical damaging of the can.

2.4 MICROBIAL SPOILAGE

There are a large number of microorganisms which are regularly found on foodstufs. These can be harmful to humans in several ways.

A number of microorganisms are disease-causing (**pathogenic**). For instance, meat from a sick animal can make people ill. It is also possible for food to be contaminated by people during preparation. In such a case the food is infected and we speak of **food infection**.

A number of microorganisms form poisonous products (**toxins**) during their metabolism. The toxins can make people ill. In this case we speak of **food poisoning**. Besides microbial food poisoning, chemical food poisoning also exists. This occurs if food contains too many toxic materials, and this leads to serious incidents. Examples of chemical food poisoning are the Chinese restaurant syndrome after consumption of large amounts of taste enhancer, and the contamination of food with heavy metals such as mercury and lead which have come into the products by means of air or water.

Microorganisms multiply if their growth requirements are satisfied and they have the time to do so. Some of these growth factors and requirements are as follows.

— *A particular quality of foodstuff* Some microorganisms need an organic nitrogen source such as amino acids, others grow only if sufficient glucose is present.
— *The water activity* (a_w) This indicates the quantity of water which is available for microorganisms, the activity being rated by numbers between 0 and 1. If the water activity is high, the bacteria come to full development. In products to which water-binding compounds have been added, less moisture is available, for example jam and cheese. All microorganisms need water. At a water activity lower than 0.6 no microorganisms should grow.
— *The acidity or pH of the foodstuff* Most microorganisms grow less quickly in acid surroundings. Compare the speed of the spoilage of milk (pH neutral) and yoghurt (pH low). A strong alkaline environment also restricts the growth, for instance the use of a solution of soda as disinfectant.
— *The temperature* The temperature range within which microorganisms can grow ranges from $-10°C$ to $+80°C$. On this basis, microorganisms are classified into three groups (see Fig. 2.3). The spoilage-causing microorganisms belong mostly to the mesophile group.
— *The presence of oxygen* Aerobic bacteria only grow with oxygen. Anaerobic bacteria cannot grow in the presence of oxygen. Facultative anaerobic bacteria are aerobic bacteria, which can exist for a shorter or longer period without oxygen. Micro-aerophilic bacteria grow best at a considerably lower oxygen concentration than that of air.
— *The presence of anti-microbial compounds* These work by restricting the growth of microorganisms. In fresh milk and eggs these may occur naturally. In the food industry, preservatives like benzoic acid and sorbic acid are sometimes added; these have an anti-microbial action also.
— *The growing-rate* In optimal circumstances, some microorganisms double their numbers in 20 minutes; others need 24 hours. When microorganisms arrive on a new substrate, they need an acclimatization period before they divide. This period, which takes on average 4 to 5 hours, is the delay or 'lag' phase. After that, a period follows which is called the growth or 'log' phase. Following this, the microorganisms maintain themselves for a shorter or longer period in the medium and finally decline (see Fig. 2.4). By influencing the growth requirements the lag phase can be prolonged. This is used in preservation.

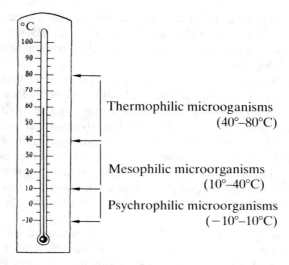

Fig. 2.3 — Optimum temperatures for microorganisms.

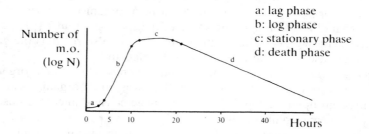

Fig. 2.4 — Graph of a growth curve (m.o. = microorganisms). (Source: Klingeren, drs B. van, *Microbiologie voor medische analisten* deel 1, Stenfert Kroese b.v., Leiden.)

— *Spore formation* Some microorganisms can form spores, when the environment becomes less favourable. In this way they can maintain themselves for a shorter or longer period, though they do not muiltiply. If the circumstances again become favourable, they return to their original form and can once more grow and multiply. Most spores are not destroyed by most food preservation methods.
— *Accompanying microflora and contaminants* In foodsuffs there are usually several kinds of microorganism present at the same time. These can influence each other in positive or negative ways.

Microbial spoilage can be caused by:
— bacteria
— moulds
— yeasts
— viruses.

Bacteria
There are a number of kinds of bacteria which cause spoilage.

— **Lactic acid bacteria** develop in foodstuffs which contain a small quantity of one or more sugars. The sugar is converted into lactic acid by the bacteria. Through this, the smell and taste of the foodstuff change. This is not harmful to humans. The bacteria are facultative anaerobes and do not form spores. In the food industry lactic acid bacteria are used positively, for instance in the preparation of yoghurt and sauerkraut.
— **Acetic acid bacteria** develop in foods which contain alcohol. The alcohol is totally or partly converted into acetic acid. Because of this the smell and taste change. The bacteria are aerobic and non spore-forming.
— **Proteolytic bacteria** develop in and on foods which contain proteins and which are not acid. The proteins are broken down, during which toxins and unpleasant smelling gases can be formed. The food becomes sticky and/or slimy. The toxins are sometimes harmful to humans, depending on the amount present.

Lactic acid, acetic acid and proteolytic bacteria cause perceptible spoilage. A number of different bacteria cause, by their number or their toxins, food infections or food poisoning respectively.

— ***Salmonella* bacteria**. There are at least 1300 types of *Salmonellae*. They are facultative anaerobes and develop quickly at luke-warm temperatures. Food can be contaminated through unhygienic preparation, but industrial slaughtering of pigs, calves and poultry is also an important cause of contamination (see Fig. 2.5). The *Salmonella* bacteria become a danger to health only when they multiply considerably. They may cause diarrhoea and, in more serious cases, fever or illnesses such as typhoid or paratyphoid. *Shigella*, related to *Salmonella*, causes dysentery. By heating above 80°C *Salmonella* is destroyed.
— ***Staphylococcus aureus*** occurs on the mucous membranes of the nose and throat in humans and animals. They develop in a protein-rich environment and produce intestinal toxins. Heating kills the bacteria, but not the toxin. A few hours after contaminated food has been consumed, symptoms such as sickness, vomiting, stomach cramps and diarrhoea occur. The patient has no fever. The microorganism is a facultative anaerobe and does not form spores. The toxin production is eliminated by quick cooling. When food is kept cool (+5°C), toxin production is nil.
— ***Campylobacter jejuni*** causes food infection. It is a facultative anaerobe and does not form spores. The symptoms of the infection are diarrhoea with a fever. The main source of contamination is poultry.
— ***Clostridium botulinum*** is a spore-forming anaerobe. It produces toxins, which act as a nerve poison. Food poisoning from these toxins (botulism) can be very

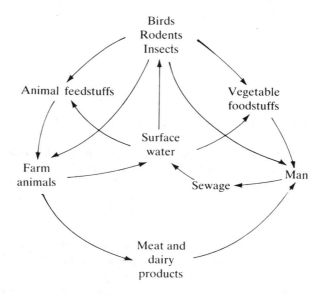

Fig. 2.5 — Possible sources of contamination with *Salmonella*. (Source: Beckers, H. J., *Besmetting van ons voedsel*, cahier Biowetenschappen en maatschappij, 10e jaargang, Leiden, June 1985.)

dangerous and is often fatal. The symptoms are: a dry mouth, double vision and breathing difficulties. Spores of *Clostridium* occur all over the world in the soil and in dung. They arrive in the food through contamination from the soil. Growth of the microorganism takes place in foodstuffs with a pH>4.5, especially in meat and fish. The spores are destroyed by heating at 121°C for 3 minutes (the 'botulinum-cook'). Toxin formation can take place in cans of products rich in protein, which have been heated to an insufficiently high temperature. Gas is also formed which causes the can to bulge (bacteriological blowing). This danger is very serious in domestic methods of preservation (bottling) of low acid products, pH>4.5.

— **Clostridium perfringens** occurs on meat and poultry which has been contaminated by the contents of the intestine of the slaughtered animal. The bacteria is anaerobic and forms spores, which can survive high temperatures. This microorganism is sometimes found in large joints of meat, which have not been cooled fast enough after processing. The anaerobic conditions within the large mass of meat favour its development, with resultant stomach ache and diarrhoea, though no fever.

— **Bacillus cereus** forms spores which are heat-resistant. The *Bacillus* lives as a facultative anaerobe in the soil. When dishes with rice, pastries or pudding are kept for a long time at room temperature, they begin to grow and form toxins in the food. Sickness symptoms can be: diarrhoea and vomiting, but without signs of fever.

— **Listeria monocytogenes** has come to the forefront only recently. It is a dangerous bacteria, which is destroyed by pasteurization. Food poisoning outbreaks have been associated with soft cheese and paté.

Moulds
Moulds develop on most foodstuffs, with a preference for a moist environment. They are aerobic. The spores are visible as a woolly fur, varying in colour from white, pink yellow, green, brown to black. Mouldy foodstuffs smell and taste musty. Some moulds form mycotoxins which cause food poisoning or other illnesses; for instance: *Aspergillus flavus*, which can occur on peanuts, forms aflatoxin.

Yeasts
To develop, yeasts prefer slightly acid sugar-containing foodstuffs. The sugars are changed to alcohol and carbon dioxide by the yeasts. The fermentation can continue until the alcohol content reaches about 15 per cent. Then the yeast cells cannot work any more because the medium has been, as it were, 'poisoned'. Yeasts are facultatively anaerobic. The fermented products have a characteristic smell and gas bubbles may be visible. They are not harmful but will cause spoilage in some products, e.g. yoghurt and coleslaw.

Viruses
These are living organisms which often cause illness. Foodstuffs can often play a part in the transmission, but no increase takes place. Heating kills most viruses, but cold can preserve a virus. That is why ice-cream is a very suitable medium. Concrete examples of illnesses are often difficult to give, because after the event it is difficult to check if a contamination by a virus has gone via food or the direct route (human infects human). Probably yellow jaundice (hepatitis-A) and the poliomyelitis virus are transmitted by food which has been unhygienically prepared.

2.5 SPOILAGE BY PARASITES

Parasites are organisms which live in or on other kinds of living organisms (host), from which they extract food. Sometimes parasites need two or more different hosts for their development (alternate hosts). A number of parasites need, as an intermediate host in the larval stage, animals which serve as food for humans. Foodstuffs of animal origin can be infected by parasites and can cause illness in the consumer. Below are a number of infections which can occur.

Tapeworm infection
This complaint reaches humans via contaminated beef or pork. The bladderworm, a development stage of the tapeworm, nestles itself in the muscle tissue of the fattening animal. The animal is the intermediate host. Meat infected with bladderworm is called 'vinnig' in the Netherlands. In the human body the bladderworm can grow into an adult tape worm, after consumption of raw or insufficiently heated meat.

Trichiniasis (muscle infection)
Raw or insufficiently heated pork can contain trichina or hairworms. A slight infection in humans can pass without clear sickness symptoms. A serious infection results in muscle illness (trichiniasis). The trichina develop in the intestines and from there penetrate the muscle tissue, where they encapsulate (cystformation). When the larvae penetrate the muscles the patient reacts with fever, aching muscles and allergic symptoms.

Herring worm illness
Raw herring can contain live larvae from the herring worm (*Anisakis marina*). The larvae can cause mild or acute gastro-enteritis in humans. The compulsory killing of the larvae, for instance by deep-freezing, strong brining or marinating, has led almost to the disappearance of herring worm illness in humans in the Netherlands.

Toxoplasmoses
This infectious illness is caused by a one-celled parasite (protozoa). The parasite occurs in almost all kinds of birds and animals. The infection rate in humans correlates with eating habits, hygiene and age. The parasite has a complicated life-cycle. The parasite can only mutiply in the small intestines of felines. The cats excrete eggs which can only cause an infection for a few days.

Humans can be contaminated before birth, because active parasites can pass from the mother via blood and placenta and so reach the foetus. This can lead to premature birth or congenital defects. After birth, contamination can take place via the mouth, by the eating of raw or insufficiently heated meat, or by contamination of hands, food, drinking or swimming water. Symptoms of the illness, such as lymph gland infection, excessive tiredness, depression or a slight fever, follow sporadically.

2.6 SPOILAGE BY VERMIN
A not inconsiderable part of the total produce in agriculture is lost by unprofessional storage, allowing gnawing by rodents and attack by insects. The loss can be as much as 25 per cent.

Gnawing by rodents
Damage can be done to crops in the field by animals living in the wild, for instance rabbits or wild pigs. During storage rats and mice especially attack the stocks. They also contaminate the stores with their droppings and can transmit illnesses. Well-constructed storage chambers are a good preventative. To obtain professional control of rodents, the help of the local council's pest control service will have to be summoned.

Voracity of insects
The development stages of insects can consist of three or four phases: egg, larva, sometimes a pupa, and the adult insect. During this life-cycle the insect lives parasitically. It can also function as a disease vector.

Some insects occur frequently.

- *Ants* are especially attracted to sugar-containing food stuffs.
- *Cockroaches* have a preference for warm and moist places. They are frequently noticed in large kitchens. They feed on all sorts of material.
- *Beetles* and *weevils* hollow grain or pulses out and deposit their eggs. The larvae further hollow the grain and form pupae in it.
- The lavae from *moths* spin sticky threads by which means clumps are formed in the product. Some frequently occurring are the larvae of the flour moth, the cocoa moth and the vegetable moth, which occurs in dried vegetables and semi-tropical fruits.
- *Mites* are spider-like animals. They have an egg-shaped shield with legs sticking out on the edges. They lay eggs from which larvae come. They occur in cheese and in flour. In the end, cheese can be completely pulverized and flour gets a sweetish, honey-like smell. Flour with mites gets a woolly surface.
- *Flies* contaminate the food with their hairy legs, to which dirt and bacteria stick. They also regurgitate their stomach contents before they eat new food. They lay eggs from which maggots quickly develop. Good hygiene and regular control of insects can overcome the problem.

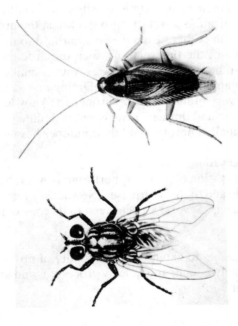

Fig. 2.6 — Cockroach and housefly.

3

Methods of preservation

3.1 INTRODUCTION

Foods will, when kept in their natural state, mostly stay acceptable only for a short time. A process of deterioration will start, through which the product first becomes less attractive and later can perhaps even become harmful to health, depending on the cause. By preservation, the timespan in which the product stays qualitatively good and acceptable can be lengthened. Various ways to make foods keep longer have been known for many hundreds of years. Drying and salting especially were often used, though in a very primitive way. Through the knowledge acquired about the causes of deterioration and through advances in techniques, methods of preservation have been continually improved and their number has increased.

The importance of preservation
In former days the preservation of food was performed out of the sheer necessity to bridge periods of scarcity and to store surpluses. Nowadays preservation is used for a number of different reasons. Several aspects need to be considered.

Economic aspects
By preservation it is possible to extend the availability of products. In the harvest periods they can be cheaply bought and quickly processed, and are then available at any time during the year.

Sensory aspects
Preservation techniques can also alter the taste, smell and consistency of a product with advantage, so that the final product is more valued; making sauerkraut or smoking fish are examples of this.

User's convenience
By some preservation techniques the product is so heated through that it will be totally or partially cooked. A reduction of volume can also result through shrinkage,

as illustrated by spinach or endive and dried soups and coffee powders. As a consequence of this, the products become easier to use or are cheaper and easier to transport.

Methods of preservation
Loss of quality and deterioration are brought about through physical, chemical or enzymatic reactions, or by microorganisms. For each product, certain specific causes of spoilage will be indicated. By preservation, the causes of deterioration can be overcome in four ways. One can:

— stimulate the activity of specific microorganisms and enzymes (fermentation), as a result of which the medium becomes unfavourable for the bacteria which cause the spoilage;
— remove pathogenic and spoilage microorganisms;
— retard the activity of enzymes, pathogens and microorganisms which cause the deterioration;
— kill pathogenic and spoilage microorganisms and inactivate enzymes.

See Fig. 3.1 for a systematic summary of methods of food preservation.

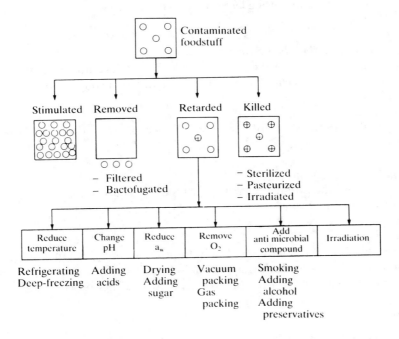

Fig. 3.1 — Summary of preservation methods.

In discussing the different forms that food preservation takes we shall consider for each in turn:

— the technology
— the principles of extending the keeping qualities
— the sensory changes, which the product undergoes as a result of the preservation method used
— the storage and storage time
— specific aspects if any.

3.2 STIMULATION OF THE ACTIVITY OF ENZYMES AND MICROORGANISMS

Technology
This uses the activity of microorganisms and enzymes by adding specific cultures or stimulating the course of various enzymic reactions. This process is called fermentation.

Some samples of fermented products are: sauerkraut (made from white cabbage using lactic acid bacteria), vinegar (made from alcohol using acetic acid bacteria), yoghurt (made from milk using lactic acid bacteria) and wine (made from fruit juice using yeasts).

Principles of extending the keeping qualities
The medium becomes specifically unsuitable for spoilage microorganisms because the pH is lowered or because a reasonable quantity of alcohol is formed.

Sensory changes
Taste, colour and texture of the food often change and all sorts of aromatic compounds are formed.

Storage and storage time
These vary with the product from a few days for yoghurt, for example, to a year or longer for vinegar or wine.

Specific aspects
Sometimes a product is fermented only to promote the development of aromatic compounds. The taste which is appreciated in coffee, tea and cocoa is specifically added by fermentation of the fresh product before further processing.

3.3 REMOVAL OF PATHOGENIC AND SPOILAGE MICROORGANISMS

Technology
Microorganisms are removed by special bacterial membrane filters (microfiltration). This is used for fruit juices because the smell and taste would be adversely affected if other preservation techniques, such as heating, were used. It is also possible to

remove microorganisms by centrifugal force from, for example, milk, which is indeed done in cheese-making (bactofugation). Here again this is done because the retention of the taste and special desirable characteristics is important.

Principles of extending the keeping qualities
The pathogenic and spoilage microorganisms are removed.

Sensory changes
No noticeable changes.

Storage and storage time
These vary for each product. Moreover, further processing plays an important role. Filtered fruit juices under refrigeration may only be kept a few weeks.

3.4 RETARDING THE ACTIVITY OF ENZYMES AND MICROORGANISMS

Under this heading come a number of preservation methods such as:

— lowering the temperature
— changing the pH
— reduction of the water activity
— removal of oxygen
— addition of antimicrobial agents.

Lowering the temperature
The temperature is lowered by *chilling* and by *deep-freezing*.

(i) Chilling
Technology
The temperature of the foodstuff is lowered to be within the range of $+10°C$ to $-1°C$. This can be done simply by putting the product outside in the cold, or with melting ice, as is used for fresh fish. Or an artificial system can be used whereby the pressure on a gas compressed to a liquid is released and the liquid gas evaporates. The heat needed for this evaporation is drawn from the surrounding space, which as a result becomes cooler (refrigerator). Freshly harvested products picked in a warm atmosphere are often immediately cooled in cold-water baths or sprinkled with water. The evaporation which then takes place uses the heat in the product, which is cooled as a result. (This technique is used on greenhouse lettuce and fruit.)

Principles of extending the keeping qualities
The metabolism and growth of microorganisms and also enzymatic activity are retarded.

Sensory changes
These are very small.

Storage and storage time

This varies, depending on the nature of the product, from a few days (fresh meat, fish, some kinds of vegetables and fruits) to half a year (potatoes and apples). Of course the quality of the product before it is cooled is an important factor in cold storage. Also the bacterial count is determined before the storage time begins. A high initial bacterial count will shorten the storage time considerably.

(ii) Deep-freezing

Technology

The temperature of the food is lowered to $-18°C$. Most kinds of vegetables are blanched (2–6 minutes) with steam or boiling water before being frozen.

The following are the most important advantages for blanching before freezing.

— Enzymes, which later during storage could cause loss of quality, are inactivated.
— The bacterial count is decreased.
— Leafy vegetables shrink and so they are easier to pack and to freeze.
— The green colour (chlorophyll) is fixed by the removal of air from the tissue surfaces, as a result of which, beans and especially peas from the deep-freeze have a bright green appearance.

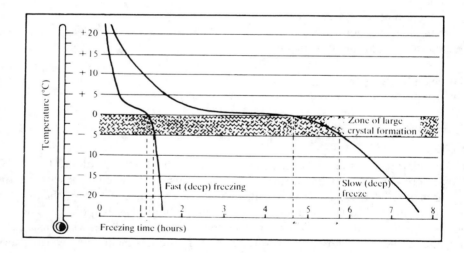

Fig. 3.2 — Ice-crystal formation during fast and slow freezing. (Source: Diepvries, Unilever, 1975.)

Freezing must take place rapidly. The temperature zone where ice-crystal formation occurs lies between $0°C$ and $-4°C$. If one passes through this zone slowly, then large ice crystals are formed. These damage the cell wall so that, after thawing, the cell contents run out and a limp, soggy product results. Fluctuating storage

Sec. 3.4] Retarding the activity of enzymes and microorganisms

temperatures (even if they stay well below zero) also induce larger ice crystals to form, which damage the cells.

Rapid freezing is usually achieved by one of two main methods. Either the packed product is held between two extremely cold plates (contact- or plate-freezing) or the product is held in a very cold air stream (blast-freezing). Another method, involving contacting the food with liquid nitrogen, is becoming more popular.

Contact-freezing
A contact- or plate-freezer is a cooled chamber with hollow metal 'shelves', through which a cooling liquid circulates. The temperature in the chamber is about $-33°C$. The products, packed in flat cartons, are placed between the 'shelves' so that they make contact above and below with them. The freezing time depends on the thickness of the packages, but is about two hours. This method is applied to flat packs of fish and fish products, meat and pastries.

Blast-freezing
Small loose products such as peas, beans or shrimps go on a perforated conveyor belt through an insulated tunnel. At the same time, air at $-40°C$ is blown from below through the perforations. The products are stirred and lifted by this cushion of air and so do not freeze together. Blast-freezing by this method takes 3–8 minutes. The products are then packed.

Packed products such as puddings, gâteaux or chickens go on a conveyor belt to a chamber cooled by cold air. The belt and its contents move in a counter-blast of air at $-30°C$. The speed of the belt varies with the size of the product. This method of freezing takes two to three hours. Sometimes called 'giro-freezing'.

Principles of extending the keeping qualities
The metabolism and the growth of microorganisms is greatly retarded, as is enzyme activity.

Sensory changes
These are small. The thawed product will lose quantities of cell moisture, and so become more limp. One calls this loss 'drip'. Fat-containing products may develop a slightly oxidized (rancid) taste (oxidation continues, but extremely slowly).

Storage room and storage time
If the temperature of the freezer is held constant from the start at $-18°C$ the storage time varies from six months for products with a high fat content to two years for vegetables and fruit. If storage at $-18°C$ is not possible then the storage time will be shorter.

The temperature which can be maintained in the freezing compartment of a refrigerator is categorized by a star rating system.

$$*** : -18°C$$
$$** : -12°C$$
$$* : -6°C$$

Deep-frozen products cannot be stored indefinitely, because certain reactions continue even in frozen materials, though of course at a slower rate. Some examples of this are:

- the oxidation of fats, resulting in the product becoming rancid.
- the denaturation of proteins, by water separating out. After thawing, much of this water is lost, and the product is dry and fibrous. This occurs occasionally with frozen meat and fish.
- the crystallization of lactose or sucrose in ice cream, which results in a gritty texture.

Changing the pH

Technology
A quantity of acid (for example acetic acid or lactic acid) is added to the foodstuff. The minimum amount required for preservation is 4 per cent acid. To achieve this, one must start with a higher concentration, because the acid draws water from the product and is diluted.

Principles of extending the keeping qualities
The activity of the spoilage organisms is slowed down by the low pH (acidity).

Sensory changes
The acidified product is changed in taste, colour and consistency; for instance, pickles are softer than the original product. Because the product is often unacceptable for consumption with an acidity of 4 per cent, less acid is used. The shelf-life is thereby shortened. This is overcome by combining the acidification with other preserving methods, such as the addition of a preservative, refrigeration or pasteurization.

Storage and storage time
Protein-rich foodstuffs (meat or fish) can be stored for one to two months. With longer storage, the protein leaches out and the liquid becomes turbid owing to the flakey protein deposit. Vegetables can be stored for six months to one year. Once the containers have been opened, acid products should preferably be stored in the refrigerator.

Reduction of the water activity
Reduction of the water activity can be achieved by *drying*, whereby moisture is removed from the product, or by combining the free moisture with *sugar* or *salt*.

(i) Drying

Technology
Moisture is removed from the product usually with aid of energy (heat), but sometimes by using low temperatures. The energy can be natural (heat of the sun) or artificial. Drying in the sun or in the shade is one of the oldest methods of preservation. Fruits and nuts grown in hot climates are still dried in this way.

Sec. 3.4] Retarding the activity of enzymes and microorganisms

In drying with artificial heat-sources one distinguishes between:

— drying with hot air
— drum-drying
— spray-drying.

Drying with hot air
Solid products like fruit and vegetables are placed on shelves in chambers through which hot air is blown. Often the pressure is reduced to increase the rate of evaporation. The air absorbs the moisture and is extracted. A special way of drying with warm air is the 'fluidized-bed-drying', by which the product is kept moving by blowing hot air in from underneath. In this way the product dries more evenly and faster.

Drum-drying
In this method the product is fed as a thick paste in between two very hot rotating cylinders, which drives off the moisture. The solids remain as a thin film on the cylinders. This is later scraped off, pulverized and packaged.

Spray-drying
This method is used for liquids. The liquid is sprayed from above or below into a tall tower, while dry hot air is blown into the tower. The dried product falls to the bottom and is collected. The moist air is extracted from the tower.

Drying at low temperatures can also be done naturally or artificially. Preservation of fish and meat in the icy outside air has long been known to Eskimos and Scandinavians (stockfish). Artificially, drying by means of cold can be done with cold air and freeze-drying.

Cold-air drying
Artificially cooled air is led into chambers where the products are spread out. Reduced pressure or vacuum is often used to increase the rate of evaporation. This process is used for vegetables.

Freeze-drying
The product is first frozen and then exposed to very low pressure, which causes the temperature to fall to between $-5°C$ and $-15°C$. Energy is provided which transforms the ice directly from the solid phase to the gas phase (water vapour), without first melting. This is called sublimation.

Principles of extending the keeping qualities
The metabolism and growth of microorganisms is slowed down by extraction of water (reduction of the water activity, a_w).

Sensory changes
The volume of the product is reduced and the original texture disappears. If drying has taken place with hot air, the product is possibly wholly or partially cooked.

Because enzyme activity continues during the drying, changes in colour and taste occur, especially with fruit. This can be prevented by blanching the product before it is dried. Sulphur dioxide is also used to prevent discolouration. Throughout drying, some of the volatile aromas are driven off, which results in the dried product having less flavour than its fresh counterpart. With some products, the volatiles can be trapped, concentrated and later re-added (fruit juices). Sometimes there is an advantage in letting a certain enzymic reaction occur before drying takes place, with the aim of creating 'flavour precursors'. These are compounds from which the characteristic aromas of the product are formed (for instance in the processing of cocoa from cocoa beans).

In the freeze-drying of products, the cell structure stays intact. The end product has about the same volume as the original product. It has a greatly enlarged surface area and is very porous. This has the advantage that the product quickly and easily absorbs water (80–100 per cent of the original water content).

Storage and storage time
The final moisture content determines the storage time. In sun-dried products this varies from 10–35 per cent. Freeze-dried products contain only 2 per cent moisture.

Dried products are sensitive to the effects of oxygen, light and moisture, from which oxidation, discolourization and lumpiness can result. Because of this, the packaging must be light-proof, air-tight and moisture-proof; freeze-dried products are sometimes even vacuum-packed to extend the shelf-life. Storage times vary from half a year for products with a high fat content to a year at least for other products.

(ii) **The addition of sugar**

Technology
Preservation through the addition of sugar is mainly used for fruit and fruit juices. The product is brought to the boil, which dissolves the sugar and creates a more viscous solution. The volume is also reduced during the heating process. For effective preservation, the final percentage of sugar needs to be 60 per cent at least in jam and 70 per cent in syrups. For fruit crystallization, fruit is consecutively boiled in increasingly more concentrated sugar solutions. Crystallized fruit contains 65–80 per cent sugar.

Principles of extending the keeping qualities
The sugar takes up the free water, as a consequence of which there is no moisture for microorganisms (particularly yeasts) to grow in. In addition, the sugar draws out water from the microorganisms.

Sensory changes
The consistency becomes more syrupy the more sugar that is used. The taste, and sometimes the colour, also changes. Crystallized fruit becomes slightly transparent.

Storage and storage time
Sealed from moisture and air, the storage time is six months to one year.

Sec. 3.4] Retarding the activity of enzymes and microorganisms

(iii) The addition of salt

Technology
Dry salt (NaCl) or salt solution (brine) is added to vegetables, meat or fish. For effective preservation, a final percentage of salt of at least 33 is necessary. Usually, less salt is added in practice and it is combined with other techniques such as cooling, smoking or drying.

Principle of extending the keeping qualities
Salt takes up the free water, as a consequence of which there is no moisture for microorganisms to grow in. In addition, salt draws out water from the microorganisms.

Sensory changes
The product loses some of its moisture and becomes more dense. The taste and, to a lesser degree, the colour changes. Because of the extremely high salt content the product has to be rinsed before use (for instance, salted runner beans).

Storage and storage time
Depending on the salt concentration, salted products can be kept for one to six months, without refrigeration, preferably in a dark place and sealed from the air.

Removal of oxygen
Removal of oxygen (complete or partial) can be done by *vacuum-packing* or by storing in a package in which the air composition is altered (*gas-packing*).

(i) Vacuum-packing

Technology
The fresh or prepared product is wrapped in an air-tight packaging from which virtually all the air is removed. As a result of the reduced pressure the wrapping clings to the foodstuff.

Principles of extending the keeping qualities
The activity of aerobic microorganisms is slowed down by excluding the air. In addition, biochemical reactions such as oxidation processes (for example, fat becoming rancid) are slowed down.

Sensory changes
None, the characteristic aroma is retained.

Storage and storage time
Vacuum packaging is mainly used to retain the aroma and to lengthen the shelf-life somewhat. The storage time and storage will depend on the product and its pre-treatment. For example: meat products about one to four weeks in the refrigerator, coffee about six months without refrigeration.

(ii) Gas-packing

Technology
There are two techniques. In one, the product is 'rinsed' with gas to displace oxygen; in the other, air is removed under vacuum and then a gas mixture is pumped in. The composition of the gas mixture is dependent on the nature of the product. For example, carbon dioxide at a concentration of 10 per cent reduces bacterial activity. Nitrogen is chemically inert and, because of this, particularly suitable, especially when a vacuum is not desired. In comparison with, for instance, the use of chemical preservatives, packaging under gas is an expensive method.

Principles of extending the keeping qualities
When oxygen is removed, aerobic microorganisms cannot grow any more. Furthermore, in the absence of oxygen, the oxidation of fats is sharply reduced.

Sensory changes
No noticeable aroma or taste changes.

Storage and storage time
Gas-packed products can be stored without cooling. Depending on the product the shelf-life varies from a few weeks for fruit juices for instance, to several months, for crisps and peanuts for example.

Specific aspects
The storage of, for instance, fruit and potatoes (living products) in special stores with controlled air composition, moisture and temperature is based on the same principles as gas packaging. This method is also called the CA method (Conditioned Atmosphere or Controlled Atmosphere).

Addition of compounds with anti-microbial activity
In this group of compounds are the constitutents of *smoke*, *alcohol* and *chemical preservatives*.

(i) Smoking

Technology
The foodstuff is hung in the smoke of a smouldering woodfire in a smoke-box or tunnel. One differentiates between:

— *warm*-smoking, in which the temperature can reach up to 60°C for meat and 85°C (=steaming) for fish for a short time (a few hours);
— *cold*-smoking, in which the temperature for meat is about 18°C and for fish about 30°C for a long period (12–36 hours).

The preservative action of smoking is a combination of factors:

— removal of moisture because the products to be smoked are first salted (cold-smoked products heavily, warm-smoked products lightly);

- moisture loss through drying during smoking;
- bactericidal and bacteriostatic action of the smoke constitutents (among other things phenols, formaldehyde, formic and acetic acids), which are deposited on the surface of the product.

Principles of extending the keeping qualities
The metabolism and growth of microorganisms are slowed down by the extraction of moisture. In addition, the smoke constituents have a weak preserving action.

Sensory changes
Cold-smoked products are recognizable by a strong smoked taste, a yellow-brown colour, fairly salt and dry flesh which is still raw. Warm-smoked products are cooked, juicy and taste lightly salted. The colour is yellowy and the smoked taste very weak.

Storage and storage time
Cold-smoked products can be stored for a few weeks to three months, the warm-smoked products from a few days to two weeks. Preferably they should be both wrapped and stored under refrigeration. The storage time can be extended by vacuum packaging.

Specific aspects
Certain compounds are evolved in the smoke with a toxic (carcinogenic) action, for example benzopyrenes.

Wood for construction work has often been treated with a fungicidal preservative, such as pentachlorophenol. Extremely poisonous dioxins can be formed from this during smoke generation. Woodshavings or sawdust from sawmills or furniture factories may not be used for the smoking.

Smoked meat products (raw bacon and dried sorts of sausage) during storage at high temperatures and under moist conditions can develop moulds on the surface from mould spores, which originate from the wood used for smoking. To prevent this, the sausages are treated on the outside with mould-preventing pymaricine.

Foods can also be given a smoked colour and smell by means of synthetic additives.

(ii) **The addition of alcohol**

Technology
Alcohol is added to fruit, often combined with sugar. A final minimum value of 15 per cent alcohol is necessary to ensure preservation. However, alcohol of a higher percentage needs to be used for preservation of fruit, because of the dilution effect of the water which is drawn from the fruit (for instance, alcoholic spirits such as brandy of 30–50 per cent alcohol are often used).

Principles of extending the keeping qualities
Alcohol draws water from the product and from the microorganisms. Alcohol retards the metabolism and the growth of microorganisms (compare its disinfectant use in medicine).

Sensory changes
The consistency, the colour and the taste change, depending on the quantity and the type of alcoholic spirit used.

Storage and storage time
For well-sealed products stored in the dark the storage time is six months to one year.

(iii) **The addition of preservative compounds**
One distinguishes between:

— compounds which prevent microbic spoilage (for instance sulphite (sulphurous acid), benzoic acid and sorbic acid);
— compounds which protect the foodstuff against oxidation (anti-oxidants like ascorbic acid and tocopherol (see section 4.2).

Technology
The compounds are added to the raw or prepared product in (mostly small) quantities, which are regulated by legislation.

Principles of extending the keeping qualities
The growth of microorganisms is slowed down by adding these compounds. Antioxidants prevent oxidation by being themselves more readily oxidized than the product.

Sensory changes
No noticeable changes.

Storage and storage time
Preservation compounds are often used in combination with other methods. Storage room and storage time vary with the product.

3.5 KILLING MICROORGANISMS

This can be done in two ways:

— by raising the temperature (*pasteurization* and *sterilization*);
— by irradiation

Raising the temperature

(i) **Pasteurization**

Technology
By means of steam or hot water the product is heated to 70–90°C, after which it is quickly cooled. The apparatus in which the heating takes place is called a pasteurizer and consists of metal pipes or plates which are heated. The product to be treated is run through the pipes or over the plates. After cooling it is packaged, for instance

milk in cartons, beer in patented spring-clipped stoppered bottles and fruit juice in aseptic packs. Pasteurization can also be done in the sealed package, as with canned fruit, Gelderland smoked sausage in plastic, and beer in bottles with crown caps. The main reason for pasteurization is to kill pathogenic microorganisms and to inactivate enzymes. This preservation method is also chosen if temperatures above 100°C would impair the quality of the product (for instance fruit) or when other methods are used in conjuction with pasteurization, for instance refrigerated storage or the addition of preservatives.

Principles of extending the keeping qualities
Pathogenic microorganisms are killed. Microorganisms which are not very thermo-resistant are also killed; but not all the enzymes are inactivated.

Sensory changes
Smell, colour, taste and consistency alter very slightly, depending on the time and temperature of the pasteurization.

Storage and storage time
Products which are pasteurized in the package can be held in store for half to one year or longer, often without refrigeration. The shelf-life of bulk-pasteurized products, with or without subsequent aseptic packaging, varies from a few days (milk) to a year (fruit juice). Pasteurized milk and milk products should be refrigerated.

(ii) Sterilization
Technology
The product is heated to temperatures in excess of 100°C (115–140°C) at which nearly all microorganisms and their spores are killed; certainly the pathogenic microorganisms are eliminated. The products, for example vegetables or meat, are sterilized while immersed in a liquid (brine or sauce) in their container (can or jar). Before the container is sealed, the contents are heated, which expands them and helps remove air. Then the container is sealed, sterilized and cooled, which leaves the jar or can under a partial vacuum. Another way to create a vacuum is to use a vacuum pump just before sealing the can. Then the can is sealed and container and contents sterilized under pressure (at 120°C) in an autoclave. A rapid cooling follows thereafter. A short sterilization period at a higher temperature (140°C) is the ideal method, since this gives a better quality. Complete sterilization is not achieved, the term 'commercial sterility' being used. That means that under normal storage conditions found in Northern Europe there is only a very slight chance of spoilage. If, for instance, the product is to be exported to the tropics, then the sterilization time must be extended to enable the same guarantee to be given. In the home, the most common application of sterilization is bottling. A new form of sterilization, which is being used for milk and fruit juice for example, is the UHT method (Ultra High Temperature), by which high pressure steam at about 140°C is used to heat the flowing product for a few seconds, after which it is cooled and aseptically packaged (see Chapter 9).

Principles of extending the keeping qualities
Virtually all microorganisms and their spores are killed and enzymes are inactivated. Also the product is hermetically sealed from the air so no re-contamination can take place.

Sensory changes
Depending on the duration of the process and the temperature to which it is heated, the product changes in smell, colour and taste. The consistency of solid products also alters (through being totally or partly cooked).

Storage and storage time
Sterilized products have an almost unlimited shelf-life without refrigeration. Because of taste considerations, care should be taken to use canned products within one year. The UHT sterilized products have a shelf-life without refrigeration of between three and six months.

Specific aspects
Calculations of the heating time and temperature are based on the resistance of *Clostridium botulinum*, one of the most dangerous pathogenic bacteria. The term used is the 'botulinum cook' (see section 2.4).

Irradiation

Technology
Ionizing rays (gamma rays and electron beams) are used for the irradiation of foodstuffs. Gamma rays from cobalt 60 and caesium 137 sources have great penetrating power and pass right through the product and its packaging. The advantage of this technique is that re-contamination is impossible. Electron beams, generated by linear acceleration, possess less penetrating power and are used for the treatment of products which are particularly prone to surface spoilage. The radiation dose or the absorbed dose used to be expressed in rads. The new unit is the gray (Gy). One gray is the energy absorption of 1 joule per kg tissue (1 Gy = 100 rad).

Principles of extending the keeping qualities
Ionizing rays posses a number of preserving effects depending on the dose absorbed. One differentiates between:

— the insecticidal effect: killing insects, their larvae and eggs.
— the bacterial and fungicidal effect: killing spoilage or pathogenic microorganisms. Bacterial and fungal toxins are *not* destroyed by irradiation.
— the physiological effect: inhibition of sprouting in roots, bulbs and tubers by preventing cell division or delaying ripening in fruit by interrupting metabolic processes.
— the physical effect: reducing the cooking time or drying time.

Sensory changes
A moderate radiation dose does not cause noticeable changes; a fresh product stays fresh, a raw product raw. A frozen product can be treated without prior thawing.

With high radiation doses, undesirable odour, colour, taste and consistency changes can occur, especially in products with a high protein and fat content. These changes can be prevented by irradiating the product in an oxygen-free atmosphere or in a deep-frozen state.

Storage and storage time
The shelf-life of foodstuffs with good keeping qualities can be improved by irradiation. An irradiated product is sometimes indicated by the irradiation symbol (see Fig. 3.3). Food irradiation is not yet allowed in all European countries, although there is now an EC directive permitting its use.

Fig. 3.3 — Logo indicating irradiated food.

Specific aspects
An irradiated product does not become radioactive. However there is a reluctance by consumers to accept the irradiation process. In the Netherlands irradiation of foodstuffs takes place at the *Proefbedrijf Voedselbestraling* (The Food Irradiation Test Unit) in Wageningen and *Gammaster* in Ede. Industries seeking permission to use an irradiation process for foodstuffs are referred to the *Gezondheidsraad* (Health Council) whose decision on its suitability is final. Twenty-four applications have been permitted in the Netherlands, including treatment to inhibit chitting in potatoes and sprouting in onions and to extend the shelf-life of peeled potatoes, cut endive, shrimps, strawberries, refrigerated snacks and fresh chickens. A cold sterilization is permitted for meals of special patients. There is a Green Paper, *Irradiated goods*, arising from the Goods Law. In this Green Paper, laws and administrative rules are set out. At the moment labelling using a logo (see Fig. 3.3) or a written declaration is not yet obligatory for irradiated products. Food irradiation is still not permitted in many countries, but its use is being actively reviewed worldwide.

4

Additives and contaminants

4.1 INTRODUCTION

In foodstuffs components occur which have been added on purpose during the production, distribution and storage to improve the quality (additives) or which more or less come about by accident in the foodstuffs (contaminants).

It is to be noted that the quantities of these components found are relatively small, but that some components nevertheless have detrimental effects; for instance, they can cause an allergic reaction in the consumer. These effects are also dependent on the quantity consumed per day.

4.2 ADDITIVES

Additives are useful components, which are added to edible and drinkable products, with a specific objective in mind. According to the Commission of General Guidelines for Additives of the Food Council additives are useful compounds which normally are not used as a foodstuff nor as a characteristic part thereof, irrespective of whether the additive is nutritious or not. They are added at the stage of preparation, processing, handling, packaging, transporting or storaging of foodstuffs or their basic components, with a technological or sensory purpose.

It is characteristic of additives in general that only small quantities are added. The use of additives in food preparation is not new. Dyestuffs and spices were already used in antiquity. In contrast with present usage, in those days only natural components were used, while now synthetically prepared additives are also used. The application has extended to the foodstuffs industry. Industrially prepared products have sometimes to be stored for a certain time after preparation, while preservation by deep freezing or sterilization is not always possible. This largely justifies the use of preservatives, so that the consumer buys a sound product from the microbiological viewpoint. To maintain the present-day foodstuff range in a technically, economically and medically sound way and to enlarge it, the use of additives has become necessary for a large number of products. Certain additives are permitted in the storage of basic ingredients, but not in the processing of foodstuffs. What remains (residues) of the additives passes into the end product (carry over).

The additives can be classified on the basis of the aim in view. We may subdivide into:

— preservatives
— anti-oxidants
— starches and thickening and gelling compounds, emulsifiers and stabilizers
— dyes (colours)
— flavour (smell and taste) components
— sweeteners
— proving or leavening compounds
— flour improvers
— anti-coagulants
— anti-foam components
— gases.

Preservatives
These lengthen the storage time of products. Used in well-defined amounts, they slow down the growth of microorganisms. The result is that the chance of bacterial food poisoning and the formation of mycotoxins is decreased. The following are a few examples.

— Benzoic acid and sorbic acid are added with a view to extending the storage time of, among other things, shrimps, mustard and mayonnaise.
— Biphenyl sprayed on the outside of citrus fruits slows down the growth of moulds and, because of that, the formation of mycotoxins.
— Sulphurous acid (sulphite) blocks the enzymatic processes which cause browning in pre-processed potatoes and vegetables. Otherwise, the attractiveness of the product would be decreased. Depending on the pH value, sulphite also works selectively on the slowing down of the growth of bacteria, yeasts or moulds.
— Sodium nitrite in meat products has a bacteriostatic effect slowing down, among other things, the growth of *Clostridium botulinum*. It is a bonus that, at the same time, nitrite also influences the pink-red colour of meat products (see Chapter 5).

Anti-oxidants
These additives also function as preservatives to prolong storage life but are directed at the occurrence or prevention of oxidative spoilage (rancidity of fats and decolarization of vegetables and fruits). The following are among those used.

— L-Ascorbic acid or vitamin C, a 'natural' anti-oxidant; used in vegetable preservation, meats, wheat/flour; aim: prevention of discolouration.
— α-Tocopherol or vitamin E, a 'natural' anti-oxidant; used in edible oils and fats; aim: prevention of rancidity.
— BHT (butylated-hydroxytoluene), a synthetic chemical antioxidant; used in oils,

fats, soups and odour and taste compounds; aim: prevention of oxidative fat spoilage.
— Citric acid; used in, amongst other things, margarine, low fat margarine, mayonnaise, beer and lemonade syrups; aim: prevention of oxidative fat spoilage and prevention of discoloration. This acid also functions as a buffering agent.

The food industry sometimes adds two or more anti-oxidants, which work synergistically together. **Synergism** is the phenomenon that the effect of a combination of compounds is larger than can be expected on the basis of the characteristics of the individual compounds.

Thickeners and gelling aids, emulsifiers and stabilizers
These additives have an influence on the consistency of food and drinks. Thickeners increase the viscosity of liquids. Gelling additives give a jelly-like consistency to liquid products. Emulsifiers make sure of an even distribution of two or more immiscible compounds. They lead to the formation and stabilization of an emulsion.
Stabilizers can maintain an achieved consistency over a longer period, even if after that the product gets a number of further treatments. The industry uses, among others the following.

— Lecithin, as emulsifier and anti-oxidant in for example ice-cream, margarine, dough products and broths.
— Organic acids (edible acids) which increase the stability of products by their influence on the pH value, because they stabilize the thickening or gelatinization. Examples are: citric acid in desserts, lemonades, mayonnaises and jams; acetic acid in meat products and salad creams; lactic acid in sour cream, buttermilk and meat products.
— Pectin, as a gelling compound in jams.
— Gelatine and agar-agar, as gelling compounds in puddings and desserts.
— Gums, modified starches and cellulose compounds, as thickeners in viscous products. These products ensure a more stable consistency for long storage than the starch-containing wheat flour or potato flour.

The thickeners and gelling compounds will be discussed more in Chapter 16.

Dyestuffs
The aim of adding dyes is to make a product more attractive. The aim can also be to restore the natural colour, if this suddenly changes through technological processes. Some dyestuffs have been shown to have detrimental side-effects. Their use is a matter of controversy. Dyestuffs often used are:

— carotenoids, also provitamin A, in cheese, margarine, low fat margarine, oils and beverages;
— caramel in vinegar, beer, meat stock cubes, and caramel custard;
— cochineal in beverages;
— beetroot red in some kinds of tomato ketchup, beverages and pink-red coloured custard.

Flavour compounds

Just as with dyes, aromas and flavours are added to increase the attractiveness of a product in this case through modifications.

Natural aromas are obtained by pressing or by extraction from vegetable or animal products, or from fermented products.

Synthesized naturally occurring and artificial flavour compounds are obtained by chemical means. This group of additives is very large. Some frequently used aroma compounds (essences) in desserts are vanillin and caramel.

Fig. 4.1 — Additives listed on packaging.

Taste enhancers can also be added to the flavour compounds, for instance inosinates and glutamates. These compounds themselves have no taste or smell, but increase the taste of any such present. They are especially used in soups, stock cubes, and meat products.

Sweeteners

These compounds are used to replace sugars in sweet foodstuffs such as beverages, jams and sweets, and in products for diabetics. One differentiates between energy-supplying and non- or very little energy-supplying sweeteners (see Chapter 25).

Proving or leavening compounds
These compounds have a function in the processing of products, but not in the end product. To obtain more volume and an airy structure, carbonates are used (sodium bicarbonate or potassium carbonate) in cakes, pastry, gâteau mixes and biscuits, and yeast is used in bread and rolls.

Flour improvers
By the use of these compounds, the proving process is speeded up and the resultant baked bread and pastry is improved. A frequently used compound is L-ascorbic acid, which also has an anti-oxidizing action. L-Ascorbic acid is an example of an additive with a so-called double function.

Anti-coagulants
These additives decrease the moisture absorbancy which finely divided, dry products possess to a high degree. Silicates, for instance, ensure that instant coffee, icing sugar, salt and milk powder stay suitable for sprinkling and retain ease of dissolving.

Anti-foam compounds
The use of the compound dimentylpolysiloxane prevents the formation of foam during the industrial production of, for instance, jams, lemonades and broths.

Gases
The food industry uses inert gases to replace air, to obtain a longer storage time for certain products. Examples are nitrogen gas which is added to soups in hermetically sealed packages and carbon dixoide in lemonades. In aerosol packaging propulsive gases under pressure are used. This happens especially with cosmetic products and some creams.

4.3 LEGISLATION IN CONNECTION WITH ADDITIVES
The usage of additives is controlled by the Goods Law (Netherlands) or by Food Standards (UK). There are four main categories of additives, each regulated by means of a **positive list**. These lists mention all the additives which are allowed. They are based on the EC guidelines. The permitted additives have an E-number. For instance, caramel (E150) is on the positive list for Food Colours. The four lists cover:

— colours
— preservatives
— anti-oxidants
— emulsifiers and stabilizers.

There may also be rulings which cover purity and nomenclature. There is also a section on miscellaneous additives.

For flavourings there is no positive list. The variation in aromas is too large. Any rulings normally relate to the origin of the aroma and the maximum permissable quantities.

The Sweeteners Regulation governs the use of sugar-replacing sweeteners. There is a list of permitted sweeteners in many countries.

Covered by vertical resolutions, the use of additives is worked out specifically for each product group. In some basic foodstuffs (milk, meat and fresh vegetables), no additives are permitted; in a number of others (bread, flour and cheese) their use is restricted and stipulated in the standards for these foods.

Information and research

The most important criterion for allowing additives on the positive lists is that it should not endanger the consumer's health. The judgement on the harmfulness or harmlessness of an additive is done by determining, through toxicological research, what quantity can be safely taken. This value, the so-called 'acceptable daily intake' (ADI) expressed in mg per kg body weight, is the guideline. Absolute safety cannot be assured. (This also applies to foodstuffs which are prepared without additives: some products contain naturally occurring detrimental compounds. Think, for instance, of oxalic acid in leafy vegetables such as spinach and purslane.) The toxological research is, for a large part, conducted by such organizations as the Central Institute of Food Research (CIVO-TNO) (Netherlands) or the British Biological Research Association (BIBRA). Some people build up an over-sensitivity (allergy) to certain additives (e.g. tartrazine, E102). Some organizations maintain a databank in which all the data connected with allergies are systematically collected. Anyone can obtain information here with a view to devising a diet which is as free as possible from allergy-causing compounds.

In the Netherlands, information about the use of additives is given by the Chief Inspectorate for the Foodstuffs and the Checking of Goods, from the Ministry of Well-being, Health and Culture. An additives booklet gives a summary of all the additions to foods and drinks.

To conclude, it should be noted that the addition of additives which influence nutritional value has not been discussed in this chapter. Other rules may apply for these. If a producer, with a view to increasing the nutritional value, wants to add extra vitamins (for instance, A and D in margarine) or minerals (for instance, iodine in bread) to a product, he should seek permission from the Ministry of Health in his country.

4.4 CONTAMINANTS

As a result of the industrialization in our country and in the countries surrounding us, a large number of contaminating compounds are dispersed into the environment. Some of these arrive in our foodstuffs. Also, contamination can unintentionally occur in foodstuffs or their raw materials as a result of production, treatment, preparation, packaging, transport or storage. Such chemical contaminants can be detrimental to health at a certain dosage. The total quantity of contaminants deposited on and in foodstuffs during production and distribution is of no concern to the consumer. It is the residue which is present at the moment of consumption which is of importance.

The quantity of this residue depends on the nature of the compound. Some break down completely after a period of time; some do not, or break down very slowly (persistent compounds). Also compounds can build up in live cells (accumulative compounds).

Contaminants are usually subdivided according to point of contamination and/or aim of usage. They can be classified as:

— residues from pesticides
— residues from veterinary medicines
— residues from packaging and utensils
— contaminants from the environment.

Residues of pesticides
To contain or prevent sickness and disease, agriculture and horticulture uses, amongst other means, chemical pesticides. There are substances to combat insects (insecticides), moulds and fungi (fungicides), weeds (herbicides), and rodents (rodenticides). Similar substances are also used in the storage of harvested agricultural and horticultural products (stock protection).

Residues of veterinary medicines
This group contains compounds which serve to prevent and treat illnesses and disorders in animals. Also, compounds which enhance nutrition and growth, to obtain the required slaughter weight faster, belong in this group. There are also compounds which influence the behaviour of animals, namely tranquillizers. The improvement of growth and prevention of illness is obtained by means of daily doses of antibiotics, chemotherapeutics and hormones in the food and drinking-water. For the treatment of illness, incidental high doses of antibiotics are used. Residues of these compounds remain in the animal tissue. Such residues can be detrimental to the consumer.

Residues of packaging and utensils
In the preparation or consumption of food some substances which are used in the packaging and the utensils react chemically with the food. There are also substances from certain kinds of plastics, which diffuse into the foodstuffs. Materials which occur in food in this manner are called migration residues. Plastics are more susceptible to this than the more traditional materials like paper, cardboard, wood, cans and glass. Compounds from which the plastics are derived or which play a role in plastic production, migrate reasonably easily, especially into foodstuffs rich in fats and oils. Some of these compounds are carcinogenic.

Contaminants from the environment
The term environmental contaminant is applicable to materials of different kinds, which are deposited in and on the food from the surroundings. There are:

— heavy metals (mercury, lead, cadmium) and arsenic;
— nitrates, nitrites and nitrosamines;
— polychlorobiphenyls (PCBs);

— polycyclic aromatic hydrocarbons (PAHs);
— mycotoxins;
— radioactive particles.

Heavy metals and arsenic
Heavy metals (mercury, lead, cadmium) and arsenic are the cause of toxicological risks. They accumulate in the foodchain and exercise, one way or another, a detrimental effect through the food.

Mercury is used along with other compounds in fungicides to protect stocks of grain. It also contaminates surface water from industrial waste. Seafish, shellfish and crustaceans accumulate mercury in their tissues. However, the toxicity is decreased through the presence of proportional amounts of selenium. Mercury compounds are absorbed into human blood and can cause kidney damage.

Lead is often used as a non-rusting metal in alloys (packaging) and in the production of dyestuffs. A great deal of lead is also released into the environment from car exhausts. In the air, lead becomes attached to rain and dust. It is deposited then on to plants through lead-containing rain and atmospheric precipitation. The uptake by way of the roots is practially zero. Lead interferes with the production of haemoglobin in the human body.

Cadmium appeared in the environment after the beginning of this century as a result of zinc processing. Cadmium settles on the ground as a sludge. Plants, especially cereals, sweetcorn and tomatoes absorb cadmium easily. Animals absorb cadmium from cadmium-containing vegetable animal foods, in which it accumulates. In humans, it accumulates in the kidneys and liver, which can cause kidney and liver damage. Cadmium is carcinogenic.

Arsenic is found in fungicides and is used as an additive in animal foods. Fish, shellfish and crustaceans have a relatively high arsenic content in the form of arseneobetaine, regarded so far as a harmless substance. Arsenic has a carcinogenic effect.

Nitrates, nitrites and nitrosamines
Nitrates serve as a nutrient for plants. Increased nitrate contents are found in vegetables (especially leafy vegetables and greenhouse vegetables) as a result of the use of organic fertilizer. Nitrates are not themselves poisonous. However, during storage of foodstuffs or in the human body they can be changed into harmful nitrite compounds. Amongst other properties, nitrites obstruct the oxygen supply. Also, nitrites can form compounds with amines from protein-containing food. These nitrosamines are carcinogenic.

Polychlorobiphenyls (PCBs)
These compounds are found as a cooling liquid in transformers and condensers. It has recently been decided not to use these substances as such any more in the Netherlands.

More then 200 isomers of these persistent organochloro compounds exist. The PCBs disperse themselves throughout the environment and accumulate in animal fat tissue. High contents are known in fish rich in fat, such as eel, in badly contaminated surface water. Toxic effects in humans show as swelling of the eyelids and as an increase in tear secretion. Tiredness, sickness and vomiting can also be the results of increased uptake.

Polycyclic aromatic hydrocarbons (PAHs)
These compounds develop during the processing and preparation of food, especially during the roasting or heating of coffee, cereals, vegetables and tobacco and during the smoking of meat and fish. The largest quantities develop during the roasting of food on open charcoal fires. It has been proved that several PAHs, including benzopyrene, are carcinogenic.

Mycotoxins
These poisons develop during mould growth on and in foodstuffs and their raw materials. They are the cause of serious illnesses. Known in history is St Anthony's fire (ergotism), which is caused by ergot in rye. More recently, the carcinogenic aflatoxin B1 was found for the first time in mouldy peanuts. A similar toxin, aflatoxin M1, was found in cereals which were used as cattlefood. From such a source it can appear in cows' milk.

Radioactive particles
Radioactivity in food and in the surroundings is the result in the main of radioactive precipitation. Radioactive particles appear in the environment mainly as a result of test explosions of atom bombs, from radioactive waste from atomic enegy centres, or as a result of accidents in such centres (e.g. Chernobyl).

N.B.: The irradiation of foodstuffs with a view to preservation (see Chapter 3) does not lead to radioactivity of the treated products.

4.5 LEGISLATION CONCERNING CONTAMINANTS
These are aimed at decreasing the use of additives which contaminate food and lowering their residual contents.

Regulations to minimize contaminants in foodstuffs are to be found dealing with such areas as:

— residues (Pesticides Law)
— animal feed additives
— heavy metals (such as lead and arsenic)
— antibiotics
— packaging.

Regulations to prevent the contamination of the atmosphere are to be found for (amongst other topics):

— contamination of surface waters
— air pollution
— soil protection
— waste material disposal.

5
Meat and meat products

5.1 INTRODUCTION

Meat is taken to mean the flesh, including fat, and the skin, rind, gristle and sinew in amounts naturally associated with the flesh used, of any animal or bird which is normally used for human consumption, and includes the offals. Excluded from the above are other parts of the carcase, such as horn, hooves, claws, bristles, wool and skins of animals, except pigs. In this chapter attention is focussed on slaughter animals, which is taken to mean one-hooved animals (solipeds, such as horses, donkeys and mules), cattle, sheep, goats and pigs. Poultry and game are covered in Chapter 6.

Meat from all these animals may be sold fresh (chilled), frozen or processed into a wide range of products.

Meat has been appreciated as a tasty foodstuff since the oldest times. The quantities consumed have varied considerably according to the fashion of the time and place. A hot meal without meat is, for people in many countries, difficult to imagine. The many kinds of meat with their manifold ways of culinary preparation make a very varied nutrition possible. One understands under meat as a foodstuff mostly the muscles of slaughter animals; in a wider sense beside muscle, fat and connective tissue, blood, organs, glandular tissue and skin are included. For instance, if we buy lean meat this contains in proportion much muscle tissue and only a little fat tissue; in fatty meat the quantity of fat tissue can reach half the total.

Consumption

In most Dutch families (about 92.5 per cent), meat (including chicken) is eaten in a hot meal four or more times a week; 1–2 per cent of families never eat meat on principle. In the higher social classes, meat is eaten less often and more meat-replacing products such as fish, eggs, cheese, pulses and soya products are used.

The consumption per year for the Netherlands and UK is expressed in kg per head of the population (see Table 5.1). These figures exclude poultry meat (Chapter 6).

The weight break-down according to kinds of meat consumed (in kg per person) in 1985 was as follows (UK figures from Annual Abstracts of Statistics):

	(Netherlands)	(UK)
Beef	17.3	14.3
Pork	40.6	17.6
Veal	2.2	low
Lamb	0.2	5.2
Horse	1.7	—
Offal	2.6	n/a

About 20 per cent of the total meat consumption in the Netherlands is in the form of meat products. This approaches 13 kg per person.

Composition
Meat contains:

water	50–75%
proteins	18–22%
fat	2–35%
carbohydrates	—
minerals	}
vitamins	} 3.5%
other water-soluble substances	}

Water
The water is almost completely bound to protein-like materials. During preparation, the water content reduces by evaporation and coagulation of the proteins, which shrinks the meat. Besides this, the total shrink or shrink factor is also influenced by the rendering of the fat.

Protein
The proteins are bound both intra- and extra-cellularly. They occur **intracellularly** in the muscle tissue as protoplasm-proteins. The most important are:
— actine and myosine, which in the form of actomyosine are responsible for the contraction of the muscles;
— albumin, which is soluble in water;
— myoglobulin, which gives the muscle a pink to red colour and varies in quantity with the age of the slaughtered animal, the kind of animal, the way it is reared, the manner in which the muscle has been used and the pH.

The **extracellular** or structural proteins consist of collagen and elastin, also called connective tissue. A large quantity of connective tissue looks white and is indigestible by humans. After being heated for a long time the connective tissue takes up water and changes into soluble and easily digestible gelatin. A low pH speeds up the process. Connective tissue also connects muscle, fat and bone tissue with each other.

Table 5.1 — Consumption figures for meat and meat products

Year	Meat and meat products (kg per person)	
	Netherlands	UK
1982	68.2	43.0
1983	64.7	43.4
1984	64.4	42.4
1985	64.8	42.8

Fat
The fat is stored in special fat cells, but a small quantity also occurs in connective tissues. The size and the quantity of the fat cells varies with the condition and the age of the animal, and the kind of animal. The largest number of fat cells is found around the organs and just under the skin. This is called **depot fat**. The storage first takes place in the depots, and then around and between the muscles, the **extramuscular fat**, and finally in the muscle tissue, the **intramuscular fat**. The fat in and around the muscles improves the moistness and tastiness of prepared meat. The fat varies in colour and consistency depending on the kind of animal, the method of rearing, the age and the feed.

Carbohydrates
Glycogen is the carbohydrate that occurs in the muscles. This has little importance as nutritional material; however, this muscle sugar plays an important part in the hanging of meat (see section 5.5) and the taste of fresh meat. Horse meat contains more glycogen than, for instance, beef or pork and tastes somewhat sweeter.

Minerals
Iron occurs particularly in the dark red-coloured muscle tissue, in kidney and liver. Besides iron, meat contains a considerable amount of calcium, phosphorus, sodium and potassium.

Vitamins
Vitamins A and C occur especially in the liver and kidney; the vitamins of the B-group occur in all the kinds of meat, but most of all in pork.

Other water-soluble substances
Muscle tissue also contains end products of the protein metabolism, such as creatine and creatinine. These are nitrogen-containing water-soluble compounds. The amount of these compounds increases the more intensely a muscle is used. Together with other aromatic compounds which develop during the heating of meat, these water-soluble compounds give a taste and smell, appreciated by many, to meat stocks and meat.

Possible health hazards
Bladder worms can occur in the meat of pigs and cows. They are visible as transparent bubbles and are recognizable as such during meat inspection. After consumption of contaminated meat, a tapeworm can develop in the intestines of humans (see section 2.5). The bladder worm is made harmless by freezing and storing at $-10°C$ for 10 days. Heating of meat is also sufficient.

Slaughter animals sometimes get compounds administered during their life of which residues remain in the muscle or fat tissue. Some of these **residues** cause danger to the health of consumers of the meat. Some examples of such residue sources are the following.

Growth-improving hormones (anabolic steroids)
In some countries the use of the DES-hormone (diethylstilbestrol) among others is prohibited for fattening animals. This oestrogenic hormone accumulates in the muscle tissue of the animal. From 1st January 1988 the use of growth-improving hormones is prohibited in cattle-rearing units. Only three natural hormones (estradiol, progesterone and testosterone) can be used for therapeutic purposes. This holds for all the member EC countries.

Growth-restricting compounds
Pesticides and herbicides are used in arable farming, for instance on crops, which are then used as animal fodder. The residues can accumulate in the fat tissue of the slaughter animal and transfer to the consumer, also in the fat tissue. This sometimes produces allergies.

Veterinary medicine
This is administered as antibiotics to the fattening animals to combat infections, or as tranquilizers to keep the animals quiet. The use of antibiotics is governed by legislation. They can be administered via medicated animalfood, by which the danger of large residue is excluded, or by the veterinary surgeon. He uses larger doses to fight infections. Antibiotics can also be the cause of over-sensitivity (allergy).

5.2 THE SLAUGHTER PROCESS

On arrival at the slaughterhouse, the live slaughter animals are checked. In Holland, the regulations prescribe an obligatory rest period for tired animals. Pigs, especially, become very tired during transportation. They are hosed down with luke-warm water to quieten the animals down. This also removes remains of faecal matter and other contaminations. After the animal is pronounced sound, the slaughter process starts.

Stunning and killing
Ruminants and horses are stunned by the use of a humane killer. An iron pin is shot in the brain with great force. Pigs are stunned by an electric shock. After stunning,

the slaughter animal has to be bled (drained of blood) as quickly as possible. This is done by a cut straight in the heart or by cutting the carotid artery.

The blood of pigs is collected and used for the preparation of such products as black pudding.

Skinning or depilation
The skin of ruminants and horses is removed by machine, salted and sent to the tannery. The skin of pigs is depilated and cleaned and comes on the market as rind.

Cutting up
The stomach cavity of the hanging animal is opened, and the organs are taken out. The carcase is eventually cut in two. Now the second check takes place, after which the carcases are cooled as quickly as possible.

Ritual slaughtering
According to Jewish and Muslim law, Jews and Muslims are only allowed to eat meat from ritually slaughtered animals. For this, the animal is first checked ritually, and then bound fast. With an especially sharp and long knife the throat is cut in a to-and-fro movement, through which the large throat arteries and throat veins are opened, so that the animal quickly bleeds. According to the Jewish laws this can only be done by a Jewish official, who also does the checking. According to the Islamic Law the slaughtering can be done by any practising Moslim, and the head of the animal has to point to Mecca. Since 1979 ritual slaughtering in the Netherlands is governed by the Meat Inspectorate Division of the Meat Inspection Law. There, for example, slaughterhouses where ritual slaughtering is permitted are listed.

The inspection of meat
All animals for slaughter, whether by general or ritual slaughtering, have to be examined for soundness. This is regulated by meat inspection laws. In the Netherlands this inspection is performed by Inspectors of the Government Service for the Inspection of Cattle and Meat (RVV).

A slaughter permit is given for a healthy animal. Sometimes live inspection cannot take place, for instance when an animal which has died or been killed is delivered (1–2 per cent of all slaughterings). The cutting up of these animals takes place in a special emergency slaughter place. The post-slaughter inspection is then especially rigorous.

The inspection after the slaughter
During the butchering a simple inspection of the carcase and the organs is made. The colour, smell and consistency are noted. Cuts are also made in some organs, such as lungs, liver, pancreas, kidneys and lymph glands. If there are any doubts, the animals are hung separately in a cold store and later subjected to an elaborate investigation. This investigation involves more cuts in organs and muscles, a bacteriological investigation, a parasitological investigation and sometimes even a boiling and a frying test to try to detect possible departures from normal taste. Dependent on the findings an inspection declaration is issued. In the Netherlands these are as follows.

— **Passed** if no faults have been detected. A satisfactory animal can be cut up. It is stamped in several places. The colour is violet/blue for the meat of cows, pigs, sheep and goats. Horse meat gets a red stamp to differentiate it from beef. (See Fig. 5.1).

Fig. 5.1 — Passed stamp.

— **Passed under obligation** if the meat is harmful or unsatisfactory because of the presence of signs of illness or parasites, but can be made satisfactory by being treated by sterilization (heating above 120°C) or freezing. (See Fig. 5.2)

Fig. 5.2 — Passed under requirements for sterilization or freezing.

Freezing is carried out if meat parasites have been found, for instance bladder worms in pigs. The parasites are killed by keeping the meat for 10 days at a minimum of −10°C. This meat receives a round stamp.

— **Provisionally passed on condition of retail sale under supervision** if the meat is not harmful, but has a potential for quick deterioration, for instance from rapid spoilage as a result of inadequate bleeding of the animal and a high pH of the meat. This can occur in an emergency slaughter. This meat receives an oval stamp. It is sold privately under the supervision of inspectors in portions of 3 kg maximum. (See Fig. 5.3.)

Fig. 5.3 — Passed on condition of 'retail sale under supervision'.

— **Not passed** if the meat is thought to be harmful to human health. This meat receives stamps in the shape of a parallellogram with the inscription 'not passed'. This meat goes straight to be destroyed. It is processed into non-harmful and useful products (for instance animalfood). (See Fig. 5.4.)

Fig. 5.4 — Not passed.

— Another stamp besides the passed stamp is given for export meat (see Fig. 5.5). This stamp guarantees that the slaughterhouse complies with the organization requirements laid down and the requirements with regard to hygiene.

5.3 KINDS OF MEAT

Beef

In many countries cattle are kept for slaughtering and for milk production. Meat from old cows is less suitable for direct consumption, but is used in the production of

Fig. 5.5 — Passed for export.

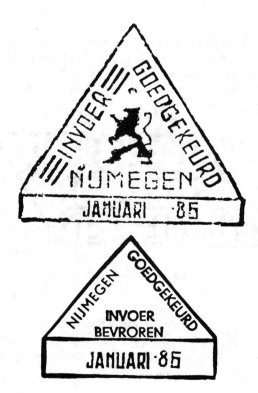

Fig. 5.6 — The meat imported into the Netherlands is rechecked. On passing, the meat receives one of the two stamps printed above.

meat products. The most palatable meat is produced by cows, steers and bulls which are specially reared for meat production.

Cows, mostly 2–4 years old, specially fattened, give an excellent quality beef.
Bulls, mostly slaughtered between 1.5 and 2 years, have tender and lean meat, with a slightly coarser fibre than meat from cows.
Steers, give fine meat when they are specially fattened and slaughtered young (1.5 years). This meat is rarely available to the trade.

Beef has a stone red to dark red colour. It is rich in smell and appetizing and contains many water-soluble compounds. The fat is light yellow and has a hard, somewhat brittle consistency. For the cuts of beef, see Fig. 5.7.

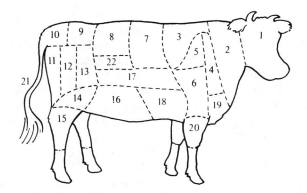

Fig. 5.7 — Cuts of beef. 1 Head. 2 Neck and clod. 3 Back rib, blade and chuck. 4. Chuck and clod. 5 Blade and chuck. 6 Top rib. 7 Fore rib. 8 Sirloin. 9 Loin and rump. 10 Tailpiece. 11 Silverside and topside. 12 Silverside and thick flank. 13 Thick flank. 14 Silverside and thick flank. 15 Leg of beef. 16 Thin flank. 17 Sirloin and flank. 18 Flank. 19 Brisket. 20 Shin. 21 Tail. 22 Sirloin.

Veal
Calves are slaughtered young. There are the following types.

— *New-born calves* These are slaughtered up to 10 days old (maximum). The meat is lean, moist, blanched (grey-pink) and not so good in taste. It is cheap and especially suitable for the meat products industry.
— *Grass calves* These, mostly male, animals are put to grass very young and get additional food of flour and perhaps of milk or skimmed milk. They are ready for slaughter at between 4 and 12 months. The meat has a light red colour, is firm and is mostly very much appreciated.
— *Fattened calves* In the Netherlands these are especially produced for export, to Italy and other countries. The animals are fed with skimmed milk and carefully formulated foods without iron. They also get little chance to move (box-calves). By these means, rapid growth is guaranteed as well as a blanched pink-coloured

Table 5.2 — Cuts of beef

Part	Application
Fore-quarter	
Neck and clod (2)	Minced meat, stewing, for soup
Back rib, blade and chuck (3)	Roasting, braising, stewing
Chuck and clod (4)	Stewing
Blade and chuck (5)	Stewing
Top rib (6)	Braising, roasting
Fore rib (7)	Roasting
Flank (17)	Stewing
Flank (18)	Boiling, braising
Brisket (19)	Braising, roasting; pickled: boiling
Hind-quarter	
Sirloin (8)	Roasting
Loin and rump (9)	Roasting, frying, grilling, braising
Tailpiece (10)	Roasting, braising
Silverside and topside (11)	Roasting, braising
Silverside and thick flank (12)	Braising, roasting, stewing
Thick flank (13)	Braising, roasting, stewing
Silverside and thick flank (14)	Braising, roasting, stewing
Leg of beef and shin (15 and 20)	Stewing
Thin flank (16)	Boiling, stewing
Tail (21)	Boiling
Sirloin (22)	Roasting

meat with little connective tissue. The meat is tender and moist. The calves are slaughtered at between 3 to 5 months.

Veal varies in colour from pink to light red. The taste is delicate, but scarcely pronounced because it contains few water-soluble compounds. It is tender and easy to digest, has a high moisture content and contains little fat. The colour of the fat is white and the consistency is fairly hard. For the cuts of veal, see Fig. 5.8.

Pork
Pigs are slaughtered young, so that they produce tender meat. The special piggeries are often completely automated (bio-industry). There are the following categories.

— *Suckling pigs* These are slaughtered when they are 5–6 weeks old. They are mostly roasted whole on the spit.
— *Bacon pigs* This special breed, with a slightly longer back produces an excellent quality bacon for export to England and America. They are about 5–6 months old when they are slaughtered.

Kinds of meat

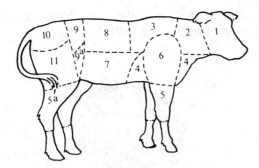

Fig. 5.8 — Cuts of veal. 1 Head. 2 Neck. 3 Best end of neck. 4 Breast. 5 Shin (shoulder). 5a Shin (leg). 6 Shoulder. 7 Breast. 8 Loin. 9 Loin. 9a Loin. 10 Leg. 11 Leg.

Table 5.3 — Cuts of veal

Part	Application
Fore-quarter	
Neck (2)	Stewing
Best end of neck (3)	Braising, roasting
	Chops: frying, grilling
Breast (4)	Stewing; stuffing and roasting
Shin (5)	Boiling (soup)
Shoulder (6)	Braising, roasting
Hind-quarter	
Breast (7)	Stewing; stuffing and roasting
Loin (8,9 and 9a)	Braising, roasting
	Chops: frying, grilling
Leg (10 and 11)	Frying, grilling, boiling, braising, roasting

— *Slaughter pigs* There are several breeds which produce good quality meat. They are slaughtered at an age of about 6 months. The taste and the firmness of the meat is dependent on the breed, but can be influenced by the feed (for instance cornflour gives flabby meat; fishmeal gives a bad taste).
— *Free-range pigs* These free-range pigs are reared according to the guidelines laid down by the appropriate legislative body. Some difference in colour (darker) and taste (stronger) may be expected in free-range pork in comparison with pork from the intensive pork industry. This meat also contains less moisture so is more substantial. The guidelines concern the production (housing, care and feeding), processing and sale of the less intensively kept pig.

Butchers who sell free-range pork in the Netherlands are allowed to use the logo of the Interim Commission (see Fig. 5.9).

Fig. 5.9.

Pork varies in colour from pink to light red, possibly in the same muscle. The meat is very finely veined with fat, which is sometimes visible and contains very few water-soluble compounds. The fat is white in colour and the consistency is soft. For the cuts of pork, see Fig. 5.10.

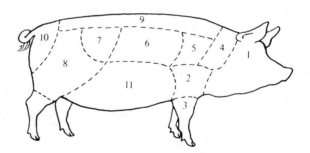

Fig. 5.10 — Cuts of pork. 1 Head. 2 Hand. 3 Trotters. 4 Spare rib. 5 Blade. 6 Loin. 7 Loin. 8 Leg. 9 Loin. 10 Tail and tailpiece. 11 Belly. 4,5,6 and 7 Chops.

Lamb and mutton
In the Netherlands, sheep-farming takes place especially in North Holland and on the 'Wadden' islands. The consumption of lamb and mutton is very slight in the Netherlands, and because of this the availability is often poor. The distinctive flavour of this meat could be a reason for its lack of popularity. The special lamb and mutton from Texel is very much appreciated by connoisseurs. The salt taste of the grass produces meat with a fine taste. It is for the most part exported to France.

Depending on the age of slaughtering the following varieties are available.

Kinds of meat

Table 5.4 — Cuts of pork

Part	Application
Head (1)	Roasting as boar's head; salted: boiling
Hand (2)	Roasting; salted: boiling
Trotters (3)	Boiling (soup)
Spare rib (4)	Roasting, frying, grilling
Blade (5)	Roasting
Chops (6 and 7)	Frying, grilling
Leg (8)	Roasting; salted: boiling
Loin (9)	Roasting
Tailpiece and tail (10)	Roasting
Belly (11)	Roasting; salted: boiling

— *Suckling lamb* This is the meat of lambs of about 10 weeks old. There is a demand, especially from Jews around Easter.
— *Lamb* This is meat from lambs slaughtered between 3 and 9 months. It is available fresh in the Netherlands between the end of May and the end of September. It is imported deep-frozen from Australia and other countries. Lamb can be expensive.
— *Mutton* This is the meat of sheep of 9 months and older. It has a stronger taste than lamb, contains a fair amount of fat, and is lower in price.

Lamb and mutton have a colour from light to dark brown-red. The taste is distinctive, for which caprylic acid, which occurs in the fat, is mostly responsible. The fat is white in colour, the consistency is very hard and crumbly. For the cuts of lamb, see Fig. 5.11.

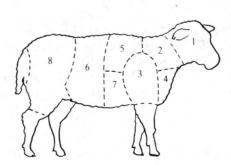

Fig. 5.11 — Cuts of lamb. 1 Head. 2 Middle neck and scrag. 3 Shoulder. 4 Breast. 5 Best end of neck. 6 Loin. 7 Breast. 8 Leg.

Table 5.5 — Cuts of lamb

Part	Application
Middle neck and scrag (2)	Stewing
Breast (4 and 7)	Stewing; stuffing and roasting
Shoulder (3)	Braising, roasting
Best end of neck (5)	Braising, roasting; chops: frying, grilling
Loin (6)	Braising, roasting; chops: frying, grilling
Leg (8)	Boiling, braising, roasting

Goats' meat
In the Netherlands goats' meat is little eaten, in contrast to in Spain and Indonesia, for instance. There, the meat of young goats is particularly relished.
  Goats' meat looks and tastes a little bit like mutton, but is less fatty.

Horse meat
On the continent, to slaughter horses a butcher needs a special business licence. Formerly, horse meat was only sold at the specialist horse butchers. Now horse meat can be sold alongside other kinds of meat in the normal butcher's shop. For a layman it is hard to distinguish horse meat from beef.
  On the Dutch market the supply of horses for slaughter is small. They are imported from the East European countries. Young meaty animals are preferred. In many countries, including the UK, horse meat is currently not very popular.
  Horse meat has a dark red colour, which becomes somewhat black as the surface dries. The taste is slightly sweet, because the muscle tissue has a higher glycogen content. The fat is bright yellow to orange in colour and has a soft consistency.

5.4 ORGANS
Heart
The heart is a hollow muscle, which consists of very fine fibred tissue. The top is often covered with hard fat. The heart has some different properties from the rest of the meat of slaughtered animals; it contains very many water-soluble compounds and connective tissue and is very lean. However, many consumers have a psychological objection to using the heart, and because of this it is especially used in meat products, mince or pet foods.

Brains
Calves' brains are seen as a delicacy by some gourmets. They are grey-coloured, lobe-shaped organs, which feel soft to the touch.

Liver
The liver contains about 20 per cent protein, 5 per cent fat and a high amount of iron, and vitamins A, C and B. The organ is divided into lobes, the number and the shape

of which are specific for the species of animal. The consumer shows a preference for pigs' and calves' liver. They are softer and tastier then ox liver. The meat has a brown-red colour and feels firm. Pigs' liver is recognizable by the light rings on the outside.

Lungs
Lungs consist of sponge-like tissue. They are not as such eaten, but are processed into meat products.

Spleen
The spleen is a long organ, which has the same structure as liver. The colour is also brown-red. The spleen is mainly processed into meat products.

Kidneys
Kidneys are smooth and firm to the touch. The weight varies between 150 and 500 g. The colour is brown-red. The smooth pigs' kidney (bean-shaped) and the lobed calves' kidney are the most popular. Ox kidneys are larger, darker in colour and less tasty. They are especially used in sausage-making. Kidney has a high nutritional value, 17 per cent protein, a high content of iron, and vitamins A, C and B, and with only 6 per cent fat is one of the leaner kinds of meat.

Tongue
Tongue consists of loose muscle tissue, with a textured, tough skin. Ox tongue, varying in weight from 2 to 4 kg, is sold fresh; but it is also processed into meat products (smoked, cooked tongue). The cooking liquor of the tongue makes a tasty stock, particularly if the throat piece has been used as well. Calves' tongue, with a weight of 0.75 to 1 kg, is especially sold fresh. The taste is more delicate than ox tongue. Pigs' and sheeps' tongues are eaten fresh, and also processed into meat products (for instance in tongue sausage).

Sweetbread
The sweetbread of calves are also sold. They are classified as neck sweetbread, which is fairly large and oblong in shape, and heart sweetbread, which is smaller and fatter. Calves' sweetbread are white in colour, soft and delicate in taste, and classified among the lean kinds of meat.

5.5 HANGING AND MATURATION

Meat consists of muscle, fat and connective tissue. The muscle tissue is built up of muscle fibres, which are surrounded by connective tissue. The fibres lie grouped in parallel into muscle bundles. Several muscle bundles are in turn bound by connective tissue, the muscle sheath, into a muscle. The muscle sheath changes at both ends into tendons. With tendons, the muscles are attached to each other or onto the skeleton. A thicker connective tissue develops in an active muscle than in a little-used muscle. This is sometimes clearly visible as so-called sinews. Much-used muscle tissue also becomes more substantial, coarser and more fibrous. The muscle bundles out of which a muscle is built extend the whole muscle length. If the butcher cuts the meat along the length of the muscle, 'with the grain', then the long muscle fibres will make the meat stringy and tough.

Meat has no flavour immediately after the slaughter of an animal. The muscles are still supple and elastic and react to stimuli. They also still contain fuel in the form of glycogen and the energy-rich phosphates adenosine-tri-phosphate (ATP) and creatine phosphate (CP). The pH in the muscles has a value of about 7.2.

After death, first the energy-rich phosphates are broken down into creatinine, ammonia and several nucleotides, such as adenosine-di- and- mono- phosphate (ADP and AMP). These compounds are an important part of the flavour components and therefore the taste of the meat.

The stores of ATP and CP can vary greatly, depending on the function of the muscle and the state in which the animal is when it is slaughtered. Muscles which at any given moment and in a very short time can do large quantities of work have a large store of ATP and CP and little glycogen, for instance the muscles in the legs. Muscles which are in action for a long period, or even constantly, but do less heavy work contain more glycogen and less ATP and CP, for instance the heart muscle and the jaw muscles.

Rigor mortis
As soon as the energy-rich phosphates are broken down, *rigor mortis* (stiffness) sets in. The time can vary from between one and twenty-four hours after the death (depending on the species of animal), and develops gradually over a period of two to three hours. An important factor which can delay or speed up the onset of *rigor* is the temperature. The lower the temperature, the slower the chemical reactions proceed, with the result that *rigor* starts later and develops slower. The carcases are cooled extremely fast in large slaughterhouses, where many animals are processed one after another, to about 1°C, to prevent growth of microorganisms as much as possible. The meat shrinks severely through this fast cooling, sometimes even to half the original length, and becomes tough. This shrinkage, which results from rapid processing is called cold shrinkage or 'cold shortening'. During the shrinking, much energy is used (ATP and CP), so that *rigor mortis* starts quickly. With the development of the *rigor* the muscle contraction stays. This muscle contraction (a reaction between actin and myosin to form actomyosin) also takes place in the muscles of the living animal. Then, however, after each contraction a relaxation follows, because oxygen as well as energy is available. After the energy-rich phosphates have been broken down the breakdown of glycogen follows. By way of a number of steps, with the help of enzymes, the glycogen is anaerobically changed into lactic acid which lowers the pH in the muscles from about 7.2 (at the moment of death) to circa 5.6. This takes about six hours in normal circumstances; at this point the stiffness of the muscles becomes apparent. The *rigor* state lasts a number of hours and is related to the speed at which it developed. After this, the post-*rigor mortis* period follows, in which the muscles gradually become supple again. The contraction is stopped by the action of proteolytic enzymes. This period (maturing or ripening) can also last a few hours or a few days, depending on the temperature.

Alkaline and acid *rigor*

The circumstances under which contraction of the muscle takes place can differ. In very tired animals, especially with pigs and bulls, the supply of energy-rich phosphates and glycogen can be completely used up before death. Therefore no lactic acid can be formed. The *rigor* sets in quickly with a high pH, which stays high (alkaline *rigor*). The meat is dark in colour, feels dry, is not supple, and tastes and smells bad.

During transportation and before the slaughter, disturbances in the muscle metabolism of the animals (especially in pigs) can occur by which the glycogen is already transformed into lactic acid before death. The energy-rich phosphates are also for the most part broken down. Furthermore, through this rapid breakdown the temperature in the meat increases. The *rigor* sets in quickly (within approximately one hour) with a low pH (acid *rigor*). The meat is very pale, nearly white in colour, and feels wet and soft.

Meat which undergoes an acid or alkaline *rigor*, keeps less well and will be mostly sold as low-grade meat or will be used in meat products.

Quality aspects of fresh meat

Proper hanging or maturation is not the only way by which to bring a tender piece of meat for the table. An appropriate method of preparation (which again is dependent on the kind and the age of the slaughtered animal and the structure of the muscle) is of equal importance. Furthermore, a number of treatments can be applied to the meat to tenderize it, which will, among other things, decrease the cooking time. These treatments include:

— some mechanical treatment, such as making the meat smaller by mincing it or by beating it flat (as with beefsteak), so that the connective tissue fibres are reduced;
— changing the pH, for example by marinating before cooking or by adding acid during the cooking;
— electrostimulation, used on meat which has become tough as a result of cold shrinkage. By means of electric shocks of 300 V, which increase the activity of enzymes which tenderize meat, this toughness can be decreased;
— a treatment with papain, an enzyme which breaks down proteins, used in some countries (but not in the Netherlands).

5.6 QUALITY DETERIORATION, SPOILAGE AND STORAGE

Meat is a product which quickly starts to spoil. Because of this it must be stored for only a short time and in the right manner. Packaging and cold storage will retard quality deterioration and spoilage, but will not prevent it. The most notable causes of quality deterioration and spoilage are described below.

Incorrect storage of meat may cause it to **dry out**. The meat discolours and looks unappetizing, but is not harmful to health.

Meat can be infected by the **blowfly** and other flies. They are disease vectors (germ carriers). The fly can also lay eggs on the meat, from which in one day maggots can develop, which use the meat for their growth and development.

Meat, which is stored too long, even in cold storage, will start to decay. This phenomenon is called **autolysis**. Dead cells are being broken down by their own enzymes. In a living cell, the working of the enzymes is coordinated; chemical synthesis and breakdown takes place without affecting the cell. After death, the lytic enzymes (proteolytic and the lipolytic enzymes) are liberated and only breakdown takes place. The breakdown is speeded up by enzymes of bacteria and fungi. We speak of **rotting** if protein-containing material is being broken down by proteolytic bacteria.

Bacteria which are found on meat and can cause food infections and/or food poisoning are *Salmonella, Staphylococci, Clostridium perfringens* and *botulinum* and *Campylobacter*.

The storage of meat is normally done with refrigeration and deep-freezing, sometimes combined with vacuum- or gas-packaging.

Refrigeration
Fresh meat has to be stored refrigerated, in the home as well as the shop. The optimum temperature is 2–4°C, which increases the keeping qualities by a few days. The storage and display of fresh meat in shops has to be done according to legal rulings (legislation). For example, in the Netherlands, pre-packed fresh meat must be stored at a maximum of 4°C and non-pre-packed meat at a maximum of 7°C.

Deep-freezing
A much lower temperature is necessary if meat is to be kept longer than a few days. The keeping quality of frozen meat is dependent on the packaging and the storage temperature. Storage at $-30°C$ increases the storage time by a factor of two in comparison with storage at $-20°C$. In the home deep-freezer, storage takes place mostly at about $-18°C$. The storage time for lean meat is then about one year, fat meat about six months and prepared meat one to four months.

With regard to the storage, defrosting or purchase of deep-frozen meat, a number of aspects are of importance for the consumer or buyer. Among other things, packaging will limit the evaporation from ice crystals, so that denaturation of the proteins during frozen storage will occur less quickly.

Defrosting of deep-frozen meat must be done slowly, and without the wrapping, in a cool place (e.g. the refrigerator). Thawing in this way will slow down drip formation, and also the growth of microorganisms.

In catering and large kitchens, portioned or portion-controlled meat is used. Each portion has exactly the same shape and the same weight. The meat for this is flaked at $-4°C$ to $-5°C$. These flakes are pressed in such a way that identical pieces of meat are produced. These portions are afterwards frozen by the shock-freezing or giro-freezing system, in which the temperature of the meat reaches $-25°C$ within six minutes. Each single portion is then wrapped in foil and packaged. After thawing, this meat is indistinguishable from 'butcher-fresh' meat (24 hours in a cold store at about 2°C).

Vacuum-packaging
This is used with fresh meat in combination with refrigerated storage. This technique is used in the slaughterhouses to ripen large pieces of meat more slowly and by doing

Sec. 5.7] **Colour formation and preservation in meat and meat products** 85

so to store it longer. Beef and horse meat are then kept at a temperature of 0–1°C for four to five weeks or at 5°C for about three weeks. Pork, lamb and veal can be kept at 0–1°C for a maximum of two to three weeks. In restaurant kitchens, portioned vacuum-wrapped meat is often used. The colour of such meat in the packaging is a little darker than fresh meat because of the absence of oxygen (myoglobin). The meat will slowly become a lighter red after reopening of the package, through the formation of oxymyoglobin (see section 5.7).

Gas-packaging
The meat is packaged with selected gas mixtures so it retains the colour of fresh meat in the package. Mostly a gas mixture of carbon dioxide, nitrogen and oxygen is chosen. The carbon dioxide and nitrogen reduce the bacterial growth; the oxygen ensures that the clear red meat colour is temporarily preserved (see section 5.7).

5.7 COLOUR FORMATION AND COLOUR PRESERVATION IN MEAT AND MEAT PRODUCTS

The colour of meat is determined by the myoglobin, a muscle-protein with a purple-red (for beef) to red-pink (for pork and veal) colour. Myoglobin is related to the blood pigment, haemoglobin. Both fulfil a role in the body for the oxygen supply. Haemoglobin takes care of oxygen transport in the blood; myoglobin ensures that a small stock of oxygen is present in the muscles.

Myoglobin can react in two ways with oxygen. Both in live tissue and in that of slaughtered animals, oxygen can combine with myoglobin to form oxymyoglobin, which is bright red in colour. In the live animal myoglobin stores oxygen for some time and then releases it again afterwards; so oxymyoglobin is not a stable compound. That is why it is not referred to as oxidation, but rather addition of oxygen to myoglobin (oxygenation). After some time the oxymyoglobin breaks down because the muscle uses oxygen, or because the compound does not survive longer (about 30 minutes). As a consequence of this, the meat loses its nice bright red colour.

Besides addition of oxygen, oxidation reactions are also possible. In this case, the iron ion of the myoglobin changes from the ferrous form into the ferric form and the brown-coloured metmyoglobin develops. This oxidation takes place slowly during storage of meat and is only possible where there is contact with oxygen. In minced meat this reaction will take place faster because of the larger surface area. Metmyoglobin cannot spontaneously change back into the reduced form (myoglobin). The occurrence of an unappetizing brown colour in fresh meat can be prevented by the use of a reducing agent. At the butchers, ascorbic acid is used. This oxidizes in preference to myoglobin. L-Ascorbic acid can also reduce a small part of the already formed metmyoglobin into myoglobin. If meat is exposed to oxygen still longer, then the brown colour eventually will develop, through the action of several enzymes.

Meat keeps its red colour in preserved meats or sausages by the addition of nitrite. Nitrite decomposes to give nitrogen monoxide. This reacts with myoglobin to form nitrosomyoglobin, a fairly stable compound which is red in colour. Nitrosomyochromogen, a pink-coloured insoluble protein, is formed on heating. Because the nitrogen monoxide formed from nitrite is detrimental to health, its use is only permitted in the form of nitrite pickling brine (sodium chloride with a maximum of

0.6 per cent sodium nitrite). Nitrogen monoxide attaches itself more quickly to myoglobin than to oxygen, in human blood also. Too much added nitrite (pickling salt) could endanger the oxygen supply. By combining it with sodium chloride the addition of large quantities of nitrite is prevented: the product would become too salty.

5.8 MINCED OR CHOPPED MEAT

The mincing of meat has as its most important advantage for the consumer the decrease in preparation time.

Minced meat is obtained by chopping or mincing. Fine mincing makes the type of meat and the cut of meat which was used hard to distinguish. The description 'minced meat' should be accompanied by the name (names) of the species of animal(s): for example, minced beef, minced veal or mixed minced meat (60 per cent pork and 40 per cent beef). The fat content may amount to 35 per cent. If the minced meat contains less fat, then the description 'lean' may be used (maximum 20 per cent fat). 'Tartare' is minced beef or (in some countries) horse meat, with a maximum of 10 per cent fat. 'Minced beefsteak' is chopped horse (in some countries) or beefsteak and contains a maximum of 6 per cent fat. A small percentage of breadcrumbs, egg white and salt may be added to minced kinds of meat for the taste. To maintain the fresh red colour L-ascorbic acid or citric acid may be added to raw minced meat. Fine mincing introduces a lot of oxygen into the ground product and the colour changes which normally occur after the cutting up of meat will appear more quickly. At the point where 50–60 per cent of the myoglobin has been converted into metmyoglobin, consumers find the colour unattractive.

Spiced minced (sausage) meat should contain 90 per cent minced meat. Besides that, herbs and spices, glutamates and a maximum of 6 per cent starch are added. Spiced minced meat which has been put into a casing is described as 'fresh sausage', 'sausage' or 'frying sausage'.

Filet Americain contains a minimum of 70 per cent minced beef with mayonnaise or salad cream. The total fat content is a maximum of 25 per cent and no breadcrumbs may be added.

Economy or composite meat products contain, besides minced meat, a stated percentage of non-meat protein (soy-protein or milk-protein). Composite meat products contain a minimum of 80 per cent (spiced) minced meat and one non-meat protein; for example, minced beef with 15 per cent soy-protein.

(Note: Exact compositions may vary from country to country.)

5.9 MEAT PRODUCTS

All kinds of prepared and preserved meat come under the **Meat and Meat product Regulations** under the name 'meat product'. The majority of meat products are prepared by the meat processing industry.

In most countries meat products are mainly used as sandwich fillings. Some kinds, such as liver sausage, pâté or smoked tongue, are also served with a drink. Ham or black pudding can be a part of the main course, while other meat products can be used in first courses, lunches or in soups.

Meat products vary in price, taste, nutritional value and keeping quality; these all depend on the materials used and/or the preparation and preservation techniques used. The nutritional value of meat products is in the main equal to that of fresh meat, and prepared meat.

In the report *Normalization of diets*,[†] meat products are subdivided according to fat content (see Table 5.6). The fatter meat is used particularly in sausage preparation.

Meat products should contain 90 per cent meat (muscle, fat and connective tissue) and 10 per cent other substances, such as binding agents, water, spices, salt and other edible products. Because of the differences in the basic materials (kinds of meat) and additions, an exact analysis of nutritional products is only possible for each separate type of meat product. But in general they contain approximately:

water	35 – 70%
protein	9 – 20%
fat	5 – 40%
carbohydrates	0.5 – 4%

The energy value is in practice dependent on the quantity of the meat product (slice size and portion size) which is used per slice of bread.

Meat products are subdivided according to the treatment applied to the kinds of meat (basic product) and the additions. These are divided into:

— prepared and preserved meat products
— kinds of sausage.

Prepared and preserved meat products
Prepared meat products are scalded, fried, roasted or boiled after the addition of salt and/or spices; for instance, sliced roast beef, fricandeau, roulade or cooked liver.

Preserved meat products keep longer than prepared meat products when one or more preserving techniques are used. Meat products can be salted, smoked, dried, marinated, and pasteurized or sterilized.

Salting
The addition of salt reduces the water activity, which in turn reduces the growth of microorganisms. Strongly salted meat can be kept for two to three months, as long as it is kept hanging and reasonably dry in a cool place (salted bacon and raw ham). Mostly, meat is lightly salted and vacuum- or gas-packaged and/or refrigerated to increase the keeping qualities.

† This report was written by the commission *Normalisatie en Organisatie* of the *Voedingsverzorging in Instellingen* (NOR).

Table 5.6 — Weight of fat per 100 g of meat products

Kind of meat	Up to 6 g	From 6 to 16 g	From 15 to 25 g	From 25 to 35 g	35 g and more
Beef	Unlarded liver Smoked meat Roast beef	Boiled tongue Fried lean minced meat Lean salted meat Roulade	Larded liver Corned beef Salted meat		
Veal	Unlarded liver	Boiled tongue Fried lean minced meat Roulade	Liver sausage		
Pork	Unlarded liver	Gammon *Casselerrib* Boiled tongue	Raw ham Cooked ham Shoulder ham Brawn Larded liver	Sandwich sausage Hot dogs Haagse liver sausage Saksisch liver sausage Berlin liver sausage Hausmacher liver sausage Liver pâté Eel sausage	Cut sausage Tea sausage Farmhouse sausage Salami Cervelat sausage Smoked sausage Eel sausage Pâté Bacon Breakfast bacon Stuffed breakfast bacon Gelders bacon Gelders smoked sausage Beer sausage Black pudding Tongue sausage
Other	Smoked horse meat Cooked smoked horse meat	Fried chicken roulade Fried turkey roulade			

Smoking
Nowadays meat products are smoked mainly to obtain the smoky taste. Depending on the temperature of the smoke, the consistency sometimes changes and the product is either partly or totally cooked or stays raw.

Cold smoking of meat, as is used for instance with smoked meat and Ardennes ham, is combined with other techniques like heavy salting and drying. Warm smoking is also combined with light salting and cooking, such as sandwich sausage and *Casselerrib*.

Drying
Drying of meat products is always combined with other techniques, especially with salting and/or smoking; for example, raw ham, saveloys and smoked beef.

Marinating
Marinated meat products can be prepared in two ways. Meat can be cooked in acid, salt and spices, poured into a mould and cooled down, as in the preparation of brawn. Or a cooked meat product (*rolpens*) can be laid in a spiced vinegar solution. These acid products can be kept for about six to eight weeks if they are kept at about 10°C.

Pasteurization and sterilization
Pasteurized meat products are heated in cans or glass containers to about 70°C (measured in the centre). The product obtained has a limited shelf-life and has to be stored under refrigeration (canned ham). In complete sterilization and preservation, the product is heated to above 100°C at the centre. Treated like this, the product has a nearly limitless shelf-life, but the prolonged, high heating makes the meat drier and more fibrous (corned beef in cans). During storage these meat products will slowly deteriorate in colour, smell and taste. A loss of consistency will also occur.

A short description of commonly occurring kinds of meat products now follows.

Fried 'roastbeef', 'fricandeau' and **'roulade'** come from cuts of meat which are salted and spiced and, after frying, sold sliced.

Fried minced meat consists of more or less finely minced meat with salt and spices. It is fried in a large piece or first cooked in moulds and then fried.

Cooked liver (calves' and pigs' liver) is sold, larded or unlarded, in its original shape, and also pressed in a square mould.

Balkenbrij is made of stock, diced fat, minced pigskin and a large percentage of buckwheat flour. The *balkenbrij* is grey or red in colour (blood); raisins may be added and different herbs and spice mixtures may be used, depending on the area where it is sold.

Pressed tongue is brined, smoked, cooked and skinned. It is sold sliced, sometimes as smoked ox tongue.

Preskop is made from pigs heads, trotters, skin, and stock. The result is a clear jelly through which coarsely minced skin and coarse pieces of meat are dispersed.

Hoofdkaas (brawn) consists of finely minced cooked pork and skin, with salt, spices and vinegar to taste. The puree is allowed to solidify in a mould.

Rolpens consists of coarsely minced spiced meat, encased in sausage skin and cooked. The sausages are put into vinegar and can be stored for eight to ten weeks.

Smoked beef is made of the fillet or the flat rump cut of beef. It is brined and smoked, and sold raw, sliced very thinly. The colour is an even red.

Smoked horse meat is made like smoked beef. The colour is darker and the meat is drier, less aromatic and often saltier. The price is lower.

Salt beef meat is made from beef breast, which is brined or injected with brine and then cooked.

Corned beef consists of pieces of brined beef with fat and gelatine. This is canned (sterilized).

Raw ham comes from a thigh-cut from a pig. It is brined or dry-salted and smoked. Raw ham is made according to specific recipes in some areas. The name of the area is given to the ham. For example: Ardennes ham, Parma ham and Coburg ham.

Cooked gammon is made of pieces of meat from the pig which are brined, cooked and sometimes lightly smoked.

Shoulder ham is the cooked whole shoulder of the pig, which is lightly salted and sometimes lightly smoked. The cooking is done in a so-called ham-cooking apparatus (canned or rolled) for six hours at about 75°C.

Bacon is made from boned rib and loin chops with the fat not removed. It is dry-salted or brined, then smoked and sold raw sliced.

Procureurspek comes from the boned neck and shoulder chops. The meat is brined, smoked and sold raw or cooked.

Breakfast bacon, back or streaky bacon, is salted and sold sliced raw, sometimes smoked.

Cooked breakfast bacon is salted and cooked streaky bacon, sometimes lightly smoked and rolled and sold as rolled bacon.

Katenspeck is streaky bacon which is salted in a spiced brine bath and then hot-smoked (steamed).

'Bladder ham' is made out of two boned rib or loin chop pieces encased in a bladder. The meat is brined, lightly smoked and cooked.

Casselerrib is the boned rib chop. This is brined and then lightly smoked and sold raw by the piece. Often it is cooked and sold sliced.

Lachsschinken consists of the long fillet of the loin chop, around which a very thin layer of bacon is wound. It is brined, smoked and cut raw.

Kinds of sausage

Sausages are prepared from finely or less finely minced meat, organs, fat, skin or blood. By adding salt, herbs and spices, a paste or 'farce' is made which is put in a casing. After that the sausages are then cooked or not, dried and/or smoked. Depending on the basic materials used and the treatments used, the sausages are divided according to their keeping qualities into **perishable** and **keeping**.

Perishable sausages are described according to the main constituents used (meat, liver or blood).

The farce of **meat sausages** consists of the raw basics (meat and fat) which are partly very finely minced. On the addition of moisture, water-binding additives, starch, salt, herbs and spices everything is kneaded to a homogenous mass. This farce

Meat products

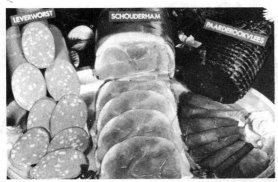

Liver sausage
Shoulder ham
Smoked horse meat

Cervelat sausage
Thuringer black pudding
Country minced sausage

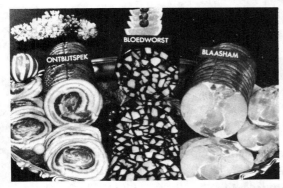

Breakfast bacon
Black pudding
Bladder ham

Salami
Liver pâté
Tongue sausage

Fig. 5.12 — Various meat products.

is pushed into a casing. The sausages are, if necessary, dried a little, sometimes warm-smoked and cooked. They are also described as cooking sausages (e.g. sandwich sausage, luncheon sausage, hunt sausage and eel sausage).

In **liver sausages**, meat, bacon and offal are used, as well as a certain percentage of liver. In the sliced varieties, the farce consists of raw basic materials, which are cooked in a mould or an intestine case (Berlin liver sausage). The farce of the semi-sliceable kinds consists completely or largely of cooked basic materials (Haagse liver sausage, Hausmacher liver sausage, fried liver sausage). The spreadable sausage (*Saksische* and *Brunswijker* spreading liver sausage) consists of cooked or thoroughly scalded raw materials. After filling, all the varieties are cooked, and after cooling either smoked or not. The smoked kinds are yellow to yellow-brown on the outside; the unsmoked are grey.

In **black pudding**, bacon, skin and offal (liver, tongue and kidneys) are used besides the main constituent of blood. The farce consists mainly of cooked ingredients. After cooking, the sausage can be smoked; the skin acquires a nice gloss (tongue sausage, blood sausage, frying blood sausage).

A short description of frequently occurring kinds of perishable sausages follows.

Sandwich sausage is finely minced meat and bacon, with an added quantity of bacon pieces. The farce is pushed into an intestine and then cooked and perhaps smoked or sterilized in cans.

'Hunt sausage' contains more lean pork and less bacon compared with sandwich sausage. The constituents are minced more coarsely. It belongs to the best quality cooked sausage.

Ham sausage is similar to hunt sausage, but the constituents are cut into small cubes. They are bound together with a small quantity of basic paste.

Knakworst is a very thin sausage with a maximum weight of 50 g. It is smoked and cooked. The meat and bacon is processed into a very fine paste.

Smoked sausage is produced in many different qualities. Since 1982, a quality ruling for smoked sausage exists in the Netherlands. The smoked sausage with the label 'Extra quality' consists of a maximum of 35 per cent fat, 13 per cent protein and 48 per cent moisture. In sausages without a label, the composition is not known. Smoked sausage is available raw as well as cooked. The ingredients can be coarse or finely minced.

Berlin liver sausage is a sliceable liver sausage, which was originally encased in pigs' gut. These days it is also encased in artificial gut, the fat tissue of the gut being imitated with a thin layer of fat. The colour is light pink because, besides liver, a large amount of pork is used.

Tea sausage is a small meat sausage of 100 to 200 g; the contents are spreadable. It has a lightly smoked taste.

Pâté is made of pigs' liver and bacon. It is spreadable and is made with all sorts of additions like mushrooms, peppers, cream or onion and herbs. It is covered with a layer of bacon fat to prevent it from drying out.

Liver cheese has about the same composition as Berlin liver sausage. The fine paste is put in a square mould, the inside walls of which are covered with thin layers of bacon fat.

Liver sausage is a cheap, dark-coloured, fairly dry liver sausage (mostly ox liver is used). With visible little cubes of fat, this is also sold as Haagse liver sausage.

Hausmacher **liver sausage** used to be made after domestic slaughtering of a pig. All the ingredients, coarsely or finely ground, are from the pig. It is grey brown in colour.

Frying liver sausage looks like ordinary liver sausage, but the composition is slightly different. This is necessary to enable the slices to be fried.

Brunswijker **spreading liver sausage** consists only of pigs' liver and bacon. The scalded ingredients are finely ground and kneaded into a soft paste-like mix. The sausages are sold cased in commercial gut.

Saksische **spreading liver sausage** is prepared as described for *Brunswijker* sausage. The constituents are not exclusively from the pig; other meat is also used in it.

Black pudding is made from blood, skin and cubes of bacon fat in a ratio of 20:25:50. It also contains salt, spices and starches. After filling, the sausages are cooked and sometimes smoked.

Tongue sausage is made by adding brined pigs', calves' or ox tongues to a blood sausage mix. After casing, the sausage is cooked and sometimes smoked.

Tongue pâté is made of the same paste and filling as tongue sausage, but cooked in a bacon-lined can.

Kidney loaf is similar to tongue pâté, but, instead of tongue, brined and pre-cooked pigs' kidneys are put in the paste.

Frying blood sausage is suitable for frying, because ryeflour is added to the paste. It is not suitable for sandwiches.

The paste (or farce) of the **keeping sausages** consists of raw minced ingredients, especially pork but also beef. No moisture is added. The minced ingredients are kneaded together with salt, herbs and spices to form a homogeneous paste. After the sausage has been filled, drying follows and sometimes cold-smoking (farmers' or country minced sausage, polony sausage, cervelat sausage, coarse sliced sausage). A ripening or fermentation occurs during drying, through which these sausages obtain a typical structure, colour, smell, and taste. Amongst other things, lactic acid is formed. This influences the keeping quality in a positive way. To prevent mould forming on the outside, the use of a mould inhibitor (pimaricine) is permitted.

Country or farmers' minced sausage is a raw sausage made from lean pork and fatty bacon. Then the sausage meat mix is pushed into casings made from the small intestines, bent into a horseshoe shape and dried. We differentiate between finely and coarsely minced farmers' or country minced sausage, and smoked and non-smoked varieties. The diameter of the slices is about 5 cm. Slicing sausage, cervelat sausage and polony sausage consist of lean meat. Pork as well as beef can be used and fatty bacon, which is minced finely or not so finely. Cervelat sausage is always minced finely and often contains white peppercorns. Polony sausage has a coarser paste, the fat is not minced, but cut into small cubes. These sausages, after drying (for three to five months), are sold thinly sliced. The slices have a diameter of about 8 cm.

Salami originates from Italy or Hungary. This keeping sausage looks most like cervelat sausage. The minced meat is mixed with garlic among other things. The skin is mostly painted white, but it is also made of white plastic. The drying time is longer than is usual with the Dutch variety (eight to twelve months). This is reflected in the price. The ripening or drying time is shorter for less expensive salami.

Diet meat products
The following types are produced.

— Meat products for a **low-sodium** diet: These are prepared without the addition of cooking salt and sodium nitrite. The unsalted meat products will, because of this, have a different taste and colour. They are also more moist, because the moisture-binding ability is less because of the absence of cooking salt. The industry sometimes uses other salts and additives to obtain an acceptable end product. To improve the insipid taste, potassium or ammonium chloride is added. To improve the colour, instead of sodium nitrite, potassium nitrite is added.
— Meat products for a **low-fat** diet: The leanest existing meat products are chosen for these.
— Meat products for a **low-sodium/low-fat** diet: Only lean meat products, prepared without cooking salt and sodium nitrite, can be chosen for this.

Quality deterioration, spoilage and storage
The durability of meat products depends upon the processing and preserving techniques used and the storage circumstances. Especially as result of poor storage, meat products can deteriorate in quality and/or spoil. Possible forms of deterioration in quality and/or spoilage are:

— drying out
— discolouration
— sweating
— souring through the effect of lactic acid bacteria
— mould
— oxidation of fats
— break down of proteins under the influence of proteolytic and/or putrefying bacteria.

Sliced meat products show these defects sooner. They also quickly lose smell/aroma and preferred colour. Good packaging and refrigerated storage lengthen the possible storage times (see Table 5.7).

5.10 COOKED PRODUCTS AND GELATINE

Cooked products legislation may mention a number of convenience food products. Some are based on meat and others are not. The legislation is appropriate for all foodstuffs which are partly or completely cooked, warm or cold, which can be sold and consumed straight away. For instance, a warm sausage roll bought at the bakers,

Table 5.7 — Storage times of sliced meat products well packaged and stored under refrigeration

Product group	Examples	Storage time
Prepared meat	Cooked liver, 'roast beef', fried minced meat	1–2 days
Cooked, lightly smoked meat products	Cooked ham, Berlin liver sausage	3 days
Raw strongly smoked meat products	Smoked meat, raw ham, country minced sausage	7 days
Marinated meat products	Brawn, *rolpens*	4–5 days
Salted dried meat products	Breakfast bacon, sliced sausage	4–5 days

snacks from automatic machines or the chip shop, pre-fried chips, and prepared but still-to-be-cooked 'croquettes' sold in take-aways come under this legislation. 'Croquettes' and other flour products, whether with or without meat filling (*loempia, nasibal,* sausage roll, pasty), are subject to rules concerning their preparation, but not their composition. For a number of meat-containing products, the composition (in the Netherlands) is governed by the Meat and Meat products legislation.

Gelatine
This is prepared from collagen from the bones of slaughtered animals. Gelatine is used as the binding agent in the preparation of jellies, aspic and puddings. It is available in the trade as white and coloured leaf and powdered gelatine (see section 16.4).

5.11 FLAVOURS, SAUCES AND SOUPS

Meat extracts, flavours and stocks
These products are used in the home or in catering establishments because they are convenient for preparing quickly and simply a stock, soup or sauce, or for improving the taste of more complicated dishes. Stocks have an appetizing effect. This depends partly on the amount of water-soluble compounds. The compostion is laid down in the meat extract regulations. The basic materials for the preparation of these products include beef meat and/or bones (also from other slaughtered animals), spices, herbs and salt.

The following forms are available:
— meat extract in liquid form or as paste;
— various liquid flavourings;
— various powdered flavourings;
— stock concentrated in liquid, cube or powder form;
— soup flavouring in liquid, tablet or powder form;
— meat consommé.

The concentrated and/or dried products have a high salt content.

Gravy, plant extracts and spicy sauces
With aid of these products a quantity of gravy or sauce can quickly be made or a poor gravy or sauce can be improved. The composition is regulated by the gravy regulations (Netherlands). There are available: gravy cubes, tablets, granules, powders, liquids or pastes. The term 'meat' may be used for a gravy if the quantity of meat proteins contained in the product is above a stipulated minimum. Also in these products the word 'gravy' is exchangable with the word 'sauce'.

Plant extracts are obtained by the processing of vegetables, spices and yeast; they help to improve the smell and taste of dishes. The term 'soya' can be used if soya beans have been used as a basis (soy sauce).

Spicy sauces are obtained from plant and/or spice extracts with perhaps other additions. These products can be used to give a spicy taste to dishes (tabasco, tomato ketchup, Worcestershire sauce).

Soups
Soups contain food solids as well as flavouring substances. The name of the soup gives an indication to the consumer of the soup's composition; for instance, tomato soup, vegetable soup. Sometimes a more exotic name may be used; for instance, minestrone soup. Most soups contain binding agents and other additives.

Convenience soups are sold in dried or liquid form. The latter have been preserved by heat treatment or by deep-freezing. The specification 'concentrated' on liquid soups means that to prepare it at least an equal amount of liquid has to be added. In the Soup Regulations (Netherlands) added substances are regulated.

5.12 STANDARDS BY LAW

Meat inspection laws lay down standards on the hygiene of slaughterhouses, the transport of meat and the inspection of meat. The inspection is usually done by the government service for the control of cattle and meat. The supervision of the processing of meat into products destined for internal consumption is usually under government control.

There are Agricultural Quality Laws which formulate, in conjunction with the regulations of the EC member states, amongst other things, quality regulations for products which are to be exported (Regulation meat products and Regulation bacon). The control of these rules is done by the General Inspection Service.

Meat and meat products regulations set out other standards with regard to the labelling and composition of meat, minced meat, spiced minced meat, meat products and prepared meat products.

6

Game and fowl

6.1 INTRODUCTION — GAME

By game we mean animals used for consumption which are obtained by hunting. In earlier times, traps and bows and arrows were used for hunting and the aim was to obtain food. In the Middle Ages the ruling class started to reserve the hunting rights for themselves. The eating of game became a status symbol. Roast game was the highlight of the festive table, symbol of riches and excess. Later the hunt was also viewed as sport and not solely to obtain meat. Game is, in general, scarce nowadays, something for special occasions. Hunting is practised by about 36 000 licence holders in the Netherlands.

Hunting comes under the jurisdication of the Ministry of Agriculture and Fisheries, which may regulate the pursuit of the hunt for a number of reasons:

— to limit damage to agriculture, horticulture and forestry from an excessive game population;
— to protect certain kinds of game;
— to regulate the hunt as an active but not destructive recreation by setting out by law the hunting seasons and the requirement for possession of a hunting licence.

Consumption
The quantity of game eaten in comparison with meat is small and difficult to estimate. Only a small part is traded through the poulterers.

Fresh game is only available during the hunting season, which is between August and January. In this period game tastes best. Outside the hunting season game is available deep-frozen or imported. Destructive game, for instance rabbits, can be hunted the whole year through. Consequently these are constantly available. However, the taste is less good during certain periods, among others the breeding season.

Composition

The muscle tissue and the organs of game have about the same composition as those of slaughtered animals. However most kinds of game have a higher content of water-soluble compounds and a lower fat content.

Paunching

To avoid spoilage and to cool the meat as quickly as possible, the organs and intestines are removed from large game straight after the animal is killed. In hunters' jargon, this is called 'paunching'.

With water game, for instance duck, only the intestines are removed.

Small game such as hare and rabbit are not immediately paunched, because the risk of contamination and infection is then larger than if these animals are kept closed.

Hanging and gaminess

In contrast to what is usual for slaughtered animals, game is not bled after killing. Game also keeps its fur or feathers until its preparation for consumption. The typical game taste is caused by the mode of life and the feed of the animal. This taste can be accentuated by lengthening the hanging period, after which the game is said to have become **high**. The changes which take place in the meat during the hanging process will be more visible the longer this process continues. To be hung, the game has to be laid or hung (unskinned, unplucked) in a cool, dry, well-ventilated place for one to seven days.

Hung game is recognizable by the strong smell and a blue green discoloration of the stomach of the animal. The meat is more tender.

6.2 KINDS OF GAME

The kinds of game consumed differ from country to country. But the following division may be made:

furred game
- coarse game, (red deer, roe deer, and wild boar);
- small game, (hare and wild rabbit);

feathered game
- water game (wild duck and snipe);
- forest and moorland game (pheasant, partridge, black and red grouse, pigeon and woodcock).

Furred game
Red deer

Red deer belong to the coarse game and are protected in many countries, where the season for them is always closed. But the meat is exported by some countries, including Germany. Red deer meat from adult animals (35–50 kg) looks like meat from a calf or young cow. If the animals are heavier than 60 kg the meat looks much

Fig. 6.1 — Game being hung.

like beef. The meat is in general very tender. The best parts are the back (saddle), the loin and the fore and hind leg. From the rib, chops are cut, and steaks from the hind leg. The best time to consume red deer is during November and December.

Roe deer
A roe deer is a small type of deer. In the Netherlands, fresh roe deer meat is rarely available (but deep-frozen is). Judged on taste, roe deer belongs to the finest kind of game. Roe deer up to three years are tender meat; old animals are tough. The weight varies from 10 to 20 kg. The cuts of the animal are the same as for the red deer.

Wild boar
The hunting of wild boar is prohibited in the Netherlands. When slaughtered, in contrast to what is customary with the pig, they are skinned. So this meat lacks crackling. Only young specimens are suitable for consumption. The weight varies from 30 to 40 kg. The cuts are as with red and roe deer, as is the preparation. The nicest piece is the back, sold as saddle or chops. The best time for consumption of fresh (imported) wild boar is November and December.

Hare
According to habitat, hare may be divided into heather, dune and clay hare. The heather and dune hare produce the nicest meat. Clay hares are heavier in weight and have darker-coloured meat. A young hare is called a leveret. Hares vary in weight from around 3 kg to 6 kg, depending on age. You can buy hare blood at the same time as the hare to make a rich gravy (jugged hare). Hare is also sold cut up into pieces: hare back, hare fillet, and front and hind legs. The head is added loose for identification if a hare is sold in pieces. This is also suitable for the preparation of game stock. The best quality meat comes from hares which are three to eight months old. In young hares, the ears are easily torn and the nose bone is soft. The meat is

lighter in colour and is tender; the older hares have darker and drier meat. Hares are hung unskinned by their feet for a few days to become high. The best time to eat fresh hare is from October to December.

Wild rabbit
A wild rabbit is somewhat smaller than a hare. The weight is about 1.5 kg. The meat is white, has a strong taste and is a bit dry. In the Netherlands, the season is open through the whole year because it is classed as a pest. Rabbit is available whole or cut into pieces (as for hare). It is hung as for hare.

Tame rabbit is the only type of furred game which is specially bred and can be obtained tame. It is for sale fresh during the whole year, but the best time to buy a good quality rabbit is from September to November. In comparison with wild rabbit, the meat has a less delicate, sweetish taste, is lighter in colour and is mostly fatter. Cultivated rabbit usually weighs about 1.75 kg.

Feathered game
Wild duck
There exist, even within Europe, many kinds of wild duck, such as the shoveller duck, the pintail, the widgeon and the teal. For marketing, they are mostly caught in duck cages, the advantage being that they stay undamaged (no shot wound). The most appreciated is the teal, a small, fairly fat duck with a weight about 400 g. Because it feeds on seeds and small invertebrate animals, the meat is more delicate than the meat of the widgeon (about 0.75 kg) and the large shoveller (about 1.5 kg). Older ducks, especially, often taste fatty. The best time for consumption is November–December.

Snipe
These live in bog areas and are easy to recognize by their very long beak. The bird weighs around 125 g; the meat is fairly fat. The open season is from August until January.

Woodcock
This bird occurs especially in dune areas and in broad-leaf and coniferous forests. Its beak is long, but less long than the snipe's. The weight is about 325 g. The best consumption time is November and December. The woodsnipe is only rarely available fresh. In the Netherlands this bird is protected.

Pheasant
They occur in the wild, especially in shrubberies or coverts with arable land and meadows surrounding them. The males are the most handsomely marked and have ornamental feathers. An adult cock (oven-ready) weighs about 1 kg, while a hen weighs about 800 g. At the poulterers, the birds available are mainly specially reared.

Young birds are recognizable by the only slightly developed spurs and the pliability of the breast bone; older ones have fully grown, pointed spurs on the legs.

The meat is fairly dry. The most suitable time for consumption is the month of December.

Partridge

The partridge, 30 cm long, is a short, compact bird. It lives especially in meadows and woody terrain. There is no noticeable difference between the cocks and the hens.

A lot of partridges are imported (wild or bred specimens).

Young partridges have a yellowy beak and yellow feet; old ones have a grey beak and grey feet. Both young and old weigh about 300 g.

Black grouse

Hunting of black grouse, a bird which looks like a large chicken, is not permitted in the Netherlands, but they can be imported. They occur in heather and peat areas and in birchwoods. The taste of the meat is described by connoisseurs as extremely fine. The weight of an adult specimen is about 1.25 kg. (Grouse shooting in the UK refers to the shooting of driven red grouse, which opens on the twelfth of August.)

Pigeon

In comparison to other feathered game, pigeon is little appreciated and eaten. There are many kinds, amongst others the destructive wood pigeon and the turtle dove. The weight is respectively 600 g and 150 g. The meat is very dark in colour and is mainly on the breast. The best consumption time is in the period April–October.

Quail

It is forbidden to hunt quail (migratory birds) in the Netherlands. They are either imported or specially bred for consumption. They are regarded as a particular delicacy. The weight is about 100 g. Quail eggs are also considered a delicacy by connoisseurs.

6.3 INTRODUCTION — POULTRY OR FOWL

To this group belong all specially reared birds such as chickens, tame ducks, turkeys, tame geese, tame quails, and guinea fowl. In the poultry industry, production is done through an integrated chain, which means that the butchers sign contracts with the producers. The chain connects up specialized breeding firms, propagating firms, and hatcheries to the poultry farm. These last firms rear the birds up to saleable, ready-for-slaughter products.

The distribution firms take care of the slaughtering and sale. The largest market share consists of roasting chickens.

Consumption

From a recent survey it appears that in one third of Dutch families poultry is eaten at least once a week. If we look at the figures for the years 1980 to 1984, we can see a clear increase in the consumption of slaughtered fowl.

Composition

The meat of poultry is of the same composition as 'red' meat from slaughtered animals, only very tender and less fatty. It is a good alternative to lean red meat.

Table 6.1 — Consumption of poultry in the Netherlands. (Total UK figures are also quoted for comparison)

Description	1980	1981	1982	1983	1984
			(kg/head of population)		
Roasting chickens	7.56	7.85	8.83	9.46	9.86
Chickens and cockerels	0.81	0.54	0.91	0.99	1.20
Turkeys	0.53	0.57	0.52	0.56	0.77
Ducks	0.03	0.04	0.05	0.04	0.04
Others	0.00	0.00	0.01	0.01	0.01
Total Netherlands	8.93	9.00	10.32	11.07	11.89
Total UK	13.40	13.30	14.40	14.70	15.40

Sources: Dutch Marketing Board for Poultry and Eggs; Annual Abstract of Statistics (UK figures).

Slaughtering

At the slaughterhouses, the birds are slaughtered, defeathered and eviscerated (which means the intestines and organs are removed), and after this the meat is quickly cooled (the fresh meat is extremely sensitive from the microbial point of view).

The cooling is done with water and ice (the so-called wet-cooling), or with the aid of cold air (dry-cooling). Wet-cooling is cheap, but has the disadvantage for the consumer that the moisture content increases (5–10 per cent) during cooling. In the EC, maximum values have been set for this added moisture. Wet-cooled birds are often frozen; dry-cooled birds are mostly sold fresh.

Birds are hung like the meat of slaughter animals. The time, however, is shorter.

Inspection of poultry meat

Poultry is usually inspected by an official veterinarian or an assistant (a poultry inspector) who is in the service of the government for the inspection of cattle and meat before and after slaughtering.

For the inspection, which is the same for all EC member states, regulations for the inspection and trade transport of fresh meat from poultry apply. The animals for slaughter are inspected alive a maximum 24 hours before slaughtering in the poultry farms or in the slaughterhouse. Only healthy animals are allowed to be slaughtered. The inspection after slaughtering consists of viewing the slaughtered animals, if necessary handling and cutting. Sometimes further investigation follows after an abnormality in consistency, colour, smell and taste.

After having passed, the separately packaged slaughtered animals or parts of the slaughtered animals in small packs are supplied with a small-size label (see Fig. 6.2) on the packaging. Poultry meat in large packs is supplied with a large oval inspection label (see Fig. 6.3) easily visible on the outside packaging. (The above refers to Netherlands practice.)

```
NE
OOO
EEG
```

Fig. 6.2.

Fig. 6.3.

6.4 KINDS OF POULTRY

Chickens, turkeys, domesticated ducks, tame geese, tame pigeons, tame guinea fowl and tame quails are produced and consumed.

Chickens
Roasting chickens are subdivided into

— spring chicken (poussin), which after 4 to 8 weeks have a slaughter-weight of 300–500 g;
— roasting chickens (cockerels and hens) with a slaughter-weight of 500–1200 g after 8 to 10 weeks;
— roasting chicken (poularde) with an age of 6 to 8 months and a weight from around 1200 g.

These chickens at the time of slaughtering are not yet fully grown and the point of the breast bone has not yet hardened.

Most of the roasting chickens come from intensive poultry farms. This means that they are reared in cages of 16 by 60 metres, in which about 20 000 chickens roam free. They are supplied with good food and water, the cages are continuously lit, the temperature, humidity and ventilation are completely automatically regulated. The meat of these roasting chickens is beautifully white, but dreary in taste and not very firm.

If roasting chickens differ greatly from those of such intensive poultry farms in their manner of rearing, kind of feed and environmental conditions, they are allowed to be recognized as such. Examples: *Bresse poularde* (France), Brussels *poularde* (Belgium), *Poularde de Stiermarken* (Austria) and *Poularde Den Dungen* (the Netherlands). These *poulardes* (specially reared roasting chickens) have a slaughter-weight of 1.5–2 kg and are expensive in comparison with intensively reared poultry.

Boiling chickens (broilers or stock chickens) are 'old' laying chickens (about 1.5 years). The meat is fairly fat, less tender (tougher) and very suitable for making stock or soup. The slaughter-weight varies from 1.5 to 3 kg. These chickens are, at the moment of slaughtering, fully grown and the point of the breast bone has become bony.

Turkeys

Turkeys are not eaten very often in the Netherlands, though they are becoming increasingly popular in the UK. They are available the whole year, in portions as well as whole specimens. Around Christmas the demand and availability are greater. A division into two types may be made.

— Turkey chick: a young turkey with a slaughter weight of 2–3 kg.
— Turkey: slaughter turkeys with a weight of 3–7 kg. They are not allowed to be older than one year. The hens are more tender than the cocks. The meat of cock turkeys is white to pink and somewhat dry, especially the breast meat. There are strong tendons in the legs, and these may be removed before preparation.

Tame ducks

Fattened ducks are, in comparison with the wild kinds, rather more fatty in taste, while the meat is lighter in colour. The young ducks from about 4 months old (duckling) weigh about 1.3–1.7 kg; the slightly older duck weighs 1.8–2 kg.

Tame geese

Fattened geese are eaten especially around Christmas time. The meat is fatty. A young goose weighs up to 4 kg, and an older one 4–6 kg at 8 to 9 months old. The liver is a delicacy and is especially used for the preparation of goose liver pâté. That is the reason why the 'Strasbourg' goose is fattened in a special manner, so that the liver is considerably enlarged.

Tame pigeons

The quality of tame pigeons is better than that of wild ones. This is because, in tame pigeons, a thin layer of fat forms just under the skin. The meat of young pigeons (250–350 g) is soft and white in colour, more like chicken meat; older specimens (400–500 g) have somewhat darker meat.

Guinea fowl

The bird gets its name from the purl-shaped white spots on the grey feathers. This fattened bird is, as far as taste is concerned, most similar to pheasant. The weight varies between 700 and 1000 g. Fresh guinea fowl is available throughout the year. The colour of the meat is darker than that of chicken, but lighter than that of pheasant.

Tame quails
See quails in section 6.2.

6.5 QUALITY DETERIORATION, SPOILAGE AND STORAGE

The same standard should be set for the meat of furred and feathered game as for the meat of slaughter animals.

The fresh meat should have have a fresh smell and colour, and should be pliable and shining. Hung game has a typical smell and a blue-green coloured stomach wall.

The availability of fresh game coincides with the hunting seasons. Outside these periods it is imported or deep-frozen.

Fresh poultry has to comply with a number of requirements. The birds must:

— be well drawn;
— look sound, which means be without bruises;
— be freshly slaughtered;
— have a good colour and smell;
— have firm meat.

In contrast to older birds, young birds have a pliable breast bone, a soft beak, and feathers which are not deeply implanted.

Fresh game and fowl have always to be stored cool, preferably at 1–3°C, wrapped in greaseproof paper or foil and not longer than one day. It has to be frozen if longer storage is necessary. Industrially deep-frozen poultry usually contains about 10 per cent extra moisture (for the water-cooled product). So, during thawing, a lot of moisture will come free (drip formation). After thawing, these kinds of meat have to be quickly prepared. The forms of spoilage described for meat in section 5.6 are liable to occur here as well. Furthermore, raw chicken especially but also other fowls can be contaminated with *Salmonella* bacteria and *Campylobacter*. A careful regard to hygiene is especially needed in the buying and preparation of these types of meat. The following are important guidelines.

— Keep raw poultry well separated from other foodstuffs.
— Thaw deep-frozen specimens in the refrigerator (wrapped). Catch the melted water, throw it away in the sink, and rinse well with hot water straight away.
— Wash well and immediately kitchen equipment which has been in contact with this meat; do not use dish cloths, but paper, and so avoid cross-contamination.
— Take care that the meat is well cooked before it is eaten.

6.6 PRODUCTS OF GAME AND FOWL

The offering by the trade of prepared and/or oven-ready meat will stimulate the consumer to use more chicken and other poultry at meal times. The treatment and processing of the meat of poultry and game can be divided into four groups: freshly slaughtered meat; pre-treated and/or prepared meat; preserved game and poultry; and meat products.

Freshly slaughtered meat
Fresh fowl is for sale as:

— whole specimens, drawn, but with the head and feet, oven-ready with or without giblets;
— halves or quarters;
— parts with bone; for instance: wings, backs and necks, breasts, legs, parts of legs (thighs) and lower legs (drumsticks);
— boned meat;
— liver;
— edible offal such as stomach, heart and wing-tips.

In some countries the meat of game and poultry, stored frozen, can be sold as fresh after it has thawed out.

Pre-treated and/or prepared meat
The availability of pre-treated and/or prepared meat is controlled by the demand of the consumer whose choice is influenced by the convenience aspect. Some examples of such convenience food are:

— oven- or grill-ready prepared whole specimens which are trussed and/or barded;
— roasted or grilled chickens or chicken pieces;
— bread-crumbed or marinated parts of poultry or game.

Preserved game and fowl meat
By preserved game and fowl meat is understood: meat (prepared or not) which has been given a better keeping quality by means of heat treatment or freezing, or in other ways, for instance by vacuum packing.
   Sterilized products are preserved by means of heat treatment (full preservation). Some examples: stock with meat in cans or jars, stews of poultry meat or game, and meat products in cans.
   Deep-frozen products are pre-treated and completely or partly prepared; then deep-frozen and packaged. Some examples: chicken schnitzels, chicken croquettes.
   Vacuum-packed products are mostly prepared before packaging or are first processed into meat products. The keeping quality is limited, but is increased if the product is also stored under refrigeration. Some examples: roasted half-chickens with gravy, sliced meat products.

Meat products
The preparation and the keeping qualities are as described under meat products of slaughter animals. It should be noted that meat products prepared with meats from poultry or game mostly have a lower fat content.
   The following are some examples.

Chicken and turkey roulade This prepared meat product consists of spiced and roasted meat from chickens or turkeys with a fat content of around 5 per cent.
Turkey ham This prepared meat consists of lightly spiced, lean turkey breast meat. The meat is smoked for a short time, pressed into block form and then cooked. The fat content is a maximum of 2 per cent.
Turkey in aspic Cooked turkey meat is mixed with gherkins, mushrooms and paprika. This is all held together in a spiced aspic. The fat content is a maximum of 2 per cent.
Game galatine This exclusive meat product is prepared from finely ground game meat, prepared to taste with smoked tongue, ham, truffles and madeira or cognac. This mixture is put in a mould and then cooked.
Game pâté A meat product prepared from ground game meat and liver.
Goose liver pâté This is a delicacy, which is prepared, with or without the addition of truffles, from the livers of fattened geese. The livers sometimes swell to a weight of 2 kg (normal: about 100 g). This is caused by the feed and the way the geese live. The goose liver pâté is most readily available canned.

6.7 LEGISLATION

Hunting Laws may regulate the hunting seasons for the different kinds of game and the supply of hunting licences or permits.

Quality rules and Labelling rules for slaughtered poultry, game and rabbits set out rules concerning, amongst other things, the labelling and the storage temperatures for slaughtered meat. Quality and Labelling rules state that slaughtered poultry have to be classified in a quality class (A, B or C); they also regulate the mandatory labelling on pre-packaged slaughtered poultry.

There may also be legislation for the Regulation Inspection and Trade of fresh meat from poultry to regulate the inspection, labelling and grading and lay down rules concerning the hygiene for transporting poultry meat, and for factories where fresh poultry meat is slaughtered, processed or prepared.

7

Fish

7.1 INTRODUCTION

Fish belong to the vertebrates. They breathe mostly through gills. The generalized structure of a fish is shown in Fig. 7.1.

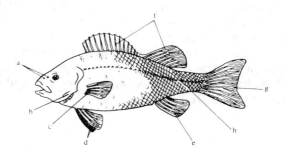

Fig. 7.1 — Generalized structure of a fish. a Nose opening: b Gill cover: c Pectoral fin: d Pelvic fin: e Anal fin: f Dorsal fin: g Caudal fin: h Lateral line.

Both male and female fish are eaten by humans. During the spawning season (hard)roe develops in sexually mature females and (soft)roe in mature males. The reproduction or spawning season occurs for most fish from the month of April until the end of June. For herring, this season occurs in October and November. There are more than 20 000 different kinds of fish known, of which about 5000 live in fresh water. Fish can be grouped in different ways, according to different criteria.

From the point of view of their **habitat,** salt-water fish and fresh-water fish can be distinguished. Some kinds live in salt water as well as in fresh water, for instance eel and salmon. These are classified as migratory fish.

From the point of view of their **build**, roundfish and flatfish can be distinguished. Most roundfish live high up in the water. Some kinds, such as herring, mackerel and sardine, which are also called pelagian fish, move freely at different levels in the water, high up as well as on the bottom. The flatfish live on the bottom. Flatfish are born as roundfish but, by adaptation to life on the bottom, the flatfish change in build and appearance. One eye moves to the other side of the head, so that both eyes are on the back. The skin on the back is camouflaged, the stomach side is white.

According to the **fat content**, one differentiates between oily and non-oily kinds. With the pelagian fish the fat content varies with the reproductive stage. They contain a lot of oil before spawning (about 20 per cent); after this they slim down considerably and contain only about 1 per cent fat.

Various **fish families** are also known, such as the herring species, the cod species, the salmon species and the flatfish. Fish which belong to the same family have, for instance, the same shape of tail or the same number of dorsal and ventral fins.

Consumption

In 1983 in the Netherlands 13.6 kg of fish was consumed per head of the population. It was divided as follows:

— 8.6 kg sea fish, of which 2.8 kg were herring;
— 3.3 kg crustaceans and molluscs;
— 1.4 kg preserved fish (canned);
— 0.3 kg fresh-water fish.

Total UK consumption was 7.1 kg/person/year, in the same year.

The consumption of fish has increased slightly in the Netherlands compared with a few years ago (1972: 11.6 kg per person; 1981: 13.5 kg per person), whereas in the UK consumption has fluctuated slightly over the same period. If the consumption figure is compared with that of meat (1983: 64.7 kg per person) then the usage of fish in the Netherlands is seen to be low. Reasons for this may be:

— the poor availability of fresh fish;
— the typical smell in the cleaning and preparation;
— the bones found while eating the fish, especially fresh-water fish;
— the less savoury taste and the lack of gravy;
— the low satiation value of cooked fish;
— unfamiliarity with the preparation of fish dishes;
— the contaminants in fish, such as mercury, cadmium and PCBs.

Composition

Fish meat contains on average the following compounds:

water	65–80%
protein	15–20%
fat	1–20%
carbohydrates	traces
minerals	iron, iodine and calcium
vitamins	A, B complex, C and D

Proteins
Fish protein is of a high biological value. In contrast to meat, fish contains a low percentage of connective tissue, which is scarcely developed. Because of this, fish meat is less compact in structure. It is very white because no large blood vessels run through the muscle tissue.

Fat
The fish oil is found in the muscle tissue, in the liver and just under the skin. A large part of the fatty acids are polyunsaturated. The linoleic acid content is about 20 per cent. In some fish, a high content of cholesterol is found, for example in eel and in the soft and hard roe of herring.

Carbohydrates
There is a very small amount of glycogen in the muscles.

Minerals
As a consequence of the smaller quantity of blood, the iron content is lower than in meat. Sea fish contain in proportion large quantities of iodine.

Vitamins
The vitamins A and D are found in the flesh of oily fish and in fish liver. The livers of halibut and cod particularly contain a lot of vitamin A. The soft and hard roes and the liver contain vitamin C. In addition, the muscle tissue contains come vitamins from the B-group.

Possible health hazards
Fish can contain contaminants, absorbed from polluted water. These are the heavy metals (mercury, lead and cadmium) and PCBs. PCBs are especially found in high quantities in fish which live on or in the bottom. Eel is an example of this. Eels are being used to check how bad the pollution of the water is. Standards now exist for the maximum permissible amount of PCBs (for eel: 5 mg PCBs per kilogram) (Netherlands).

Fish can be contaminated with pathogenic bacteria, for instance *Clostridium botulinum* and *Salmonella*. During preparation, fish can also be contaminated with pathogens such as *Staphylococcus aureus* and *Clostridium perfringens*.

In some fish, parasites are found. The most well-known parasite is the *Anisakis marina,* also called the herringworm (see Chapter 2). All herring which are to be consumed raw must be frozen for 24 hours at $-20°C$ to eliminate this worm. This is a legal requirement in some countries. Other methods of killing the worm are heavy salting for 10 days at 0°C in a salt concentration of 20 per cent NaCl, or marinating for 10 days at pH 4.2.

7.2 PRODUCTION AND DISTRIBUTION

Fishing

Fishing has been an important source of income for many Northern European communities, for centuries, especially for those living near the coast. Since the discovery of curing, by Willem Beukelszoon van Biervliet (circa 1400), the Netherlands has played a large role in herring fishing, whereas its fresh-water fishing is less important.

Sea fishing

The Dutch fishing fleet fishes mainly in the North Sea. Along with the Netherlands, England, Scotland, Norway and Denmark play a part here. To prevent the extinction of fish species the EC has set up catch quotas or even imposed bans. In 1977, for instance, the Dutch fleet was forbidden to catch herrings. This bar was lifted in 1983. In the meantime herring were imported from Denmark. At the present time (1985) the Dutch fleet has a catch quota for all roundfish and for some flatfish (sole and plaice). Sea fishing is subdivided into small and large sea fishing.

Small or **coastal fishing** takes place near the coastline. The fishing is done in cutters which are fitted with trawl-nets. These have the shape of a large pointed bag, or bow-net, and are pulled by steel ropes over the sea bottom from portside or starboard. A different method of catching is applied by beam dredge-net fishing. The fishing is done with two nets which are set out one on either side of the boat. Because the beam is pulled over the bottom, the fish are disturbed and are chased into the nets. The catch from coastal fishing consists mainly of flatfish (such as sole, flounder, dab and plaice) and shrimps. Some roundfish are also caught, such as herring and mackerel. The catch is immediately sorted on board, and often brought still fresh ashore and traded at a fish auction.

Large or **deep-sea fishing** takes place far from the coast. The fishing is done with freeze stern-trawlers. These do not fish over the side, but pull the net in over the stern. The fish is then processed below deck, thus avoiding interference from adverse weather conditions. Because the ships often stay at sea for weeks on end, the fish is preserved at sea. The catch is mainly mackerel, herring and halibut.

Fresh-water or inland fishing

On rivers and other inland waters, fishing takes place with cod-end nets (bag-shaped nets which are pulled by two boats), blow-nets and angling rods. Some kinds of fish are reared in special breeding ponds. Some examples of this are salmon, eel and trout. Sometimes the fish are released into inland waters after they have reached a certain length and weight. These released fish stay there for some time before they are caught. But mostly the fish are sold straight from the breeding pond.

A recent development in fish farming is the use of warm water ponds (25°C) for the breeding of sheatfish.

For sport, angling fish are also released, for example carp and hake. The Dutch Fishery Law (1963) stipulates a 'closed season' during which it is not permitted to fish. A fishing licence is also necessary to take part in angling.

Stripping, washing and cooling

The fish is killed after catching. The artery behind the head is cut so that the fish is quickly drained of blood. It is then, in most cases, 'stripped', which consists of cutting open the stomach from the head to the anus, so that the intestines are removed. This has the advantage of also removing intestinal bacteria, while autolysis by digestive juices is also prevented. Then the fish is rinsed with cold water and stored in crushed ice. Sometimes salt is added to the ice to lower the temperature a little more. The ice surrounds the fish and will melt as it extracts heat from the fish, which will be cooled down. Residues of dirt and bacteria will rinse away with the melting ice at the same time.

A small number of fish are not gutted. The 'closed' or 'stiff' fish are bled empty, but all the organs remain in the stomach cavity. In the fresh state they are very easily spoiled. When these fish are later smoked, the presence of the digestive organs has a beneficial effect on the taste. Often these fish also contain soft or hard roes.

The process of opening the stomach cavity in herrings is called **gutting**. This differs from stripping in that not everything is removed from the stomach cavity. The pyloric appendage of the stomach has to remain because it contains enzymes (trypsin) which play a large part in the maturation of the herring, so that they can be eaten 'raw'. With anchovies, the 'emptying' process is called **heading**. This is the same as gutting, except that here the head is also removed.

Post-death processes
The after-death processes follow roughly on the same lines as those for meat. *Rigor mortis* sets in within a few hours and ceases after one day. Then a ripening process starts during which the glazed effect disappears and flavour components are formed. The final pH value which is reached lies around 6.2. Higher final pH values (pH 7) do occur if the fish became very tired while being caught (for instance if it was dragged in the trawl for a long time and the energy store totally used up). This has later consequences for the storage of the fish. Not all fish go through the *rigor* stages. Fish can also be killed and straight away be prepared and eaten.

Distribution
After arrival in the harbour the boats unload their fish in plastic crates packed with ice. In the auction hall the fish is sorted by size and quality. The sorting by quality is done by smell, colour, condition of the mucus layer, and possible bruising.

Then the fish are auctioned in lots to the wholesaler, the retailer or restauranteurs. More expensive fish such as sole and turbot are auctioned by the clock; the cheaper kinds are sold by the auctioneer in the hall. Because of temperature considerations, auctioning always takes place early in the morning. Often, the wholesale trade has rented room in the auction hall where they fillet the fish and distribute it. There are also some in the auction hall whose only function is filleting fish. Some filleting machines can process more than 1000 fish per minute. Wholesalers can also sell processed fish to retailers. For transport to the retailers, the fish — filleted or whole — is packed in ice in crates in refrigerated trailers. Fresh fish is sold to the consumer in the shops, in markets or from street stalls.

7.3 KINDS OF FISH

A much-used classification consists in dividing fish into fresh- and salt-water fish. A short description of some common kinds of fish follows.

Salt-water fish

Anchovy
A small fish (length 9–15 cm), which looks like herring with a pronounced upper jaw. The colour of the skin is silvery with a blue-green back. Anchovy is headed, filleted and preserved in a large amount of salt.

Anchovy has to be de-salted before use, but, for the consumer, lightly salted anchovy is available. This is de-salted and canned in oil after the salting and fermentation process. The end product has a limited shelf-life. Anchovies are much used in anchovy sauces and pastes.

Flounder
A flatfish (length 30–40 cm) similar to plaice. The body is slimmer than that of plaice, and the back is brown-grey with yellowy spots and bumps. Along the fins are sharp prickles. The flesh of flounder is firmer and finer in taste than that of plaice.

Dogfish
The dogfish, which can reach a length of 1 metre, belongs to the cartilaginous fishes. The body is elongated with teeth-like scales. On the back are two dorsal fins, each with a thorn-like projection. The skin is grey-blue. Skinned dogfish is also sold smoked, and is very similar to smoked sea-eel.

Brill
A flatfish, 30–60 cm in length, with a smooth yellow-grey skin. The flesh is slightly blueish in colour.

Herring
A silvery fish with a dark blue back and a pronounced lower jaw. The tail is divided. There are several species, such as North Sea herring with a length of 35 cm and Channel herring and Norwegian herring which can both reach a length of 50 cm. The *Hollandse nieuwe* or *matjes* herring is caught from May to the end of September and also sold in these months with the description *nieuwe*. In this period the herring does not contain soft or hard roe. The fish flesh is white, fatty and tastes good. *Matjes* are traded gutted and lightly salted. 'Roed' herring containing soft or hard roe, caught in the autumn, and spent herring, caught after the spawning season, are strongly salted, smoked or marinated. These herrings are tougher, less fatty and less tasty than the *matjes*.

Halibut
The largest of the flatfishes, with a maximum length of 3.5 metres. The body is long and slim with a sleek brown-grey skin. The weight can be up to 250 kg. As a rule, the fish is sold in cutlets.

Scad
This fish belongs to the cod family. It lives in large schools near the coast. Scad is one of the fish caught in the North Sea. It can reach a length of 40 cm. The flesh is fairly fatty. Scad is reconizable by a pronounced series of large hard gleaming scales along the backline and by the prominent large eyes.

Cod
A large fish (length 80 cm) with a large head and a chin barbel. The back is olive-green with brown spots. On the side of the body is a white lateral line. Cod liver oil is sold in capsules.

Coley
This fish looks like cod and is about the same length. The back is coloured dark green to black. The lower jaw is pronounced. The inside of the mouth is black. Coley fish (sometimes called saithe) are occasionally dyed, smoked and sold as a substitute for smoked salmon.

Mackerel
A medium large fish (length 40 cm) with a pointed head and a slim body. The back is a metallic colour with black stripes across. The tail is sickle-shaped. Mackerel is often steamed and sold whole or filleted.

Red gurnard
This fish with its large head and reddish back with spiny fin rays is also called sea-cockerel or grunting cockerel. The name comes from the fact that the gill covers push over each other and make a grunting noise when the fish is lifted from the water. The fish can reach a length of 55 cm.

Sardine
A small (10–25 cm) silvery fish, similar to herring, with a pronounced lower jaw and large scales. Sardines are mostly caught in the Atlantic Ocean and preserved canned in oil. Large specimens are called pilchards.

Dab
A small flatfish (30 cm) with a brown-yellow skin rough to the touch, a fat tail and a curved lateral line at the height of the breast fin.

Haddock
This fish belongs to the cod family, but is smaller (50 cm). The head is small with a very short chin barbel, the back is darker, and on both sides is a black lateral line from the head to the tail. At the height of the first dorsal fin just behind the gill covers is a black spot, also called the St Peter's thumb mark. The liver of haddock is canned and sold.

Plaice
This flatfish (20–30 cm) can be recognized by its brown-grey smooth skin with orange spots, also called 'rust spots'.

Sprat
A small herring-like fish (about 10 cm in length) with sharp stomach scales. Sprats are often smoked and sold in bunches.

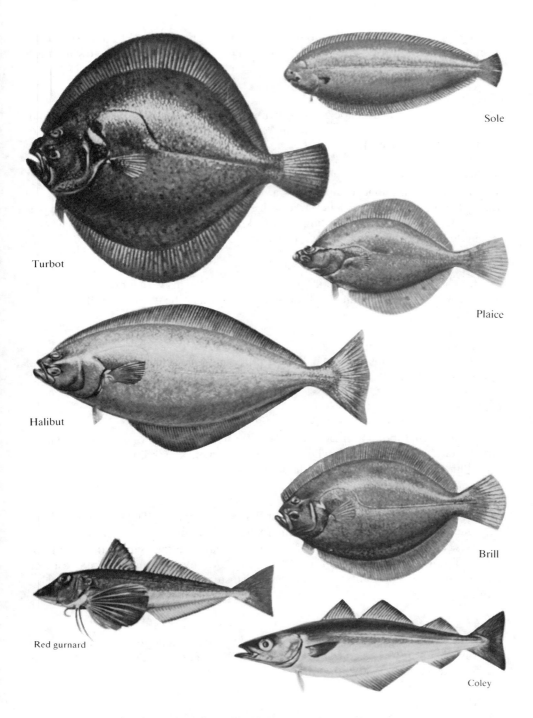

Fig. 7.2 — Several kinds of salt-water fish.

Turbot
The body of this flatfish is diamond-shaped with a grey-brown knobbly skin. The diameter of the turbot can be up to 1 metre. Turbot is regarded as one of the finest kinds of fish.

Sole
This flatfish has an oval shape and a brown-grey skin. The head is small in relation to the body. Sole reaches a length of about 40 cm. Sole, like turbot, is regarded as one of the finest kinds of fish. In the last few years sole has become a very expensive fish.

Tunny fish or Tuna
This fish looks like mackerel and has a dark blue back with grey flanks. The size varies from 50 to 200 cm. The largest size can weigh up to 300 kg. The meat is coloured red to white depending on the kind. In some countries, tuna is only aviable in cans (with vegetables, with sauce, or in oil or brine).

Whiting
Whiting belongs to the cod family. It is a small fish (40 cm) with a pointed head and light brown skin. The flesh is a little glassy and less delicate in taste than cod or haddock.

Conger eel
This salt-water fish looks like an eel, but is fatter and less delicate in taste. Conger eel is sold skinned and is prepared as for eel.

Fresh-water fish
Perch
A rapacious fish (about 40 cm long) with hard scales and many bones. The head has a large mouth with many teeth. The body has a green back with lines across. The first dorsal fin is full of spines. The caudal fin is red.

Trout
Trout belongs to the salmon family. The fish is 25–40 cm long. The colour and shape vary according to the kind (mountain, river, or rainbow trout). A thick layer of mucus is characteristic. Most trout are bred in special trout farms in basins of running water. Trout is preferably sold alive to fish restaurants. Killing just before preparation improves the quality, the looks and the taste of the fish. Trout is also available smoked.

Carp
A medium-sized fish (60–70 cm) with a flat high body with large scales and sensory barbs on the mouth. The colour of the back is olive green. Young carp is released in Dutch waters and in many other places.

Sheatfish
This fresh-water fish originates from the tropical waters of Africa and Asia. The fish has a remarkably large head and long barbs around the mouth. The colour of the skin

is grey black. Sheatfish is bred in warm-water basins in the Netherlands on a small scale. The consumption of sheatfish is still very small. This fish is sold fresh and also smoked.

Eel
Eel is a long cylindrical fish with grey-green to olive-brown back, and a yellow stomach. Microscopically small scales are hidden in the thick skin. Eel can reach a length of 1.5 metres. The eel goes to sea to breed. The larva or glass eels take several years to reach the coast, then swim up-river via the river mouth. After a certain time they go out to sea again to spawn. It is during this migration that the eel is caught. In some places, eels are bred. Eel is sometimes available fresh, but mostly steamed. Small eels are called 'elvers'. Jellied eel is sometimes available.

Pike
This hunting fish has a large flat mouth with teeth. The back is black on the flanks with grey-green bands on the sides. The length is 60–80 cm. Pike is also bred to be released in the open waters.

Sturgeon
The sturgeon is a large kind of fish, the female of which can reach a length of 5 metres and a weight of 150 kg. The body has a torpedo shape with a brown-green back. The sturgeon swims in Russian waters. The fish is well known because the roe is processed into caviar.

Salmon
This medium to large fish (60–100 cm) has a slim, silvery body. In front of the caudal fin is a so-called adipose fin. This is a small fin without fin rays. There are many kinds of salmon, coming from several countries, such as Russia, Norway, Japan, and Alaska, with varying weights (from 2 to 20 kg), varying taste and varying colour (from white to pink and red). Salmon is born in fresh water but travels later to the sea. The fish swim back up the rivers when they are ready to spawn. They are caught during this voyage. The quality of the flesh is then very good. Salmon is also bred and later released in inland waters. Salmon is mostly sold fresh, smoked or canned.

Salmon trout
This is a cross between salmon and trout. The fish has got pink-coloured meat and is very tasty. The body is more elongated then the body of the salmon, the mouth is rounder and smaller. The back is silver-white with black flanks. The salmon trout measures about 80 cm, and has a little fat fin like salmon.

7.4 FISH PRODUCTS

Some fresh fish is industrially preserved. Preservation is of great importance to the trade. It also offers advantages for the consumer such as convenience and variety in taste. Preservation also permits a country's fish products to be exported and foreign preserved fish products which either are rare or do not exist here to be imported into that country.

Fig. 7.3 — Some fresh-water fish.

Fish products

The methods of preservation which are used for fish are:

— deep-freezing
— salting
— drying
— smoking
— marinating
— sterilizing
— packaging in cans, glass or plastic foil
— irradiation.

Deep-freezing

After stripping, the fish is filleted and cut into pieces. These are pressed into blocks or thick layers, sometimes covered with breadcrumbs and pre-fried, deep-frozen and packaged. Some fishes are frozen whole, for instance salmon and trout. Because the chance of drying out through the evaporation of ice crystals from the relatively large surface during storage is great and because this will cause protein denaturing, the fish is sometimes glazed. To do this deep-frozen fish is for a moment dipped in cold water, by which a very thin layer of ice forms. This layer will partly prevent evaporation. The storage time for deep-frozen fresh oily fish is about half a year, for leaner kinds up to one year, for prepared kinds a few months. The storage time is given on the packaging. Some examples of deep-frozen fish:

— salmon and trout in stripped, non-filleted form;
— fillets of cod, haddock, coalfish and whiting;
— breadcrumbed, fried pieces of fish (cod, coalfish or haddock) of regular size for fishfingers, for example.

Sometimes it states on the packaging that the fish has few or no bones.

Salting

This preserving method was frequently used in former days but has been replaced by deep freezing. At present salting is mainly used for herring and anchovy. There are two methods.

Dry-salting

The fish is sprinkled with salt. If the moisture, which is extracted from the fibres by the salt, can drain away, a drying of the fish takes place at the same time.

Brining or wet-salting

In this case the fish is put in the brine, or is sprinkled with salt, and the moisture is not drained. The brine solution has to contain at least 20 per cent NaCl to preserve. This is called 'heavy salting' in the herring fisheries. If less salt is used, combined with refrigeration, the fish is suitable for immediate consumption. De-salting is in the latter case not necessary.

The storage time is, depending on the salt concentration, the desired structure and the fat content, 1–6 months, often combined with refrigerated storage (0–1°C); with heavily salted anchovies it may even be several years.

Some examples are given below.

— *Matjes* herring, a gutted lightly salted herring (2–3 per cent NaCl), without soft or hard roe.
— Salt herring, a non-gutted herring with hard or soft roe and heavily salted (20 per cent NaCl). The final salt value in the fish is about 12 per cent. Before consumption, the salt herring has to be de-salted in cold water or cold milk.
— Anchovy, a little fish which is headed, filleted and packed with a lot of coarse salt. After one year's maturing, the fish is tasty and soft. After several years, the fish liquifies and is processed into paste or mixed with butter into anchovy butter.
— Caviar or salted roe from the sturgeon. Roe from other fishes (among others, salmon and tuna) is available as coloured fish roe, lightly salted (3–3.5 per cent NaCl) or salted more heavily (12 per cent NaCl).

Drying
The drying process can take place in two ways; a natural drying by the wind and/or freezing, or artificially in wind tunnels.

Mainly lean fish is dried. Fat fish gives a bad product, because it can become rancid. During drying, the moisture content of fish drops from 80 per cent to 15 per cent. The storage time for dried fish is about one year, if well packaged (vacuum-packed).

Examples:

— Dried cod and haddock (stockfish). The fish has to be soaked before use; sometimes it is sold pre-soaked.
— From several kinds of dried fish, fishmeal can be made, which can be processed into other products.

Smoking
There are two methods used.

Cold-smoking
After a heavy salting the fish is strung onto skewers. The smoking time can vary from 12 to 27 hours at 30°C.

Warm-smoking
The smoking process takes several hours at 80–120°C. It is important for the keeping quality that the temperature in the centre of the fish has been at least 65°C. If this temperature is not reached, then several original fish enzymes are still active, amongst others fish phosphatase, which can cause spoilage.

The consumer can learn to distinguish between cold- and warm-smoked fish. The summary shown below gives a review of the differences in looks, taste, keeping qualities and uses of cold- and warm-smoked fish.

Cold-smoked fish
— The skin is smooth around the fish and is difficult to remove.
— The flesh is not cooked and is attached to the bone.
— The flesh is dried through, reasonably salt to the taste and has a strong smokey smell and taste.
— It is easy to slice the flesh thinly.
— Storage time: several weeks in the refrigerator.
— Examples: herring, salmon, sprat, trout.

Warm-smoked fish
— The skin bulges and is loose on the fish.
— The flesh is cooked and is easy to remove from the bone.
— The flesh is moist, lightly salty in taste and has little smoke smell and taste.
— The flesh is not easy to slice thinly.
— Storage time: several days in the refrigerator.
— Examples: herring, mackerel, eel.

Marinating
A mixture of vinegar, herbs and salt is added to fresh, salted, deep-frozen or baked fish. Marinated fish can be stored under refrigeration for 1–2 months.
 To increase the keeping quality a preservative can be added or the product can be pasteurized.
 Examples:

— Herring fillets, *rollmops* (herring fillet rolled around a gherkin), fried herring.
— Jellied eel.

Sterilization
Fish is mostly cut into steaks for canning, and is sometimes skinned and boned. Before sterilization the can is sometimes topped up with water, oil or sauce.
 To prevent the fish sticking to the packaging, the fish is pre-cooked or the can is lined inside with paper. Pre-cooking is also done with oily fish, to lessen their charcteristic oily taste (mackerel). They are afterwards sterilized with another (better tasting) oil, sauce or vegetable mixture. If a smokey taste is appreciated then the fish is smoked before canning. These preserves have nearly an unlimited keeping time as a result of the sterilizing process; but because of quality deterioration (especially loss of consistency), it is not advisable to store them longer than about one year.
 Examples: salmon, tuna, herring, sardines, mackerel, haddock liver.

Packaging in cans, glass or plastic foil (without heat treatment)
Airtight and possibly vacuum packing in cans, glass or foil takes place after pre-treatments such as smoking, salting or marinating. These semi-preserved products can be kept for about 1–2 months if stored in a refrigerator.
 Examples: smoked eel, *matjes* herring, anchovies in oil, sliced smoked salmon, steamed mackerel.

Irradiation
Fresh and fried fish can be treated by irradiation, to lower the bacteriological count. If stored under refrigeration (4°C) the storage time is lengthened by several weeks. Example: several kinds of fresh fish and shrimps.

Diet products
Fish and fish products in cans or deep-frozen are available for a sodium-restricted diet.
There is a choice of several lean kinds of fish for a low-energy diet.

7.5 QUALITY DETERIORATION, SPOILAGE AND STORAGE

The quality of fish depends on, among other factors, the time in the season of the catch, the fishing grounds, the age and the spawning time. Also, freshness is an important factor. All fish which are brought in are divided into three freshness classes (Extra, A, and B), as laid down by the member states of the EC. The freshness class depends on smell, colour, damage to the fish such as light bruising or scratches, and the condition of the mucus layer.

Fresh fish can be recognized by:

— *the smell* Sea fish especially contains urea and trimethylamine oxide. The last-mentioned compound is after death broken down by bacteria into tri- and di-methylamine, which spreads a disagreeable smell. Fresh fish should smell fresh. The longer the fish is kept (even refrigerated), the stronger the typical fish smell becomes.
— *the skin* This is covered with a protective mucus layer, which should be undamaged. The skin should look moist and glossy. If there are scales on the skin, they should be securely fastened.
— *the flesh* This should feel firm to the touch and be attached to the bone. The colour is a transparent blue.
— *the gills* These should be red and moist and should smell fresh. One exception to the red colour are the gills of the mackerel, which are coloured brown.
— *the eyes* Fresh fish has spherical, somewhat protruding eyes, which are clear white with gleaming black pupils. Through storage on ice they become flat and turbid.
— *the stripping cut* This should have firm, fresh edges.
— *the stomach cavity* This cavity contains the stomach meat, which should look black and glossy.

Fish spoils faster then meat. Reasons for this can be:

— the chance of contamination from the intestinal flora during stripping is high;
— the pH of fish is higher than that of meat, which causes spoilage to begin sooner;
— the fish flesh is and stays more moist, also while kept on ice;
— fish is by nature contaminated with psychrophilic bacteria which even while stored under refrigeration can still cause spoilage.

Forms of quality deterioration resulting in spoilage are, among others, proteolysis, oxidation of fat or mould growth, especially with smoked fish. It is advisable to prepare and eat fresh fish on the day of purchase. It can possibly, if well wrapped, be kept for one to two days refrigerated (0–5°C). If the fish is to be kept longer, then storage at deep-freeze temperature will be necessary.

7.6 LEGISLATION

The fish regulations demand exact labelling and specify the composition of preserved products.

Hygiene regulations set forth guidelines with regard to hygiene in the rooms where the fish is traded.

The herring regulation 1984 (Netherlands) defines the obligation to deep freeze, to salt heavily or to marinate herring in connection with the herring worm. The regulation also contains rules about the organization of the fishing fleet, the way in which the fish is supplied (fresh, lightly salted, heavily salted or deep-frozen) and the processing of the same.

Laws on agricultural residues and surface water contamination try to prevent, as far as is possible, the contamination of fish by toxic chemicals.

8

Crustaceans and shellfish

8.1 INTRODUCTION

Crustaceans and shellfish are often mentioned together. According to most Fish Regulations and in the Fisheries Law they are classed as fish. But there are large differences.

The **crustaceans** belong to the arthropods without an internal skeleton, but with a stout calcareous armour by which the body is protected. The body consists of:

— the head, often with stalked eyes and long antennae;
— the thorax with walking appendages;
— the abdomen with appendages which facilitate swimming.

Shrimps, crabs and lobsters are typical crustaceans.

Shellfish are molluscs which live in one shell, or in two shells which close together, in water or on land. The calcareous shell is outside in most cases, with the exception of the squid, which has an internal shell.

Mussels, oysters, scallops, cockles and snails are typical shellfish.

Consumption

In 1983 the consumption of crustaceans and shellfish in the Netherlands was 3.3 kg per head whereas in the UK the figure was below 0.3 kg per head. Because of availability, consumption is highest in coastal areas. Fear of contamination by coastal pollution may affect consumption figures.

Composition

Crustaceans and shellfish contain roughly the following compounds:

        water   80–85%
        protein   6–18%

fat   1–2%
carbohydrates   0–4%
minerals   calcium, iodine and iron
vitamins   A and B complex.

Possible health hazards
Because most crustaceans and shellfish are cooked alive, the chance of deterimental effects from microbial infection is very small. Cooking alive is necessary to inactivate the very active proteolytic enzyme system, which attacks the meat straight after the death of the animal and makes it soft and crumbly. With crustaceans, the incurved tail is a sign that they have been cooked alive.

It is important with shellfish, such as oysters which are eaten raw, that the water in which the animals live and are caught is free from pathogenic microorganisms. This is strictly checked by the authorities.

Shrimps imported from Asiatic countries are irradiated on arrival in Holland to prevent contamination with pathogenic bacteria.

Mussels can absorb a neurotoxin from plankton. After consumption of such mussels paralysis can occur in humans. The protein of shellfish, especially mussel protein, can sometimes lead to allergic reactions.

8.2 KINDS OF CRUSTACEANS

Shrimps
Shrimps are somewhat grey, a little transparent and elongated (see Fig. 8.1). They

Fig. 8.1 — A shrimp.

live at the bottom of the sea. They are caught by shrimpers throughout the year with special shrimp trawls. To prevent spoilage, they are cooked on board straight after being caught. They are then quickly cooled, often, however, with sea water, from which subsequent contamination can occur. As a result of the cooking, the shimp changes in colour and shape. The pink–red coloration is caused by carotenoids which were previously bound to proteins, but are released on heating. Also as a result of heating, the live shrimp body curls.

Most shrimps are brought ashore unpeeled. The peeling is done as a home industry. This is a not unimportant source of income for the fishing villages. Home peeling of shrimps is permitted under the regulations of the **Shrimp Regulations (Goodslaw)** (1984) (Netherlands).

To improve the keeping quality of the shelled shrimps, preservatives such as benzoic acid may be added or they can be irradiated.

Kinds of shrimp
The **Dutch** or **North Sea shrimp** (*Crangon crangon*) occurs along the coasts of the whole of western Europe and in the Mediterranean.

The colour of this shrimp varies from brown to grey, depending on the colour of the sea bottom. In England they are called brown shrimp, in France '*crevette grise*'. The size of the unpeeled shrimp is 5–7 cm, peeled 1–2 cm. They are brown–pink after being cooked.

Next to these another kind of shrimp is found in smaller quantities in the shrimp trawls: the **red sturgeon shrimp** (*Pandalus borealis*). This shrimp is slightly different in colour and shape, and larger in diameter than the *Crangon*.

The **Norwegian** or **North Atlantic shrimp** is caught in the cold North Atlantic waters. This shrimp can be 10 cm unpeeled, and peeled about 3 cm. After cooking, the shell and the meat on the outside become red; inside the meat stays white.

The **small Asiatic shrimp** comes from the warm tropical and sub-tropical waters in south-east Asia (such as Indonesia and the Phillipines) and West Africa (such as Senegal). These shrimps are the same size as the Norwegian, but they have more white divisions.

The **large Asiatic shrimp**, also **tropical** or **Chinese shrimp**, called **Gamba** or **oedang**, is recognizable by its large tail. They sometimes are incorrectly called scampi, which is the name of a very small lobster.

The **small Asiatic fresh-water shrimp** is bred in Bangladesh in rice fields and ponds. They look very much like the Asiatic fresh-water shrimp, and are imported into North America.

Most of the Dutch and Norwegian shrimps are sold cooked and peeled, perhaps deep-frozen. The small Asiatic kinds are cooked, peeled and deep-frozen in the country where they have been caught, and are then exported. The large shrimps are mostly deep-frozen uncooked, often only the tails. Because of the unreliability of the water, shrimps from Asiatic countries are irradiated in the deep-frozen state directly after arrival in the Netherlands to kill pathogenic germs.

Crabs
These crustaceans have a flat oval-shaped body with a scarcely visible little tail and a very large combined head and thorax. Ten legs are attached to the body. The front pair of legs are supplied with claws and contain the nicest meat. This meat has got small pieces of cartilage in it. Compared with the meat from the crab claws, the crabmeat from the head and thorax is less red from the outside and inside pink-white. Underneath the carapace is the less valued grey-coloured meat and the liver, which has a sandy or grey–green colour.

Fig. 8.2 — A crab.

Kinds of crab

In the Netherlands, the **North Sea crab** is found, which is recognizable from the stone-red-coloured carapace with black pincer points. The carapace of an adult animal is about 20 cm wide.

In France, other types of crab occur, for instance the **Sweet Spidercrab** with longer legs and smaller body.

In Asia, Russia and Canada, the **Giant Crab** (*kamsjatka* King Crab) is found. The name relates to the legs which can reach a length of 1.5 metres, while the body is about 50 cm.

The North Sea crab is sold alive or cooked. The meat of the giant crab is imported deep-frozen or canned. From Japan come crabsticks, which contain a minimum of 10 per cent crabmeat with the rest other fish meat, for instance white fish. White fish is a collective name for carp-like fish with flakey white flesh.

Lobsters

The body is long and narrow. In the front, two appendages are developed into large claws. The shell varies in colour, and depending on the sea bottom may be brown, black, green or red. When the lobster is cooked it turns red as a result of the presence of carotenoids. The liver is thought to be a delicacy. Females (hens) may contain an orange–red roe (coral), which in restaurants is used as decoration or is used in the preparation of a sauce.

Kinds of lobster

The **Sea Lobster** is caught near Scotland, round the Scandinavian countries and near Canada and other sites. This kind of lobster can reach lengths of up to 80 cm with a weight of 700 g. The shell is deep blue in colour and heavy as a result of chalk

Fig. 8.3 — Lobsters.

deposition. One of the two claws has large teeth, while the other is supplied with small sharp teeth.

The **crawfish**, **langouste** or **spine lobster**, is also a salt-water lobster, and is distinguished from the sea lobster by its red–brown shell and the back being not smooth but notched. The crawfish has no claws but very long antennae on the head. Weight for weight it yields more meat than the sea lobster because the shell of the langouste is less heavy. After cooking, the shell of the langouste is coloured orange. This kind of lobster is imported, for example from France, Yugoslavia and South Africa.

The **langoustine** or **Dublin Bay Prawn** is a small sea lobster with kidney-shaped black eyes, a pink shell and slim claws with very sharp teeth. The claws supply hardly any meat. The total length of the langoustine is 30 cm. They live in the depths of the North Sea, the Mediterranean and the Adriatic Sea and in the Atlantic Ocean. Langoustines are also sold under the name **scampi**.

The **crayfish** (*ecrevisse*) is a small (15 cm) slim lobster which lives in fresh water. The shell is brown–black and colours red on cooking. They are found in small rivers in the Ardennes, in South Europe and in North America.

Lobsters are sometimes brought in alive and kept in aquaria. The claws are bound, to prevent nipping. They are cooked alive, after which the meat can be removed from the claws, the shell and the tail. From the langoustines (scampi) sometimes only the tails are sold.

8.3 KINDS OF SHELLFISH
Mussels
The mussel is a mollusc which lives in a bivalvular shell. The shells are oblong, the colour outside is blue to black and inside white to light blue. The mussel has a 'foot' of muscle and glandular tissue with which it can move itself a little and from which the beard or byssus threads, with which they attach themselves, are splayed out.

Mussels occur in the North Sea, the English Channel, the Atlantic Ocean and the Mediterranean. They are also bred in Belgium and in the Netherlands. The breeding season is from the month of April until the end of June; the secretion of the mussel spat takes place. This is fished up out of the sea and (in the Netherlands) reared further in waters leased by the Dutch Government in Zeeland and the Waddenzee. After one and a half to two years the mussels are caught, separated from any attached shells, 'cut' (byssus threads or beard removed) and then washed clean, packaged and sold refrigerated.

Fig. 8.4 — Shellfish (oysters and mussels).

There is no official division according to varieties with mussels. A differentiation can be made according to the circumstances in which the mussel grows (naturally or specially bred). The bred mussel is of a better quality than the naturally grown kind.

Mussels are brought in alive throughout the year with the exception of the months of April, May and June. Breeding takes place in these months and the mussel is less fit for consumption.

When alive, the shells are closed or close after touching. If they stay open then the mussel is dead. Fresh mussels spoil quickly. It is advisable to keep them refrigerated, perhaps under water so that the mussels can rinse themselves clean. Before cooking, any dead or mud-filled mussels must be removed. During cooking, the shells open, so that the mussel can be removed.

The colour of cooked mussels depends on the thickness of the shell; through a thick shell little light permeates so the mussel stays white. A thin shell gives mussels an orange colour after cooking. However the colour of the mussel does not interfere with its quality. Cooked mussels have a short storage life (only 1 day under refrigeration). Sometimes they are sold cooked, crumbed, and deep-fried or marinated.

Oysters
The oyster is a bivalvular shellfish with a rough shell of which the lower half is waved and the top half, which carries the annual rings, lies on top like a lid. The two shells can move by means of a strong contractor muscle.

The oyster is an hermaphrodite and the larvae develop in the shell. After about three months the young oysters are expelled. They attach themselves to suitable places, after which the shell formation starts. They are ready for consumption after about six years. Previously, oysters were only gathered on oyster beds which had been formed in a natural way. Now they are often farmed on special sites. In the Netherlands, chalky rooftiles or empty mussel shells are used, to which the oyster attaches itself. The oysters are gathered when they are full-grown and are stored in oyster pits until they are transported. The water where the oysters are kept has to be free of pathogenic bacteria. There is a continuous bacteriological supervision by government officials. When the oysters are transported, a certificate is enclosed which declares that the water was of the required standard.

Kinds of oyster
The **Zeeland oyster**, is cream-white in colour and is grown in the Netherlands (East Scheldt) and Belgium. As well as the flat kind, which is ready for consumption only after five years, a rounder kind is grown in the East Scheldt, which has been developed from the spat of Japanese varieties. These oysters grow very quickly and are ready for consumption within two years.

The round oyster, which contains less meat and is cheaper than the flat oyster, is sold under the name *Creuses de Zélande*. The largest and best kinds of the flat oysters are sold as *Zeeuwse Imperialen*. The largest *Imperialen* are labelled with six noughts, corresponding with the number of annual rings on the oyster shell.

The **French oyster** is a grey–green colour inside, because of algae and diatoms which are especially grown in the storage places and which give a specific taste. They are bred along the whole west coast of France. They look like the Zeeland oyster.

The **Portuguese oyster** is more oblong in shape, less expensive and less delicate in taste than the Zeeland oyster.

The oyster is sold alive from the month of September until the end of April, packaged in baskets which are filled with seaweed. A live oyster has a well-closed shell. When closed, these can live out of water for about 14 days without oxygen if stored under refrigeration. If the oyster is to be eaten raw then it is necessary to cut the contractor muscle with a special knife. Sometimes they are delivered ready for the table, which means with the shell opened and the oyster loosened from the shell.

Scallops (*Coquilles Saint Jacques*)
This bivalvular shellfish lives in the Mediterranean and gets its name from the pilgrims who went on pilgrimage to the shrine of St James in the northern Spanish town of Santiago de Compostela. An empty scallop shell was used as a begging-bowl. On their return home, the pilgrims used the shell as decoration on their coat or as a head cover, as a sign that they had completed the pilgrimage.

The scallop shell is light in colour, sometimes white and towards the edge brown–orange; it is fan-shaped and grooved. The size is about 12 cm. In addition to the mollusc, sometimes an orange–red roe is found inside, and a little bag with black liquid, which is not edible. Before consumption, the beard has to be removed. Where fresh scallops are not available the choice is between deep-frozen and canned. The empty shells are often used as 'plates' to serve salads, etc.

Cockles
This bivalvular shellfish occurs on the sandy grounds of the North Sea and the Mediterranean. The cockle digs itself in and lives in the sea bottom. With the aid of a pressurized water technique the cockle is stirred up and spouted into a suction basket. The cockle shell is about 8 cm long and ribbed. The colour is white to yellow with brown stripes. In the Netherlands, cockles are sold fresh on the spot. Cooked and deep-frozen they are exported to Spain and other places. Cockles are also put in vinegar.

Snails
Snails belong to the univalve molluscs and live in a curved turret-like shell, some on land and some in the sea. Only a few are acceptable for consumption.

Fig. 8.5 — Apple or vine snail (*escargot*).

Kinds of snail
The **apple** or **vine snail** (*escargot*) is especially bred in areas with vineyards in France and Belgium. In the autumn this land snail digs itself in to hibernate and closes the shell with a white chalky lid. It is at this stage that they are collected.

The **winkle** is a sea snail which lives in a small bullet-shaped little shell of a few centimetres length. The shell is closed with a horny lid. They are caught in Zeeland and Belgium with special dredge-nets. Just like the oyster, the winkle can live for a short time outside the water.

The **whelk** is a large sea snail with a spiral-shaped pointed shell 10–15 cm in length. The shell is grey to orange in colour, with dark stripes on the spirals. Sea snails are sometimes sold fresh near the coast; mostly they are sold cooked or canned.

(Also mentioned here are **frogs' legs** and **turtles** because, in regard to availability, storage time and method of presentation, they fit into this group.)

Frogs' legs

The thighs of certain types of frogs are regarded as a delicacy in French and Chinese kitchens. The meat is tender and white–pink in colour. Before cooking, the legs are skinned and put in ice-cold water for several hours to keep the meat white and to let it swell a little. They are imported deep-frozen from East Europe, France and Asiatic countries. As with shrimps, frogs' legs from Asiatic countries are also irradiated after arrival in Holland.

Turtle

The turtle belongs to the reptiles, which live on land or in the sea. The meat from one specific kind, the sea turtle from the West Indies, is used for the preparation of a delicate clear soup, turtle soup (*consommé tortue clair*). The stock for this soup is made from the meat just under the topshield, the undershield and the flippers.

Turtle soup is available in cans.

Cuttle-fish, squid (or inkfish) and octopus

These are molluscs with an internal shell. Eight or ten legs or tentacles which are supplied with suction pads are attached to the body. In the squid and cuttle-fish the small body, except the head, is covered in a coat in which the shell can be found. This shell is a chalky oval plate which is found washed up on the beach as cuttle-fish shell. The name inkfish comes from the compound produced as a defence (sepia or black ink). They are often eaten in countries around the Mediterranean.

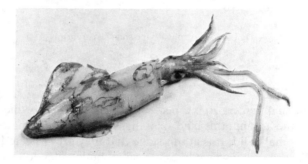

Fig. 8.6 — Inkfish or squid.

Kinds

The **arrow squid** (*calamari*) has wide fins on the sides of the body. It can shoot quickly backwards by forcefully squirting out water.

The **seacat** (*sepia*) is larger then the arrow squid.

The **octopus** (*pulpo*) has eight tentacles, in contrast to the previously mentioned kinds, which each have ten tentacles.

Squid is available fresh. The smaller species have tastier and more tender meat than the larger ones which are tougher. Squid is also sold canned, with or without sauce.

8.4 PRODUCTS OF CRUSTACEANS AND SHELLFISH

Crustaceans and shellfish have to be pre-cooked and taken out of their shell(s), or their carapace removed, before they can be processed and preserved.

Methods of preservation which are used are:

— **Sterilizing**, natural or with oil or a savoury sauce. Sometimes the product is smoked beforehand. Examples: shrimps, crab, smoked oysters in oil, squid in savoury sauce and *escargots*.
— **Marinating**, in vinegar or acid jelly. Examples: shrimps and mussels.
— **Drying**. Examples: shrimps and *trassi* (a fermented shrimp product).
— **Deep-freezing**. Examples: shrimps, with or without shells, crab-sticks, *escargots* in their shells with spiced butter, and poached scallops.

8.5 QUALITY DETERIORATION, SPOILAGE AND STORAGE

Crustaceans and shellfish spoil quickly because of the high water and protein content. Quality deterioration is quickly obvious because of a disagreeable smell, discolouration as a result of drying out, and the meat becoming tough. As with fish, shrimps, langoustines and North Sea crabs are divided into classes of freshness. In the raw state the storage time is very limited. For most types this is a few days in a cool place. By keeping them under water, the storage time can be a little extended. An exception to the short storage life are oysters and winkles, which can be stored for a short time out of water, but refrigerated. The storage time for irradiated shrimps is about four weeks at 4°C.

9

Milk and some milk products

9.1 INTRODUCTION

Milk is a product secreted by the milk gland or udder. The milk gland makes milk from the food under the influence of hormones. Cows eat grass, hay and cut maize. Besides that they also get concentrate foods. The cow chews and ruminates all this food. In the rumen (one of the four stomachs of the cow) the food is broken down by bacteria and enzymes; it is then absorbed into the blood.

Milk is really meant for the young calf during its growth. The lactation period is about 10 months. The rest of the year the cow remains 'dry' until a new calf arrives. The first milk which the cow gives after the birth of a calf is called colostrum. The composition differs from that of normal milk.

Dutch cattle are world famous for their high milk yield. The average milk yield per cow per year can be more than 5000 litres, while some pedigree cows can yield up to 10 000 litres of milk. Because a high milk yield is an inherited characteristic, cows which have this characteristic pronouncedly are registered in herd books. Such pedigree cattle are further bred very carefully to improve the milk yield still more. Besides the milk yield being high, the fat and protein content are also high, especially in the Dutch-Friesian breed (or black-spot breed), the Groningse breed (or blazed-head breed) and the Meuse-Rhine-Ysselbreed (or red-spot breed).

Sheep or goat milks are, in the Netherlands, mostly used fresh only by people who keep a goat or sheep. Sheep and goat's milk is processed into cheese. The milk yield of goats and sheep is fairly low, respectively 500 and 400 litres per year. The taste of this milk is much stronger than that of cows' milk.

The EC plays an important role in the dairy business. The price of milk can never fall below a certain minimum price through the marketing and pricing policy of the EC. The EC will intervene if the milk price threatens to come below that minimum, and buy the surplus milk as skimmed-milk powder and butter at prices determined beforehand by the Ministers of Agriculture of the EC states. Thanks to this intervention the farmers get a minimum price for their milk at all times, not only in

Sec. 9.1] Introduction 135

bad times but also when the market is already saturated. This, however, encourages over-production of milk. The stock of skimmed-milk powder and butter in the storehouses of the EC has increased because of this. To counteract this situation, the EC ministers have set up a super-levy system in 1984, to restrict over-production. This means that the farmers can produce a certain quota of levy-free milk. If they produce too much they have to pay a levy on it.

Consumption
In 1985, 134.7 kg milk and milk products were consumed per head in the Netherlands. In comparison with a number of years ago, the total consumption has scarcely changed (in 1983: 135.2 kg per person; in 1981: 131.0 kg; and in 1979: 135.6 kg). The corresponding figures for the UK are: 1979, 200.7 kg per person; 1981, 196.5 kg per person; and 1983 198.2 kg per person. When individual products are examined, a rising trend in the use of low-fat and acid milk products, such as semi-skimmed milk and skimmed-milk yoghurt products, is seen.

Composition
Raw milk contains (source: *Nederlandse voedingsmiddelentabel*, 34 (1983).):

protein	3.4%
fat	3.9%
carbohydrates	4.6%
water	88 %

Milk also contains minerals such as calcium and phosphorus, and vitamins A, β-carotene (especially in the spring), B_2 and D.

Protein occurs in milk as casein (80 per cent) and the whey-proteins, albumin and globulin (total 20 per cent). Casein is bound to phosphorus. By the addition of acid or alcohol the linkage is broken and casein separates out. This is called curdling of the milk. Casein is barely soluble. In the foodstuff industry the pure casein or caseinates are separated. These are used as stabilizers (for instance in halvarine), as emulsifiers (for instance in coffee whiteners and toppings), as foaming agents (for instance in pudding powders). The albumin occurs in a soluble form. When the milk is boiled, it separates out at the bottom of the pan. Colostrum contains much albumin, which separates out quickly. For this reason colostrum is never processed in the dairy.

Milk fat is yellowish and aromatic, and occurs in emulsified form, which means as little globules invisibly dispersed throughout the milk. The amount of fat can vary depending, among other factors, on the kind of feed and the time of year. In the spring, the fat content can increase even to 5 per cent. Milk fat, as a result of its lower specific gravity, has the characteristic of separating out. This happens when the milk stands undisturbed for a long time. The fat globules float to the top and form a layer of cream. By stirring the milk, the cream can be quickly dispersed. At the moment, a special technique is used to prevent the formation of a cream layer (see section 9.2).

Milk sugar or lactose gives milk a slightly sweet taste. Lactose can be converted into lactic acid by lactic acid bacteria. The milk becomes thicker, the casein separates out and smell and taste compounds develop. During heating, lactose can caramelize, or react with the amino acids of protein (Maillard Reaction), which results in the formation of a yellow-brown colour.

Possible health hazards
Pathogenic microorganisms can be found in milk and milk products. These can originate from a sick cow (for instance, the TB bacillus, or foot and mouth bacteria), or from humans (for instance, *Salmonella*, the TB bacillus, or paratyphoid bacteria).

Pathogenic bacteria are killed by pasteurization and sterilization. However, during packaging or bottling of pasteurized products a re-infection can occur. Also, contaminants can occur in milk and milk products (see section 4.4).

9.2 PRODUCTION AND PROCESSING

Cows are milked twice a day, early in the morning and late in the afternoon. Previously, milking was done solely by hand. The work has to be conducted hygienically because milk is an especially good growth medium for bacteria. During milking by hand, the hind legs and the tail were chained and the udder wiped with a clean cloth. The milk was first strained to remove coarse contaminants such as hair and straw. The milk churns were collected twice a day or taken to the dairy by the farmer. With an eye to hygiene as well as the scarcity of labour, the milking machine has replaced hand milking, the milk churns are rarely used. With machine milking the milk goes straight into a cooled tank (4°C). The chance of pollution from the surroundings is small. But attention has to be given to the cleaning and sterilization of the apparatus. The tank is emptied two to three times a week by a cooled tanker (RMO=motorized milk receiver), which takes the milk to the dairy. A sample is taken from each batch of milk. In the dairy laboratory the samples are analysed for protein and fat content. Besides this, every 14 days the milk is sampled at the farms and a check is made for smell, taste and quality, among other things, through a determination of the kinds and number of microorganisms present. In the dairy, after processes which involve heating, the milk is checked by means of the peroxidase and the phosphatase tests.

The **peroxidase test** shows if the milk has been heated above 85°C. Raw milk contains peroxidase naturally. Peroxide is broken down by the enzyme peroxidase, and gives off oxygen. Oxygen can be detected with a dye. In milk which has been heated above 85°C, no peroxidase is found, so no oxygen formation can occur, and the reaction with the dye will be negative in this case.

The **phosphatase test** shows if milk has been heated above 72°C, so that the TB bacillus is killed. To do this test, sodium phenyl phosphate is added to milk. This compound is broken down by the enzyme phosphatase (naturally present in milk) into phosphorus and phenol. Phenol can be detected with a colour reagent. If the milk has been heated above 72°C, then the enzyme phosphatase will have been destroyed and the reaction with the colour reagent will be negative.

If the milk complies qualitatively with the requirements laid down, it is then pumped into the dairy and undergoes one or more of the following processes.

Flash-heating (Thermization)

Flash-heating is used if the milk cannot be processed immediately, for instance milk which is delivered at the dairy on Saturdays. The milk is heated for 15 seconds at 65°C and immediately after that cooled to 4°C. The enzyme lipase is inactivated and the number of microorganisms are in a limited way reduced by the short heat treatment. Flash-heated milk can be kept refrigerated for one or two days during which taste and smell do not noticeably deteriorate.

Clarification

This is done in a cleaning centrifuge to remove coarse contamination such as dust and dirt from the milk.

Pasteurizing

In this process, the milk is heated for 15 seconds at 72°C. By this means, all pathogenic microorganisms and many other microorganisms which are detrimental to the quality and the storability of the milk are killed, and many enzymes are rendered inactive. Most pasteurization is now done in a continuous process, during which the milk flows in a thin film through a heat exchanger (pasteurizer). The milk is then heated to 72°C, kept hot, and then cooled to 4°C. Through this treatment, the milk, from the point of view of public health, is safe for consumption. An alternate process involves heating milk at 63°C for 30 minutes.

Fig. 9.1 — Louis Pasteur.

Standardizing
In a centrifuge, the milk is separated into cream and skimmed milk. After that, depending on the milk product to be produced, a certain percentage of cream or fat is again added to the skimmed milk. For instance, for milk 3.5 per cent fat is added, for whipping cream up to 35 per cent.

Homogenizing
In milk the fat is present as small emulsified globules, which tend to float to the top. The speed at which this happens depends on the diameter of the fat globules and on the presence of agglutinine (a protein) in cold raw milk. Agglutinine disturbs the working of the surface-active compounds. These compounds are situated around the fat globules and ensure that the fat stays emulsified. The fat globules join together and will float faster as a result of the disturbed working of the surface-active compounds. In the homogenizer, the milk, at a temperature of 55–65°C, is forced under high pressure (200 atmospheres) through very narrow orifices, which decreases the diameter of the fat globules and renders the agglutinine inactive. Homogenizing prevents the cream separating out, and also gives a fuller taste and a whiter colour.

9.3 MILK AND MILK PRODUCTS

Milk is sold pasteurized or sterilized, as full-cream milk (minimum 3.5 per cent fat), semi-skimmed (1.5–1.8 per cent), and skimmed (maximum 0.3 per cent fat).

Pasteurized milk is heated to 72°C. At this temperature all pathogenic bacteria is killed. The milk is packaged in cardboard packs, covered with a thin layer of polyethylene to prevent leakage, or in wide-necked glass bottles with an aluminium cover. The sell-by date is printed on the packaging. Pasteurized milk should be stored in the refrigerator.

Sterilized milk has undergone a heat treatment which is at a considerably higher temperature and often for longer than that of pasteurized milk. Sterilized milk can be divided into *conventional sterilization* and *ultra-high-temperature process* (UHT).

In *conventional sterilization*, free-flowing milk is first heated (15 seconds, circa 132°C) and then held for 10 minutes at a temperature of 110–120°C in bottles sealed with caps and after that cooled as quickly as possible.

UHT sterilization consists of a single temperature treatment, the milk is kept flowing for 1 to 5 seconds at a minimum temperature of 132°C. UHT sterilization can be done in one of two different ways: the direct method, where the milk is injected with hot steam; and the indirect method, where the heat transmission takes place via heat exchangers. After UHT sterilization the milk is aseptically packaged in sterile packaging, such as cardboard covered with an aluminium layer or in plastic bottles. (Aseptic packaging means: packaging in such a manner that no contamination by microorganisms can take place after sterilization.) UHT milk has the date of the extreme limit of storage time and storage instructions printed on the packaging. Compared with conventional sterilized milk, UHT milk has suffered minimal chemical, physical, and sensory changes (such as browning and cooked flavour intensity) in relation to the intensity of the heat treatment.

Cream is the collective name of a number of products which contain a minimum 10 per cent fat and contain no other fat than milk fat. Formerly, cream was obtained by leaving raw milk standing undisturbed for 12 hours; a layer of cream was then formed on top of the milk. This was spooned off and used as whipping cream or processed, for instance into butter. Nowadays, the dairy collects surplus cream from high-fat milk during the standardization of milk. The cream is then pasteurized or sterilized.

Half cream contains 10–20 per cent fat, coffee cream contains 20–35 per cent fat.

Whipping cream contains 35 per cent fat. Because of the high fat content and the surface-active compounds, the emulsified fat can hold air through whipping. The result is that the cream becomes stiff and about doubles its volume. Uncooled and very fresh whipping cream is less easy to whip than cooled and slightly older whipping cream. By whipping too long and too heavily the cream is beaten into butter. The surface-active compounds around the fat globules are then broken, which causes the emulsion to disappear. The fat flows together into large conglomerates and also pushes water out.

A product with less fat (25 per cent) is available as a substitute for whipping cream. This product could not normally be whipped, because of its low fat content. However, by the addition of such aids as thickener, caseinate and stabilizer this is made possible.

Imitation whipping cream powder or **topping** consists of sugar, vegetable fat, milk protein, thickener, flavour and colouring agents, and anti-oxidant. The powder can be whipped up with milk and be used as a low fat replacement for whipped cream. Compared with whipping cream, topping has a larger volume when whipped, is more stable in the whipped state and cannot be whipped up into 'butter'. Topping is also available with cocoa.

Sour half-cream, sour cream and **sour whipping cream** (*crème fraîche*) are produced by adding lactic acid bacteria to pasteurized half cream, coffee cream and whipping cream. By this, the products become sour and thicker in consistency.

Whipped cream is a mixture of cream and an inert gas (usually air), agitated to produce a foam. Sugar and stabilizer are also added. The product is available in aerosol cans. Cooling of the aerosol can and its contents is necessary to obtain the cream in an airily 'whipped' state from the can. The cream collapses more quickly than normal whipped cream.

Coffee milk is also called evaporated milk. To make this, water is removed from standardized milk. Concentration takes place at a temperature of 40 to 50°C under vacuum. The chance of taste and colour changes (cooked taste and brown colouration) is greater at higher temperatures. The milk is homogenized and sterilized after evaporation. Coffee milk is available in the full (minimum 7.5 per cent fat), the semi-skimmed (4–4.5 per cent) and skimmed (maximum 1 per cent fat) varieties. Compositional standards will vary from one country to another. Linoleic acid-rich oil is also added to skimmed coffee milk to bring it to the fat percentage of semi-skimmed or full coffee milk. In this case the product may not be called coffee milk, but it is described as 'skimmed evaporated milk with X per cent vegetable oil'.

Condensed milk with sugar is a product which is evaporated in the same manner as coffee milk but its shelf-life is increased by adding about 40 per cent sugar. The thick syrupy product is then canned. This long-keeping product which can be stored without cooling is used a lot, especially in the Tropics. After thinning with water it can be used as sweet drinking milk.

Coffee whiteners are added to coffee in powder form instead of milk, coffee cream or coffee milk. They are not dairy products. The composition varies, but mostly they contain milk protein as their only milk component (caseinate), together with maltodextrine, coconut fat or palm kernel fat. Emulsifying compounds are added to aid the solubility, stabilizers to prevent the milk proteins from separating out and anti-oxidants to prevent fat oxidation. Coffee whiteners are easy to use and have a long storage time, without refrigeration.

Chocolate milk is milk mixed with cocoa compounds, sugar, thickeners, and odour and taste additives. The product is sterilized and available in the full (minimum 2.5 per cent fat), the semi-skimmed (1.5–1.8 per cent fat) and the skimmed (maximum 0.3 per cent fat) variety.

Fig. 9.2 — Various milk products.

Buttermilk is a product which is produced in butter-making during churning. The butter is manufactured from soured cream. After churning, the butter floats on an acid liquid, buttermilk, which is drained off. This buttermilk has got a characteristic taste, and is somewhat flocculent, At the moment, buttermilk is also produced by acidifying skimmed milk with lactic acid bacteria. The drink which is produced in this manner has a refreshing acid taste and is homogeneous. The buttermilk contains between 0.4 and 1 per cent fat. Separation occurs quickly during heating. Usually it is not clearly marked on the packaging if the product originates from butter-making or from soured skimmed milk.

Kefir is a weakly effervescent drink with an acid taste which contains some carbon dioxide and alcohol. Pure cultures of lactic acid bacteria and yeasts are added to pasteurized milk. While ripening during a fermentation of 24 hours, lactic acid, alcohol (about 0.05 per cent) and carbon dioxide are formed. The product is then cooled. Kefir can also be prepared at home with the aid of dried kefir corns, which contain the cultures of lactic acid bacteria and yeasts. Kefir contains a minimum of 3.5 per cent fat, semi-skimmed kefir 1.5–1.8 per cent and skimmed kefir a maximum of 0.3 per cent fat.

Custard is prepared by adding binding agents (usually starch), sugar, colour and flavour compounds to milk or buttermilk. The mixture is heated so that the starch thickens and binds the liquid. After that, the custard is cooled and poured into some form of packaging. Custard which is to have a long storage life subsequently gets a UHT treatment. Custard contains a minimum of 2.6 per cent fat, skimmed custard a maximum of 0.3 per cent fat.

Porridge is prepared in the same way as custard. As the binding agent, wheatflour, oats, rice, barley or semolina is used. Porridge, like custard, is available full (minimum 2.6 per cent fat) and skimmed (maximum 0.3 per cent fat).

Pudding, mousse and bavarois, perhaps with sauce or cream, are prepared in the same way as custard. Larger amounts and different binding agents are used. Pudding is always packaged hot. For these products there are no composition rules, but hygienic standards may be stipulated in the regulations.

Yoghurt is made by adding an acid inoculum to pasteurized and homogenized milk. The inoculum contains thermodynamic organisms of yoghurt culture, *Lactobacillus bulgaricus* and *Streptococcus thermophilus*. About 40 per cent of the lactose in the milk is changed into lactic acid by the enzymes of the yoghurt bacteria. Yoghurt is more acid and more viscous than milk and has a characteristic yoghurt flavour. Yoghurt can be prepared in two ways.

Stir or liquid yoghurt develops after inoculating with 0.25 per cent starter in large tanks at 30–45°C. If after about 16–20 hours sufficient acid has been formed, the yoghurt is cooled down to 6°C. This stops the working of the yoghurt bacteria. The mass is stirred smooth and packaged. This yoghurt can be poured.

Stand or thick yoghurt ripens in the packaging. After the milk has been mixed with 2.5 per cent starter, plastic pots and bottles are filled. For 2.5 hours the filled bottles or pots are placed in a warm water bath (40°C), after which the yoghurt is cooled to 6°C. The end product is thick, gelatinous, not easy to pour, and more acid than stir yoghurt.

Yoghurt has a longer shelf-life than milk because of the higher acidity. It can be kept for about 14 days in a closed package. The storage quality can be extended by heating the acid product. In doing this, the yeasts and moulds are killed and the number of yoghurt bacteria are reduced so that no further acidification takes place. Depending on the degree of standardization of the milk a full yoghurt (2.95–4.4 per cent fat), semi-skimmed yoghurt (1.5–8 per cent fat) or skimmed yoghurt (maximum 0.5 per cent fat) is obtained.

Bulgarian or thickened yoghurt is made with milk which has been thickened to two-thirds of its volume, or to which skimmed milk powder has been added to increase the percentage of milk solids. Bulgarian yoghurt is often prepared according to the thick yoghurt procedure. Bulgarian yoghurt contains a minimum of 4.4 per cent fat, skimmed Bulgarian yoghurt a maximum 0.5 per cent fat.

If lactic acid bacteria other than those laid down in the Directives for yoghurt preparation are used, then the name yoghurt may not be used. The acid product appears in the trade under a different name, for example Biogarde (in the Netherlands). Dependent on the type of lactic acid bacteria, laevo- or dextro-rotary lactic acid is formed. The microorganisms of the yoghurt cultures form about the same quantities of both types of lactic acid.

If mainly dextro-rotatory lactic acid is to be obtained, then for instance *Lactobacillus acidophilus* or *Lactobacillus bifidus* are used. Dextro-rotatory L(+) lactic acid is a compound formed in the body which is liberated in the muscle after heavy work. The L(+) lactic acid is broken down again by a special enzyme.

The body does not contain specific enzymes to break down laevo-rotatory D(−) lactic acid. Because of this, laevo-rotatory D(−) lactic acid circulates longer in the blood and is eventually excreted via urine and sweat. A low pH is regarded by some people as less desirable, so these are claimed to be beneficial aspects of *acidophilus* or *bifidus* products.

Within the assortment of acidified milk products, there are also products available with exotic names like Viili, a product from the bio-dynamic milk dairy farms, and Umer from the milk processing industry (Netherlands). The consistency and taste vary, dependent on the pure cultures of lactic acid bacteria and perhaps yeasts used. Mostly, the products are produced according to the thick yoghurt procedure.

Milk (full and skimmed), buttermilk, bulgarian yoghurt (full and skimmed) and yoghurt (full, semi-skimmed and skimmed) can be mixed with fruit juice. Sugar, thickening and emulsifying compounds, citric acid flavours and colouring may also be added. The storage quality can be extended by heating the product.

Bulgarian or thickened yoghurt (full and skimmed) and yoghurt (full, semi-skimmed and skimmed) are also mixed with fruit or fruit pulp. Also sugar, and several additives are added to the yoghurt. By heating, the shelf-life of the products can be extended.

Ice-cream is a frozen, sweet, aerated, beaten custard, which is produced industrially from ice-cream powder or ice-cream mix. The mix or powder, which is also available for the consumer, is a product based on milk powder to which milk fat, sugar, thickening and gelatinous compounds, emulsifiers, colour and flavour compounds have been added. Water is added to the ice-cream powder or the ice-cream mix. The mixture is pasteurized and homogenized and cooled to 1–2°C. The liquid product is beaten during freezing, which aerates the mixture. Because of this, the ice crystals formed are small and the ice-cream does not become too hard. After this the ice-cream is packaged and stored at −18°C. Broadly speaking, the descriptions and fat composition of the various types are as follows.

Dairy ice-cream contains a minimum of 9 per cent milk fat, **milk ice-cream** a minimum of 2.5 per cent. For ice-cream the type of fat is not prescribed, although usually it should contain greater than 5 per cent.

Water ice is made from a sugar solution in water, to which smell, colour and flavour compounds are added.

Yoghurt and buttermilk ice-cream is prepared from ice-cream mix of which yoghurt or buttermilk are important components. The milk fat content of yoghurt ice-cream is a minimum of 2.5 per cent, of buttermilk ice-cream a maximum of 0.5 per cent.

Soft ice-cream contains less milk fat and more milk components like protein. This type of ice-cream is not pasteurized and is not deep-frozen. Soft ice-cream is kept at −2°C, which makes it stay soft.

Soft scoop ice-cream differs from ice-cream in composition, because different sugars, thickeners and stabilizers are used. Through this, less water freezes at −18°C, but at, for instance, −24°C it becomes hard. Soft scoop ice-cream is easy to use, even straight out of the freezer.

Local compositional requirements should be consulted for the exact requirements for individual countries.

Infant milk or **milk food for infants** or **baby milk food** is a liquid product based on cows' milk, that has been made suitable in composition to serve as a food for infants.

Milk powder is obtained by removing nearly all the moisture from milk. For milk powder for consumption, the milk is spray-dried. The most popular milk powder is skimmed, with a maximum of 1.5 per cent fat. A cheaper method of drying produces roller-dried powder. This is less easy to dissolve and is used for the preparation of milk chocolate, chocolate milk and custard.

Diet products: Milk and milk products are available for a low-sodium diet (for instance, milk and coffee milk with low sodium contents), for a diabetic diet (for instance, kinds of ice-cream and toppings) and for the linoleic-enriched diet, in which in certain products the milk fat has been largely replaced by linoleic-rich oils.

9.4 QUALITY DETERIORATION, SPOILAGE AND STORAGE

Milk and milk products reduce in quality during storage by the following processes.

— *Souring of sweet milk products* when lactose is changed into lactic acid by lactic acid bacteria. Acid milk products can also acidify too far and become tart and bitter, and the moisture separates out.
— *Fermenting*, especially of products with added fruit, sugars and/or thickeners.
— *Mould formation*.
— *Fat oxidation*, which is accelerated by exposure to light. Also, the heat-resistant lipase can cause rancidity.
— *Tainting* caused by the absorbing of strong smells from the surroundings especially by the milk fat.
— *Proteolysis*.

Vitamin B_2 can become inactivated if milk or milk products are exposed to light for a long time and the essential amino acid methionine can be converted into the inactive compound methional.

Storage advice for milk and milk products in closed packages
Store pasteurized products in a cool dark place, up to the date stipulated on the package.

Store acid milk products also in a cool dark place. The storage time is longer than for the non-acid milk products (about 2 weeks).

Sterilized milk in non-light proof packaging does not have to be cooled, but must be stored in the dark (about 1 year or longer).

UHT sterilized milk can be stored uncooled (at ambient temperature) for about 3 months.

Milk powder can be stored uncooled for about 1 year.

Storage advice for milk and milk products in opened packages
All milk and milk products after opening, regardless of the treatment, must be stored cooled and in the dark and then treated as a fresh product. Further, depending on the type of product, it should be consumed within 2 to 7 days.

9.5 LEGISLATION

The **Fresh Milk Regulations** set down regulations concerning milking, and the labelling and composition of milk and milk products.

Ice-cream Regulations contain rules for the composition of ice-cream.

Condensed Milk and Dried Milk Regulations set down regulations concerning the composition of milk powders and ice-cream mixes among others.

For the procedures (the dairy farmers and the dairy industry) there are several quality criteria concerning the production and processing of milk. The Production Board for Dairying sets a number of regulation standards for the processing and labelling of milk for consumption. By Ministerial Law in the framework of the Goodslaw (Netherlands) the fat content of full milk has been laid down as 3.5 per cent (**Law for Milk Standardization**, 1975).

9.6 CONTROL

The health of the dairy herd is supervised by the government agricultural advisory services and by veterinary surgeons.

At regular intervals a sample of the milk from each dairy farm goes to a milk control station for examination of its hygienic quality. The milk control stations are set up by the farmers and the dairies. They are under the supervision of the government.

The dairies also check the milk (its composition, and microbial count) through the milk samples taken by the driver of the tanker who collects the milk from the farmer.

The dairies and the retail trade are monitored for hygiene. The final products of the dairy are also monitored. This duty is in the hands of the Government Inspection Service (Environmental Health Officers). Besides this check, most major dairies will instigate their own quality assurance programmes.

In some countries a small levy is paid on dairy products, and this is used to fund research and development into milk and milk products.

10
Butter

10.1 INTRODUCTION

Butter is a smooth, light-yellow, reasonably firm, but still easily spreadable mass, which mainly consists of milk fat and water. The product butter has existed since man has drunk milk. From archaeological finds it is known, for instance, that butter was used by the ancient Persians and the ancient Egyptians. Butter was also known to the Greeks and Romans as a basis for ointments and beauty aids. For cooking, olive oil was used; only the less affluent used butter. The use of butter in the kitchen is still uncommon in the Mediterranean countries.

In Northern Europe the consumption of butter gained real momentum in the fifteenth century, especially among the higher classes of the population. The Netherlands began to export considerable quantities of butter, and in the eighteenth century had the monopoly position in the export of butter. However, as a result of adulterating the butter by the addition of water, amongst other factors, the Netherlands lost this position to Denmark, which produced butter of a better and more consistent quality.

Since the beginning of this century the Dutch butter trade has undergone stringent controls to put a stop to adulteration. A butter mark was put on butter which had been inspected and approved. Export of Dutch butter increased after this measure.

Another reason for the collapse of the Dutch export market for butter at that time (and for reduced consumption of butter in the present day) was the invention of margarine halfway through the last century.

Consumption

The use of butter in the Netherlands is relatively low in comparison with the consumption figures for other EC countries. Butter consumption in 1985 was about 3.5 kg per person per year (compared to 5.0 kg per person in the UK) and this accounted for 7 per cent of the total fat consumption.

Composition
Butter contains approximately:

fat	82%
water	16%
fat-free milk components	2%
vitamins	A and D and β-carotene
minerals	calcium, phosphorus and sodium.

The fat in butter consists mainly of triglycerides. The fatty acids present differ in chain length and saturation (see Chapter 13). The high content of fatty acids with short chains makes butter easily digestible. Largely half of the fat is composed of saturated fatty acids; of the unsaturated fatty acids about 70 per cent is oleic acid and only very little is linoleic acid. The fatty acid composition varies with the seasons, and the lactation period; the cattle feed also has an influence. During the lactation period the amount of fatty acids with short amd medium-length chains increases and the amount of linoleic acid decreases. When the cows graze outside in the summer, the concentration of oleic acid is higher; this decreases in the winter and the amount of palmitic acid increases. These differences have a noticeable effect on the firmness of the butter. The content of vitamins and β-carotene is higher in the summer when the cows eat mainly grass — this shows in the colour of the butter. However, the corresponding colour change in the winter months is masked by the addition of β-carotene or the colourings bixin or norbixin.

10.2 PRODUCTION AND DISTRIBUTION

The traditional preparation of factory butter

There are a number of steps in the preparation.

— Raw milk is separated in a centrifuge into cream with a fat content of about 40 per cent and skimmed milk.
— The cream is pasteurized at 60 to 90°C and cooled.
— Acid-forming and flavour-forming bacteria are added.
— For 15–20 hours fermentation takes place, whereby lactic acid and diacetyl, among other substances, are formed (the specific butter aroma is partly caused by diacetyl).
— For 30 minutes, the acidic cream is vigorously stirred or churned. The fat globules stick together and start to float as conglomerates on the buttermilk. The fat-in-water emulsion changes to a water-in-fat emulsion.
— The buttermilk is drained off and the conglomerates are perhaps mixed with salt and/or dyes and are kneaded into a homogeneous mass in a churn kneader.
— The butter is packaged and cool stored. The whole process of churning, kneading

and packaging takes place mostly continuously in a 'butter canon' (continuous butter processor).

Fig. 10.1 — A butter canon.

The new or NIZO method
In this method non-acidified pasteurized cream is used. After churning, the sweet buttermilk is drained off. Only during kneading is an acidfying inoculum (a lactic acid concentrate and a culture of lactic acid bacteria) added, after which the butter is left to ripen. The drained, sweet buttermilk is processed into other dairy products.

The two different methods of preparation have no noticeable influence on the taste and composition of the end product. The preparation of **farmhouse butter** differs in a couple of points from the dairy preparation. The cream is not pasteurized and sometimes the cream is not acidified, but the fermentation is allowed to occur spontaneously. This will give the butter a more pronounced taste of its own. (Sometimes called sweet butter.)

Price management
The high price of butter stops many people from buying it. The price is determined by the agricultural politics of the EC. In Brussels each year a guide price is determined for industrial milk. The butter producer pays this guide price and offers fresh butter for sale according to normal cost calculation. Because this butter is expensive, the producer is left with a surplus, which has also a limited storage life.

For an arranged price, the producer can hand in his surplus to the Intervention

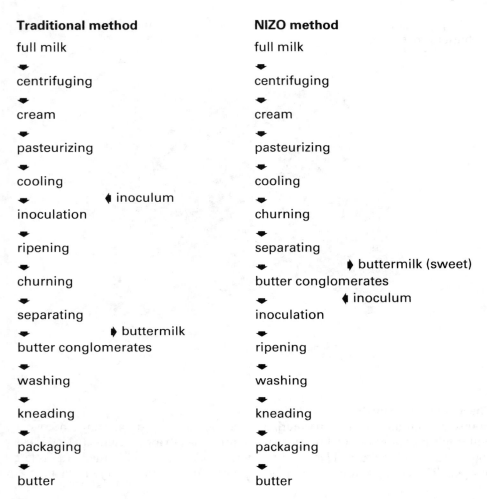

Fig. 10.2 — The preparation of butter. (Source: Karkens, F. W. P., *Traditionele boterbereiding en boterbereiding volgens NIZO-methode*, Productschap voor Zuivel, Rijswijk 1981.)

Bureau of the EC. The butter is stored at at least −15°C in cold stores. This cold store butter is sold wholesale for a lower price than usual to businesses, hospitals, bakeries or the army. This butter is sometimes offered to the consumer for a sharply reduced price as so-called Christmas butter, or is processed into baking and frying butter with about 96 per cent milk fat.

10.3 KINDS OF BUTTER
A number of different kinds of butter are available.

Unsalted butter
This is either not stored cooled or only for a very short time and does not contain added cooking salt. This butter is suitable for use in a restricted-sodium diet.

Salted butter
The group 'salted' is divided into salt butter with a salt content of 0.7 per cent minimum and 2 per cent maximum, and 'lightly salted' where the salt content varies between 0.1 per cent and 0.7 per cent.

Cold-store butter or Intervention butter
This is stored in cold-stores at a temperature of $-15°C$. The storage and the sale is organized by the government. Cold-store butter is always unsalted. It has to be mentioned on the packaging that this is butter from the cold-stores. It has got a limited shelf-life, because it has been frozen.

Herb butter
Finely chopped garden herbs, lemon juice, pepper and salt are added to the butter.

Whey butter
This is churned from whey cream, a byproduct of farmhouse cheese preparation. It does not occur packaged in the trade and is only of importance for use by the farmer-producers.

Farmhouse butter
This is prepared according to the traditional method from raw farm milk, which may or may not be inoculated. Farmhouse butter differs sometimes with regard to the moisture content (higher than is permitted by law) and the requirements concerning the microbial status.

Baking and frying butter (anhydrous milk fat)
This product is real butter, from which nearly all the moisture has been removed. Then, protein-rich whey powder and lecithin are added, to obtain browning compounds and to prevent spitting, respectively. Frying butter contains 97.1 per cent fat, 2 per cent fat-free dry matter with proteins and minerals, 0.5 per cent lecithin, 0.4 per cent moisture, vitamins A and D and a marker compound (this last to prevent the product once again being processed into butter). This product has also a limited shelf-life because it has already been frozen.

Testing
Butter is tested for quality and composition before it reaches the market. The testing is done by the Central Organization of Dairy Control (COZ) (Netherlands). It is classified on the consistency, smell and taste, appearance and moisture, based on the testing regulations for COZ butter products. The classes are 'extra quality' and 'standard quality'. On the retail packs of Netherlands-produced butter, a government 'butter mark' has to be affixed together with the quality marks 'extra' (blue) or

'standard' (green) (see Fig. 10.3). Similar grading procedures are used in other countries.

Blue quality mark

Green quality mark

Fig. 10.3 — Dutch 'butter marks'.

10.4 QUALITY DETERIORATION, SPOILAGE AND STORAGE

Butter contains, besides fats, small percentages of proteins, milk sugar and water. Because of its composition butter is a suitable substrate for microorganisms (mould) and is sensitive to fat oxidation. Butter can also dry out.

Acidified butter has such a pH that the growth of spoilage bacteria is slowed down. However, butter has a limited shelf-life, especially because of the sensitivity of the fats to oxidation. The packaging must bear a 'use by' date.

Unsalted butter can be kept in the refrigerator for about 6 weeks. It can be kept in the deep-freeze, well wrapped, for 6 months.

Salted butter can be kept in the refrigerator for about 6 weeks. It is not desirable to freeze it, because salt promotes the oxidation of fat.

Cold-store butter has already been kept for some time at $-15°C$. This butter should preferably be used straight away, It can possibly be kept for another 1–2 months at $-18°C$ if the butter is sold frozen. The process of thawing and re-freezing increases the rancidity.

10.5 LEGISLATION

The rules concerning butter are laid down in the **Butter Regulations**. Rules are set out in these concerning the basic compounds, composition, permitted additives (including the dyes carotene, bixin and norbixin), control, quality, packaging and labelling.

11
Cheese

11.1 INTRODUCTION

According to the Dutch Cheese Decree, cheese is the product which is obtained by curdling milk and/or buttermilk and removing whey. The product may or may not be ripened and/or matured. The milk can come from cows, goats or sheep, but also from a number of other animals. For instance, reindeer cheese in Lapland or mozzarella (buffalo and/or cows milk) in Italy. In Northern Europe most cheese is made of cows' milk. Cheese made with goats' or sheep's milk is called goats' milk cheese and sheep's milk cheese specifically.

Cheese has been made for centuries. This has been concluded from archeological finds in Egypt and old papers from the ancient world. Cheese-making developed when cows and goats were domesticated. Because there was a generous supply of milk for drinking, the surplus was put into containers, which were mostly dirty and full of bacteria. The milk became sour and thick. Soon people realized that the product was nice to eat. Only much later was it learned that milk can be curdled in a different way, to produce cheeses which are of a better quality than the spontaneously curdled cheese and which can also be kept longer. The making of cheese was originally done on the farm. The cheese varied in each area, for instance in shape, fat content or specific additions such as cumin or cloves. Much later on, the dairy cheese developed. There are at the moment about 4000 different kinds of cheese in the world.

Consumption

In recent years the Dutch have started to eat more cheese. Table 11.1 shows that the consumption figures expressed in kg per person per year have increased greatly in 15 years. Also shown are figures for consumption in the United Kingdom.

The quality of cheese consumed consists of about 95 per cent of Dutch cheese, of

Table 11.1 — Consumption figures for cheese (Netherlands and UK)

Year	Number of kg per person	
	Netherlands	United Kingdom
1970	8.3	n/a
1979	11.5	6.2
1982	12.5	6.2
1985	12.7	6.6

Sources: Dutch Dairy Bureau and Annual Abstract of Statistics.

which about 4 per cent is farmhouse cheese. The Dutch eat mainly full fat dairy cheese (about 83 per cent), 40+ cheese (about 10 per cent), and the rest of the cheeses account for about 7 per cent of the market share.

Composition

10 litres milk yields about 1 kg cheese. In Table 11.2 the composition is shown of 100 g full-cream milk and 100 g (mild) Gouda cheese. The figures are obtained from the Dutch Foodstuff Table.

Table 11.2 — Composition of full-cream milk and Gouda cheese

Component	In 100 g full-cream milk	In 100 g Gouda cheese (mild)
Water	88.0 g	44.0 g
Protein	3.4 g	23.0 g
Fat	3.5 g	28.0 g
Lactose	4.6 g	—
Sodium	0.04 g	1.0 g
Calcium	0.1 g	0.6 g
Phosphorus	0.1 g	0.4 g
Vitamins A	0.03 mg	0.35 mg
B_1	0.03 mg	0.03 mg
B_2	0.17 mg	0.20 mg

The differences in protein, fat, lactose and water contents in the two products are obvious. With regard to protein it should be noted that the protein in milk is present in the form of casein and whey or serum proteins; cheese has only the casein.

Through the concentration, the fat-soluble vitamins (A and D) occur in large quantities in cheese, while the water-soluble vitamins (B-complex) are partly removed with the whey.

11.2 THE PRODUCTION OF DUTCH CHEESE (COWS' MILK)

The starting point in cheese-making is milk, which is brought up to a certain fat content (standardized). The fat percentage depends on the type of cheese which is being made (for instance, 'full fat' from full-cream milk or 'low fat' from skimmed milk). Then the milk is pasteurized and cooled down to around 30°C, after which an acid inoculum, rennet, dye, calcium chloride and potassium nitrate is added.

The **acid inoculum** is a pure culture of lactic acid bacteria, which, among other effects, change lactose into lactic acid. This has a beneficial influence on the keeping quality, the taste and the consistency of cheese. Because of the disappearance of lactose a food source for a number of microorganisms is removed. The pH for this type of cheese lies around 5.0–5.1. If the acidity decreases to a value below 4.8, the cheese then becomes hard and crumbly and much too acid in taste. Also the acidity of enzymes responsible for the maturation is slowed down. The acid inoculum also has a beneficial influence on the curdling process. About 1 litre of inoculum per 100 litres cheese milk is added.

Rennet, prepared from the abomasum of new-born calves, contains the proteolytic enzyme chymosin or rennin. This enzyme is necessary in the calves' stomach to make the milk go more slowly through the intestines, which improves the digestion and the absorption. The enzyme activity disappears in older cows.

The separation of the milk proteins by addition of rennet which contains chymosin is called **curdling**. Chymosin breaks casein down into paracasein. This reacts with calcium to form paracaseinate and separates out. Fat and some of the water together with its dissolved compounds are also locked into the separated mass. A spongy substance develops known as the **curds**. The liquid left over is the **whey**.†

The cheese becomes harder or softer depending on the quantity curdled. The curdling time and the curdling temperature also influence the consistency. For a hard cheese more rennet is added and the curdling temperature is higher. Because of this the curdling time is shorter than for the preparation of soft cheese. (Compare the curdling time of Gouda cheese, which is 30 minutes, with that of quark, which is 16 hours.) Cheese makers in some countries are only allowed to use calves' rennet.

Some microorganisms produce an enzyme which can curdle the milk in the same way as rennet. If cheese is to be made and sold in this manner then permission may have to be obtained from the Ministry of Agriculture and Fisheries. In the Netherlands, Bicheka cheese is sold by the wholefood trade for vegetarians (microbial-curdled cheese). For the making of kosher cheese (white cheese) certain moulds are allowed to be used instead of calves' rennet. To give cheese a uniform colour through summer and winter a natural dye (β-carotene, annatto or norbixin) may be used.

---

† Whey is a byproduct of cheese-making, The whey (whey-liquid) contains, amongst other substances, the serum of whey-proteins, and the water-soluble vitamins and minerals which also occur in milk. The whey-protein is sometimes separated and used in products such as protein-enriched foodstuffs. The whey-liquid with the serum-proteins is used in the beverage industry. Addition of fruit juice, sugar, flavour compounds and/or yoghurt and cream, produces whey-drinks that are similar in appearance and taste to fruit juice or soft drinks or milk or yoghurt drinks. The whey-liquid, without the whey-proteins, serves as basis for a clear beverage.

To improve the curdling process **calcium chloride** ($CaCl_2$) is added. The curd formation is stimulated by calcium.

Potassium nitrate (KNO_3) plays a role in the prevention of unwelcome bacterial action, especially of the coliform bacteria and the butyric acid bacteria. Coliform bacteria can occur in the milk through unhygienic handling of the milk; spores of butyric acid bacteria can appear in the milk via silage. Both kinds of microorganisms form hydrogen gas and can cause the cheese to crack in either the early or late stages. KNO_3 acts as acceptor of hydrogen gas. It also slows down the development of the butyric acid bacteria.

The curdling process begins after all the previously mentioned compounds have been added to the milk and takes about 30 minutes. The curdled milk is cut, which causes the curd to shrink. The separation of the whey is in this way improved. The whey is drained off.

Warm water is added to the curd. This, combined with stirring, causes the curd to shrink and releases more whey. This process is called **syneresis**. After again draining off the moisture the curd is sometimes mixed with herbs or spices such as cumin or cloves and the put in a cheesemould. For each type of cheese a special mould is used, round-ball-shaped for the Edam cheese, flattened-cylinder-shaped for the Gouda. In earlier times the moulds were made of wood in which a linen cheese-cloth was laid. Now a plastic material is used with a nylon internal net. The moulds are sealed with followers (lids with holes) and placed under a press for a number of hours. By doing this, the curd loses still more whey. The cheese is taken out of the mould after the pressing-time. Through pressing, a curd edge often develops between the lid and the cheesemould. This is first cut away. Then a rest period of 6–10 hours follows, during which the cheese is regularly turned over. This is necessary because the cheese is not yet symmetrical in form. Some people refer to this period as 'walking the cheese round'.

Then the cheese is dipped in a brine bath, which contains a salt solution of 20 per cent NaCl. This takes 1–5 days depending on the type of cheese and the weight, until the cheese absorbs 1–3 per cent sodium chloride. The use of salt is essential in the making of cheese. It has a preserving action and is necessary for the crust formation and the taste. Through osmotic action moisture is extracted from the cheese, which makes the crust firmer.

Then the cheese is wiped off and covered with liquid plastic, which is gas and water permeable. To slow down mould formation during ripening and storage the crust is covered with an antibiotic (natamycine or pimaricine). Edam cheese is often covered with an air- and watertight yellow or red paraffin layer. Leyden farmhouse cheese is rubbed on the outside with a brown-red dye.

The cheese is not yet ready for consumption: the salt has not yet been uniformly distributed and the taste development has scarcely started. To mature the cheese, it is placed in storage rooms on bare wooden boards. The cheese is turned regularly during the ripening to obtain a good shape.

Some consumers can distinguish **raw milk dairy cheese** from dairy cheese, especially by the taste. In the raw bactofugated milk (which means that the detrimental bacteria have been removed by means of centrifugal force) several non-detrimental bacteria remain which would have been killed by pasteurization. These bacteria are the cause of some pronounced nuances in taste.

Farmhouse cheese: The preparation of farmhouse cheese differs in a few points from the preparation as described of Dutch dairy cheese. Raw non-pasturized milk is used, which results in a more savoury and pronounced taste. Also the use of dye and potassium nitrate is sometimes omitted. Some farmhouse cheeses have not been covered with plastic and treated with pimaricine.

Maturing and maturation times

During maturation a number of changes occur in cheese. In one of these changes, the salt diffuses in the cheese and the first crust formation takes place, so that the cheese does not lose too much moisture or dry too quickly. In long storage the weight loss (moisture loss) can be as high as 40 per cent.

The changing of lactose into lactic acid is completed. During this transformation, carbon dioxide is formed by the lactic acid bacteria from citric acid (an in-between step). This forms the holes or eyes in cheese.

The formation of flavours is started by proteolytic and lipolytic enzymes which break down protein and fat.

The consistency of the dairy product changes through loss of moisture. The cheese becomes firmer and contains gradually more dry matter. Also, the crust becomes firmer and less permeable through drying out.

Dutch cheeses are sold as mild, mild-mature, mature, extra-mature, old and extra-old depending on their maturing time (see Table 11.3).

Table 11.3 — Maturing times for cheese

Mild	4 weeks
Mild-mature	2 months
Mature	4 months
Extra-mature	7 months
Old	10 months
Extra-old	12–24 months

11.3 CLASSIFICATION OF CHEESE

Cheeses can be classified by kinds of milk used, the consistency, fat content, form and taste.

The kind of milk determines for a large part the taste, smell, colour and consistency of the dairy product. Most cheeses are made from cows' milk. Abroad, a lot of cheese is made from goats' and sheep's milk. Sheep's cheese often has a sharp, pronounced taste, and goats' cheese is often somewhat crumbly. A well-known example of sheep's cheese is Roquefort. Small French goats' cheeses are available in the shops under the general name of 'chevre'.

The consistency depends on the degree of compressibility (hardness) of the kinds of cheese. On this basis cheese can be classified as:

— hard; for instance Friesian clove cheese, Emmentaler;
— half-hard; for instance Gouda, Roquefort;
— soft or dessert; for instance Brie, Kernhemmer;
— fresh; for instance quark, Boursin.

In cheese-making, a number of factors and processes influence the consistency. Some of these are: the amount of curd, the curdling time, the curdling temperature and the final moisture and fat content of the end product.

For comparison with the preparation of Dutch cheese, a short description of the preparations of Gruyère and Emmental, Friesian cheese and Cheddar now follow.

In the preparation of Emmental and Gruyère, after draining off the whey, the curds are heated in the cheese vat to 53°C for 30 minutes. This scalding process produces more whey and the curds become drier and firmer. Only then is it put in the mould.

A different technique is used for Friesian cheese and Cheddar. After draining the whey, the cheese vat is covered, and a ripening of several hours follows. After this, the curds are ground with salt and/or mixed with spices. In this way, more whey is released. The cut curd is put in the cheesemould and put under the press. This procedure is called the Cheddar process. Friesian cheese after pressing is also submerged in hot water to be scalded. After this, another pressing follows. Further brining is not necessary. These cheeses appear in the shops after about 6 months.

In the Netherlands, the **fat content** of cheese is expressed as the percentage that occurs in the dry matter. The figure would continually change if the fat content were calculated on the total weight, because during storage the cheese loses moisture. The legalized marks on Dutch and other European cheese are:

60+
50+, 48+, 45+, perhaps with the mention 'full fat'
40+
30+
20+
10+ skimmed
skimmed

For cheeses from other countries, the fat content is also calculated as a percentage of the total dry matter; for example, on German cheese 60% *Fett im Trocken* (Fett i. tr.), and on French cheese 60% *de matières grasses* (m.g.).

The consistency of the dairy product is influenced by the fat content. The more fat, the softer the cheese. Compare mature Gouda cheese (48+) with Friesian cheese (20+): the latter is harder because of the lower moisture and lower fat content.

Classification of cheese

The shape is characteristic for some cheeses, for instance the round ball-shape of Edam, the flattened cylinder of Gouda and Friesian. Several other well-known cheeses have distinct forms, such as Emmental, Camembert and Cheddar. The shape of cheese has to do not only with tradition but also with the consistency and the maturing. It is impossible to imagine a well-ripened Brie of the dimensions of an Emmental, which can have a diameter of 1 metre and weigh about 100 kg.

Cheese can also be made in other shapes for practical reasons, for instance in small individual portions, or in a loaf shape for catering establishments. For Dutch cheeses the shapes are regulated by law (Agricultural Quality Order for Cheese products).

The taste depends on a number of factors in the cheese-making. The fat and moisture content have an influence, but by the use of bacteria and moulds, cheeses can be made with totally different appearances and characters.

The most used kinds are **lactic acid bacteria**, which are added in the form of an inoculum. Examples: Gouda, Edam, Cheddar and Emmental.

Propionic acid bacteria are added in some cases besides lactic acid bacteria. They change part of the lactic acid formed into carbon dioxide and propionic acid. The gas causes large holes to be formed in the cheese, and the propionic acid gives a somewhat hazlenut-like, sweetish taste. Examples: Emmental, Leerdammer, Westberger and Bergumer.

Red bacteria and **Corynebacteria** are active on the outside of the cheese and cause a red-orange coloured crust. A high humidity is necessary to to ensure vigorous activity of the bacteria. The cheeses are regularly sprayed with water during the ripening, which takes 4 weeks. The cheeses are recognizable by the coloured crust, but also by the more-or-less strong penetrating smell and taste. Examples: Kernhemmer cheese, Limburger, Port Salut and Saint Paulin. The crust of Gruyère is also obtained in this manner.

Blue moulds belong to the aerobic *Penicillin* species. The curd is mixed with the mould culture. The cheese is pricked through with needles, to let the necessary oxygen penetrate into the middle of the cheese. The culture develops in a humid atmosphere and at a temperature of around 7°C, as a result of which the characteristic blue-green veins appear in the dairy product and the structure becomes crumbly. Blue mould cheeses are heavily salted to slow down the activity of other microorganisms during the ripening. To prevent crust formation, the cheeses are packed in aluminium foil after 3–6 months ripening. Examples: Roquefort, Danish Blue, Blue Stilton and Gorgonzola. Some of the cheeses have in combination with a blue-green-veined dairy product a thin orange-brown or whitish crust, as a result of the working of red bacteria of other moulds.

White moulds also belong to the *Penicillin* family. These moulds grow on the surface and cause a white velvety crust. To obtain this, the cheeses are placed in rush mats and sprayed with mould spores. At a temperature of around 12°C, this ripens in 4 weeks. The mould growth influences the taste and consistency of the dairy product. With further ripening, the consistency becomes runny. Example: Camembert and Brie.

The taste of cheese can also be varied by adding herbs, spices, nuts or other foodstuffs. Cheeses with a low fat content gain more taste through this, but cream cheeses or low-sodium kinds of cheese are also spiced. Examples: Leiden cheese with

Fig. 11.1 — Some kinds of Dutch cheese.

cumin, Friesian clove cheese, sage Derby, Double Gloucester with chives and onion, fresh cream cheese with mixed herbs and low-sodium cumin-cheese.

11.4 KINDS OF CHEESE

Dutch cheese

Gouda cheese is a half-hard (48+) cheese usually with a mild creamy taste. There is also **Farmhouse Gouda** (52+) which is more pronounced in taste. Gouda is sometimes also spiced with cumin. The cheese has a flattened-cylinder shape with round sides and an average weight of about 8 kg. Other cheeses of the Gouda shape are: *Baby Gouda* which are smaller in size (200–1100 g) and are always sold young; *Cream cheese* (60+) also young and varying in weight from 4.5 to 6 kg (large type) and 200 to 1100 g (small type). *Amsterdammer* (48+) is a soft mild-tasting cheese with an increased moisture content and is only consumed young.

Edam cheese is a half-hard (40+) cheese with a more piquant taste. The cheese is shaped in a ball form with a weight of 2–3 kg. The crust is sometimes covered in yellow or red paraffin wax or the cheese is packaged in red cellophane.

Leiden cheese is a half-hard cheese (20+ to 40+) spiced with cumin. There are farmhouse and dairy Leiden. The cheese has a flattened-cylinder shape with lightly bowed-out sides, which form a sharp edge with the flat top and bottom. It weighs around 8 kg.

Farmhouse Leiden mostly has a brown-red crust and is clearly recognizable by the imprint of the two crossed keys and the words '*Boeren Leidse*'. Leiden cheese is piquant in taste but fairly dry.

Friesian clove cheese is a 20+–40+ cheese spiced with cloves or with cloves and cumin. The cheese has a flattened-cylinder shape with straight sides, which form a sharp edge with the flat bottom and a round edge with the flat top. This shape is made according to the Cheddar process (see section 11.3). The 20+ cheese is somewhat harder then the 40+. Clove cheese is usually eaten when it is half a year old.

Kantercheese is a non-spiced cheese of Friesian cheese shape (20+–40+). This cheese is, like clove cheese, mostly sold when mature (around 6 months old).

Loaf cheese in an Edam cheese in a loaf shape, 40+, with and without cumin. This shape exists mainly for restaurants and catering establishments.

Square and **oblong** (blockshape) cheeses exist in 48+, 40+ and 20+. These shapes are suitable for use in restaurants and by caterers. Often marketed without crust, or the crust is removed before use. *Crustless cheese* is a cheese of the Gouda or Edam type with a flat block shape (48+ and 40+). The cheeses are a wrapped in aluminium foil within ten days of making and then stored for at least 18 days at 5°C to ripen and are sold after that.

Maaslander is a Gouda dairy cheese (48+) with a little lower salt content than Gouda cheese. This cheese comes onto the market after two to three months. Other trade names are **Yssel cheese** and **Laaglander**.

Maasdammer is a 45+ cheese, flat-round in shape, with large holes and slightly rounded on the top. This 'Dutch holey cheese' gets its holes and typical sweet taste from the action of propionic acid which is added with the inoculum. The cheese is also sold under the names Westberg, Leerdammer, Bergumer and Gouwetaler and is for sale after a ripening time of two months.

Kollumer, **Texelaar** and **Terpcheese** are cheeses of the Gouda type (48+). They are made partly out of pasteurized milk and partly out of bactofugated milk. They are somewhat more full-flavoured than the Gouda and appear on the shop shelf after three months.

Kernhemmer cheese is a 60+ cheese with a creamy but piquant taste and a light orange-coloured crust caused by the action of the Corynebacteria. The cheese inside is light yellow. This cheese was developed in the fifties by the Dutch Institute for Dairy Research (NIZO). The ripening time is about four weeks. This cheese is suitable as a dessert cheese.

Limburger cheese is a small, square (7×7 cm) full fat (50+) little cheese with a very pronounced taste and a penetrating smell. The cheese gets the taste from the action of the Corynebacteria and a white mould on the crust.

Pompadour is a 50+ cheese, flat-round in shape and weighing up to 4 kg. There are many taste variations, including garden herbs and garlic, kummel and pepper. The colour of the crust is then also varied (green or brown).

Subenhara is a soft dessert cheese (48+), weighs about 1.5 kg, and has a flattened cylindrical shape. This cheese contains garden herbs and garlic. The crust is black. A diet variety also exists with a yellow crust and less salt.

Feta is a 40+ or 45+ cheese made from sheep and/or goats' milk. The cheese ripens in a minimum of 15 days at 10°C and is kept in a brine solution. The cheese has no crust when sold and a moisture content of around 50 per cent. Feta is soft but keeps its shape, is uniformly white in colour and has a fresh sour and also salty taste.

White Maycheese (48+) is of the farmhouse Gouda type with a weight of 2.5–5 kg. The moisture content is around 58 per cent. The cheese is brined and sold after 24 hours. It should be eaten within three days of preparation.

Ground and grated cheese: In principle, all hard kinds of Dutch cheese can be ground or grated. The skimmed and older types are easier to prepare and have more taste. They are useful for, among other things, the preparation of sauces and butters. They are packeted in small amounts related to their storability.

Cheese powder is dried and ground cheese or ground dried-up processed cheese. It is more concentrated than ground or grated cheese and keeps a little bit better. The moisture content is lower.

Other cheeses

The number of different countries is so large that it is virtually impossible to describe all existing cheeses. Table 11.4 reviews some non-Dutch cheeses, giving the most important characteristics of a number of cheeses, including the fat content and the taste characteristics. The type of milk is mentioned if it differs from the norm, which means if it is not cows' milk. Sometimes additions and particular uses are briefly noted.

Diet cheese

Diet cheeses are available for fat-restricted and sodium-restricted diets. There is a large choice of low-fat types from the 20+, 10+ and skimmed varieties. There are cheeses in the shops for the sodium-restricted diet which have been prepared with or without the addition of diet salt. During the preparation, the curd is placed in a bath with KCl instead of a brine bath with NaCl. Less crust develops and so the cheese holds its shape poorly and bulges out quickly. The storage time of this cheese is considerably shorter, because the cheese is more susceptible to all sorts of bacterial activity and also often has a higher moisture content. If the sodium restriction is not too severe, a kind of cheese with a lower sodium content can be chosen, such as the Maaslander.

Quality deterioration, spoilage and storage

Forms of spoilage which can occur are as follows.

— *Drying out*.
— *Sweating* The fat separates out as result of storing at a temperature above 20°C; full fat cheeses especially sweat quickly.
— *Mouldiness* This happens especially with young cheeses which still contain much moisture.
— *Undesired bacterial working* During the ripening, butyric acid bacteria and coliform bacteria can cause the cheese to rupture because of gas development. This is prevented by the addition of potassium nitrate (KNO_3). Continued activity of *mould cultures*, which previously have been added to give the cheese a different taste or appearance, will slowly render the cheese no longer acceptable.
— *Attack by cheese mites* A little spider-like creature burrows through the cheese during the storage of whole cheeses, which must then eventually be pulverized.

The best storage conditions are cool dry surroundings (around 12°C) and good packaging.

Cheese which is stored in the refrigerator has to be taken out well before use, so that it comes up to room temperature, which benefits the taste and smell. This also

Table 11.4 — Summary of a number of non-Dutch kinds of cheese

Country of origin and variety name	Taste	Fat content	Characteristics
BELGIUM			
Brussels cheese	Very soft and salty	Skimmed	Soft cheese, no crust, kept in brine
Remoudou	Strong, piquant, sharp	45+	Most well-known kind of Herve-type cheese, cube-shaped with red bacteria crust
GERMANY			
Limburger	Strong, piquant, sharp	20+–50+	'King of the smelly cheeses' with red bacteria crust, looks like Herve; cube or loaf-shaped
Sauermilch käse	Very fragrant, sharp, piquant	10+ or less	'Sour-milk cheese'
Tilsiter	Full, fairly strong, sour after-taste	30+–50+	Light red bacteria action in crust, small holes in the cheese, mostly in loaf shape
DENMARK			
Blå Castello	Full, creamy, savoury taste	70+	Veined with blue mould
Danish Blue	Full taste, savoury and salty	50+–60+	Veined with blue mould
Danbo	Sharp, aromatic, strong taste	10+–40+	With red bacteria crust, sometimes with added caraway seed; square shape
ENGLAND			
Cheddar	Piquant, full, somewhat dry	48+	Scottish Cheddar is coloured strongly orange, fat cylindrical shape; also farmhouse cheddar
Stilton	Very piquant, yet creamy	48+	Veined with blue mould, brown-red crust; very crumbly
GREECE			
Feta	Salty, piquant	45+–60+	Sheep or goats' milk, no crust, stored in brine, white crumbly cheese, scarcely ripened
Kassein	Salty, aromatic	40+	Cheddar-like; sheep or cows' milk
ITALY			
Bel Paese	Mild, creamy	50+	Small, irregular holes, crust formed with red bacteria action
Gorgonzola	Fragrant, creamy, sharp	48+	Blue-green mould veins with red-orange crust; or white, without blue mould
Mozzarella	Mild, sometimes smoked	45+	Buffalo or cows' milk, stringy when melted — the 'pizza cheese', used when very fresh and supple
Parmesan	Spicy, savoury	32+	Very hard, dry cheese, crumbly, especially suitable for grating
Ricotta	Mild, sweetish and lightly salty	10+–30+	From sheep's milk and whey or cows' milk and whey; fresh cheese looks like cottage cheese, crumbly

Table 11.4 — *(Continued)*

Country of origin and variety name	Taste	Fat content	Characteristics
SWITZERLAND			
Emmental	Soft, nutty	45+–48+	Large, glossy holes and hard dry crust, size to 1 m diameter, made from raw milk
Gruyère	Saltier and fruitier than Emmental	45+–48+	Small scattered holes, red-brown crust
Raclette	Mild, soft, nutty	50+	Used as melting cheese, toasting cheese
Sapsago	Sharp, herby	10+	'Schabzieger' grating cheese in the shape of hard little cones, often with Alpine herbs
FRANCE			
Bleu d'Auvergne	Salty, creamy, piquant	40+–45+	Well-developed blue mould
Bleu de Bresse	Creamy, mild	50+	Slightly developed blue mould
Brie	Creamy, fragant	40+–50+	Circular cheese, 50 cm diameter, white mould crust, cream coloured cheese which runs when ripe
Camembert	Creamy, mild somewhat mushroomy taste	45+–55+	Circular cheese, diameter 12 cm, white mould crust, runs out thick and creamy when ripe
Chèvre	Savoury, coarsely sharp, penetrating	45+	Goats' cheese with chalk-white colour, sometimes white mould crust, 'leaf-crust' or with herbs; for sale under many names in the shape of oblong little rolls or flat little cylinders
Munster	Mild, fragrant	40+	Red bacteria action gives orange crust, soft cheese, in round, small flat little shapes
Port Salut	Mild, fragrant	40+–50+	Red bacteria cheese, a Trappist cheese by origin, name protected by law, flat cylindrical, yellow crust
Roquefort	Penetrating and salty, piquant after-taste	50+–60+	White, blue mould veined cheese from sheep's milk, crumbly, name protected by law
Saint Paulin	Mild, aromatic	40+–50	Dairy version of Port Salut

applies for pre-packed sliced cheese and dessert cheese. Whole Dutch cheese can be stored for half a year or longer, pieces of cheese about 2–3 weeks, and sliced cheese about 1–2 weeks. If the cheese is vacuum-packed, the storage time is about 1–2 months.

11.5 FRESH CHEESE

In this group such products as quark, cottage cheese and fresh cream cheese are counted. They are obtained by the curdling of milk (cows', goats' or sheep's milk) with no or minimal ripening. Also no crust is formed.

Fig. 11.2 — Some small French cheeses.

Quark or 'fresh cheese'

This is a well-known cheese, and most probably the very first, most primitive form of cheese. Quark develops on its own when milk becomes sour and thickens through the presence of certain microorganisms. The consumption of quark used not to be high. However the use of quark in pastry or for mixing with syrup or fruit has increased its popularity. For example in the Netherlands in 1971, 100 g quark per person was consumed; by 1985 this had risen to about 1 kg. In other countries, including the United States, France and Germany, more quark is being eaten.

The making of skimmed quark

In sequence: skimmed milk (0.3 per cent fat) is pasteurized and cooled to 30°C, inoculated with lactic acid bacteria (starter), and rennet is added, the ratio of the added quantities in comparison with cheese-making being that more of the starter inoculum is used and less or no rennet. (Curdling time: about 16 hours). The curd is cut, and the whey and curd are separated in a separator. Sometimes the curd is mixed with salt, herbs or fruit. Finally the mass is cooled to 5°C and packaged.

Composition

On average, 100 g skimmed quark contains:

water	83.0 g
protein	13.0 g
fat	0.2 g
vitamin B_1	0.04 mg
vitamin B_2	0.30 mg

The source of the figures is the Dutch Foodstuffs Tables. Because less whey is extracted the content of vitamins of the B-complex is higher than that of young Gouda cheese (see Table 11.2).

The quark range can be classified acording to fat content, treatment and/or additives. Depending on the amount (if any) of **added milk fat**, quark gets the labelling 'skimmed', 'semi-skimmed', 'full' or 'cream quark'. The percentages of fat in the dry matter are respectively 10 per cent, 10–35 per cent, 35–50 per cent and minimum 50 per cent. The moisture contents are 85–87 per cent maximum.

Dependent on the treatment the curd receives, quark belongs to one of two types: **German quark**, which is somewhat stiff and crumbly, and **French quark**, which has been made somewhat softer and smoother by beating with extra moisture.

To vary the taste, the whole can be mixed with salt and herbs or with fruit and sugar. Available are **dessert quark**, where the fruit is in a layer at the bottom and **fruit quark**, where the fruit is mixed into the quark.

A special kind of quark is **cottage cheese**, (**Hüttenkäse** or **Kotta**) made of skimmed milk, which is soured and curdled at 32°C for about 29 hours. After coarse cutting, the whole is heated to about 50°C to kill most of the lactic acid bacteria. After this the whey is drained off and the curd washed with cold water. The mass is not homogenized, so that the crumbly structure remains. Sometimes the end product is mixed with salt and/or cream to a content of about 5 per cent fat.

Fresh cream cheese

To this group belong little cheeses with a mild creamy taste, which are easily processed into a soft paste. The basis is cream with a fat content of 10 per cent. After souring and curdling, the curds are separated from the whey, pasteurized, homogenized and brought up to taste with salt and herbs. The cheese is then aseptically packaged. The little cheeses can be kept for several months in an unopened state. Some examples: Mon Chou, Boursin, Paturain.

Spoilage and storage

Because of the high moisture content and the absence of a crust, fresh cheese has a much lower keeping quality than ripened cheese with a crust. Fresh cream cheese can be kept longer in its original package than the quark varieties. The best storage place for both kinds is the refrigerator.

11.6 PROCESSED CHEESE

Processed cheese was a discovery of the Dutchman Eyssen, towards the end of the nineteenth century. The long-keeping quality compared with traditional cheese, and also, especially, its spreadability and constant taste are responsible for its success. For the producer, the economic advantage is that imperfect, misshapen, damaged or soiled cheeses are still usable for processed cheese preparation. At the moment, because of the great demand, special cheese is made for it. The consumption per year is on average 0.5 kg processed cheese per person in the Netherlands.

Production
The cheeses are cleaned, crusts removed and sub-standard parts are cut off. The remainder is ground up. Water and other additives such as melting salts, milk powder and perhaps flavour components are added. The melting salts (phosphate salts) have the function of binding the protein and the fat with the water, so that the cheese stays spreadable and sliceable after cooling down. In this way an emulsion develops. While stirring and increasing the pressure, the whole mass is heated to 74°C, during which the cheese melts. Because the temperature is so high, the molten cheese is also pasteurized. It is then aseptically packaged.

Composition
Because of the high moisture content of the processed cheese (45–60 per cent), these products contain, in comparison with cheese, less fat, protein, B-vitamins and vitamins A and D. So, full fat cheese contains 23 per cent protein and 28 per cent fat, while full fat cheese spread contains 16 per cent protein and 22 per cent fat.

Depending on the quantity of fat, the following labelling is used, expressed in percentage fat content of the dry matter:

> 60+
> 50+ ⎫
> 48+ ⎬ sometimes with the labelling 'full fat'
> 45+ ⎭
> 40+
> 30+
> 20+
> 10+ skimmed
> skimmed

Kinds of processed cheese
The assortment can be arranged according to the moisture content, the lactose content of the end product, and the use of ingredients other than cheese, for instance herbs, ham or bacon. They are classed as:

— **processed cheese**, which can be sliced;
— **cheese spread**, which can be spread and has a higher moisture content;
— **processed cheese products**, in which the lactose content is higher than 5 per cent through the addition of a large quantity of milk constituents. Only half of the dry matter in these products originates from cheese.

The name must indicate if other foodstuffs have been added, for instance 'Processed cheese with cumin'. With the exception of herbs and spices the percentages of the characterizing foodstuff added has to be mentioned, for instance cheese spread with 10 per cent ham. Also flavour components may be used, for instance cheese spread with smoke flavour or smoked processed cheese (40+).

There are also low-sodium varieties for sodium-resricted diets.

Spoilage and storage
Through the high heating during melting, the cheese is in fact pasteurized. It is then also aseptically packaged. Because of this it can be stored for a long period (up to one year) without refrigeration. After opening of the package, processed cheeses have a limited storage life.

11.7 LEGISLATION

Legislation with respect to cheese is usually described in the **Cheese Regulations**. The Cheese Regulations stipulate amongst other requirements, the composition, the permitted additives and the obligatory labelling of the cheese. One of the obligations may be to mention the country of origin, if the cheese variety has not been developed in that particular country. Otherwise the name of the country of origin is not necessary. For instance: Emmental is by origin Swiss cheese; if it is copied in Germany, then the product has to be called German Emmental. When Gouda cheese is made in Denmark, it has to be called Danish Gouda.

Agreements about whether or not to mention the country's name were made at the Convention of Stresa in 1951 by a number of cheese-producing countries to protect the typical kinds of cheese. It is also prohibited to use some cheese names in other countries at all, for instance Roquefort and Parmesan.

There may also be regulations for processed cheese and quark.

With regard to cheese made in the Netherlands the legislation is set out in the Agricultural Quality Order for cheese products (Agricultural Qualities Law). Its contents include instructions for the preparation, the arrangement according to quality and the shapes of the various Dutch cheeses and the Government Cheese Mark.

Government Cheese Mark
The cheese checking stations have introduced the Government Cheese Mark to combat the imitation of Dutch cheese. This mark guarantees the correct composition (fat and moisture content) and quality. The stamp is made of casein and merges with the crust soon after application, so that removal is impossible. On the stamp are given the fat content, the country of origin, the producer and the checking station under which they are regulated. And for most kinds, the name of the cheese is also included.

Fig. 11.3 — Several Dutch Government Inspection Marks of cheeses.

12
Eggs

12.1 INTRODUCTION

'Egg' in everyday speech usually means a chicken egg. But eggs from other birds are also edible, for instance quail, duck and goose eggs.

From time immemorial, eggs have been used as a source of food. Around 1500 BC a start was made to domesticate poultry. The production of chicken eggs has been increased by breeding and improvement of breed selection.

Fig. 12.1 — Chicken eggs.

Consumption

In 1985 the consumption of eggs in the Netherlands was about 200 per person whereas the UK figure was 224. Two-thirds of these were used in the home. The rest was used by industry and bakeries. There has been little change in the number of eggs consumed per person in the last ten years.

Besides chicken eggs, goose and ducks' eggs are also eaten, on a very limited scale.

Composition

An egg consists of a shell with contents. The shell is not eaten. The following substances occur in eggs:

water	75%
protein	13%
fat	11%
carbohydrates	—
minerals (phosphorus, sodium) vitamins A, B_1, B_2, D	1%

The ratio shell:yolk:white is about 1:3:6. The yolk contains the largest part of the protein and the total content of fat. Besides fat, fat-like compounds also occur in the yolk, for instance lecithin, which acts as a natural emulsifier, and cholesterol, a compound which is often mentioned in connection with heart and circulation problems.

12.2 STRUCTURE OF THE EGG

An egg consists of the embryo of a bird, surrounded by a stock of food (the yolk and the albumen) and a shell. In Fig. 12.2, the section of an egg is drawn.

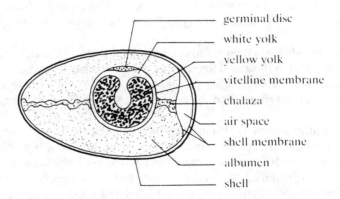

Fig. 12.2 — Section of an egg.

The **eggshell** is white or brown, depending on the breed of chicken. The colour has no effect on the taste but is genetically determined. The shell contains chalk, is breakable and porous. On the outside, the shell is surrounded by a thin egg membrane (cuticle) which feels somewhat fatty to the touch; it has a soft gloss and it protects the shell against intruding microorganisms. Sometimes it can happen that no shell is formed through shortage of chalk in the food. The resulting egg is called a soft-shelled egg.

On the inside of the eggshell two **shell membranes** are found, which lie against each other. At the blunt end of the egg, in between the membranes, an air space is formed. In a new-laid egg the air space is missing. On cooling down, the contents of the egg shrink and the air space develops. During storage, moisture evaporates out of the egg and the air space becomes larger still.

The **albumen** consists of dense albumen and thin albumen. The thin albumen is underneath the shell and close around the yolk. Through the action of enzymes, the older the egg becomes the more thin albumen develops.

Raw albumen contains avidine. This compound makes the vitamin biotin, which plays a role in carbohydrate metabolism, inactive. By boiling the egg or cooking it in any other way, avidine is inactivated.

The **yolk** is surrounded by a very thin vitelline membrane, to which the chalaza is attached. The colour of the yolk varies from light yellow to orange-yellow and is dependent on the β-carotene content of the food.

The **chalaza** ensures that the yolk stays in the middle of the egg. During ageing, the chalaza loses its firmness, through which the yolk will start to float up. The chalaza leading to the pointed end of the egg is the strongest; that is why it is advisable to store an egg with its point downwards.

12.3 PRODUCTION AND DISTRIBUTION

Egg production is carried out by specialized firms. They employ laying-batteries or deep-litter systems.

In the **laying-battery** the chickens are in wire-mesh cages, alone or with four to five chickens together. The supply of food and the collection of the eggs is a continuous system. This method has for the producer the advantage of a high production of clean, undamaged eggs. This means relatively cheap eggs for the consumer.

With the **deep-litter system**, (also called the free system) the chickens walk free in scatter-sheds with ten thousand together in one enclosure. They have some freedom of movement, but the eggs easily get damaged and dirty. If the deep-litter system complies with certain rules, then the eggs can be sold as free-range eggs (see section 12.6).

The deep-litter system is also used by bio-dynamically or ecologically orientated firms. The chickens are fed with biologically or bio-dynamically grown food, and they also have more freedom of movement (maximum 5 chickens per m^2). In some countries eggs so produced are specially marketed. For example, in the Netherlands eggs from bio-dynamic firms carry the BD label and EKO eggs get the EKO stamp (see Fig. 12.5).

However, to achieve a high egg yield the system is not the most important aspect but a good laying breed. In breeding stations breeds are developed either for a high egg yield and a good egg quality (laying breed) or for good meat production (fattening breed).

Production and distribution

Fig. 12.3 — Laying-battery.

Fig. 12.4 — Deep-litter system.

Fig. 12.5 — EKO stamp on EKO eggs.

Firms which produce the chicks obtain cockerels and hens from the breeders and produce hatching eggs, which are hatched in hatcheries. The laying chicks (poulets) are reared by the egg-producing firms and are ready to lay after about five months. They lay approximately 175 eggs per laying season of half a year. After this they are slaughtered and used as soup chickens. The eggs are collected and checked for soundness by 'candling' which means to look at in transmitted light. By doing this, bad or hatching eggs can be picked out. They are then sorted for size and packed in cartons or in boxes. This is done in registered packing stations. The eggs are traded by cooperatives such as the egg auctions or at egg markets.

12.4 EGGS AND EGG PRODUCTS

Chicken eggs are available loose or packaged in boxes, sorted according to quality and weight. For the consumer, only quality class A is for sale. These eggs, so-called fresh eggs, have to comply with certain rules, for instance the shell has to be clean and undamaged and the air pocket must not be too deep.

Eggs of class B or class C go to industry. These sorted-out eggs contain for instance haircracks in the shell (such an egg is called a rattle) or blood or meat stains in the yolk or the egg white. It can also be possible that the shell is badly damaged. The weight of an egg is expressed as a class number (see Table 12.1). Very heavy eggs

Table 12.1 — Weight classes of chicken eggs (Netherlands)

Class	Weight per egg (g)
1	more then 70
2	65–70
3	60–65
4	55–60
5	50–55
6	50–55
7	less than 45

sometimes contain double yolks. Most eggs sold weigh 55–60 g. Recipes are mostly written with this size in mind.

Ducks' eggs are on average larger then chicken eggs (about 70 g) and have a strongly yellow-coloured yolk. The shell is white or light blue-green in colour. They are often used in the rusk industry. Because the chance of infection by *Salmonella* bacteria is higher, these eggs have to be boiled for a minimum of 10 minutes or fried on both sides.

Goose eggs weigh about 150 g. The shell is dirty white to light green in colour; the yolk is orange. Because these eggs, as with duck eggs, can be infected with *Salmonella* bacteria, it is advised to boil them for 15 minutes or to fry until completely cooked.

Quails' eggs are the size of cherries. The shell is beige-brown speckled. Quails' eggs are regarded as a delicacy. They are poached and served on toast, or hard-boiled and used in salads. Sometimes the eggs are stored in a bowl with a vinegar solution, or shelled and bottled in water.

Besides fresh eggs, eggs and parts of eggs are also preserved for sale, such as frozen egg products or egg powder. Bakeries and industry especially use these products. In restaurants the 'egg by the yard' (egg-roll or long-egg) is used. This roll (approx. 35 cm long) is made from about 12 fresh eggs. The yolks and the egg whites are separately cooked in tube-moulds, pushed inside each other and cooled. The rolls are stored deep-frozen. They can be kept for about one week outside the deep-freeze. About 75 identical slices of egg garnish can be cut without waste.

Around Easter, prepared eggs are offered for sale, which can be cooked, painted and laquered, and then kept for about 6 weeks.

12.5 QUALITY DETERIORATION, SPOILAGE AND STORAGE

The quality of fresh eggs deteriorates during storage. The porous shell lets moisture evaporate through it, resulting in the contents shrinking. Also, moulds and bacteria can penetrate. The protein membrane gives protection; that is why a clean, unwashed shell is important.

In the packing stations all eggs are checked for freshness during candling. A fresh egg is translucent, the yolk is clearly visible and the air pocket is small.

In the home the freshness can be determined with the help of the following tests.

— *The floating test* The egg is put in one litre of water to which 80 g salt is added. An older egg will stand at an angle. The egg is unsuitable for consumption if it starts to float.
— *The roll test* An old egg will roll long and stable because the chalaza have weakened and the yolk will seek the middle. With a fresh egg, the strong chalaza will hold the yolk a little away from the middle so that the egg will wobble somewhat as it turns around.
— *The breaking test* A fresh egg on breaking has a round, firm yolk and a lot of thick white compared to thin white.
— *Shaking* The contents of an old egg will slosh about because it has a large air pocket, formed through evaporation of the watery part of the contents.
— *Yolk position* In a boiled egg the freshness can be determined by the place where the yolk is situated.

Fresh eggs can be kept for three to four weeks if stored cool and in the dark. It is better not to store eggs in the refrigerator when they are in regular use, because this speeds up drying out. For longer storage it is advisable to store the eggs in the refrigerator to prevent unwanted bacterial growth. In this case, the eggs ought to be turned regularly. It is not advisable to wash the eggs, because then the thin protective wax layer is rinsed from the shell. Yolks can be stored, well covered, in the refrigerator for a few days. The white of the egg can be kept in the refrigerator for about a week.

Boiled eggs can only be kept for a few days in the refrigerator. The protective effect of the cuticle has been lost in the boiling process. Whether an egg has been boiled or not can be determined with a roll test. Start to spin the egg and put a finger on it for a count of one. If the egg is raw, it will start to turn again when the finger is removed. In a raw egg the contents keep moving even if the outside has been stopped for a short period. If the egg has been cooked, it will stay still after being touched.

Eggs can be stored in cold stores at around 0°C if they have to be kept for longer periods. They can be kept there for about six months. Once out of the cool store they have to be processed within a week. These eggs never get the quality class A and the consumer cannot buy them. They are processed in industry, in bakeries and large kitchens. Eggs can be stored even longer if the contents, either whole or separated into yolk and white, are dried or deep frozen.

12.6 LEGISLATION

The EC Ordinance concerning certain trade standards for eggs contains regulations for classing them by weight and quality (see section 12.4). On packaged eggs, besides the above-mentioned labelling, there is also stated the number of eggs, the name and the registration number of the packing station and the date of packaging. Exceptions to this rule are eggs which are sold straight from the farm or sold loose in the shop. On the small packaging of class A eggs a sticker is allowed with the word 'extra' if they are very fresh eggs. The sticker has to be removed from the packaging if the eggs are not sold within seven days after packaging. Regulations with regard to quality may also describe what free-range eggs are, namely eggs from free-range chickens, which can roam in a space of 1 m^2 containing a maximum of seven chickens. One-third of the floor has to be covered with sand, straw, shavings or similar material. Free-range eggs may only be allowed to be packaged by certain firms which are members of an Association for Free-range Eggs Control (SSC). The packaging of free-range eggs can be recognized from a special label, for example the SSC label (see Fig. 12.6).

Fig. 12.6 — Labelling on the packaging of free-range eggs.

Loose sold free-range eggs may be stamped with a red logo (see Fig. 12.7).

Fig. 12.7 — Logo on loose free-range eggs (x=registration number of the packaging station).

Regulations on the delivery of chicken, duck and goose eggs require that duck and goose eggs which are intended for consumption must be labelled with the notice:

'Duck's egg to be cooked for 10 minutes'
'Goose egg to be cooked for 15 minutes'.

13
Oils and fats

13.1 INTRODUCTION

Oils and fats are collective names for a group of products which universally occur in nature. They have similar characteristics: feel fatty to the touch, are not soluble in water and are lighter than water. The main difference between oils on the one hand and fats on the other is that oils are liquid at room temperature and fats are solid. Changes in temperature can convert the one into the other. A lowering of the temperature will congeal oils (make them solid); an increase in temperature melts solid fats (they become liquid).

Consumption

In the compilations for consumption figures, a distinction is made between visible fats — such as butter, margarine, low fat margarine and edible fats and oils, whether or not added while preparing dishes or products — and invisible fats which occur naturally in products such as meat, eggs, nuts, cheese or milk products.

In 1985 the consumption of visible edible fats was about 24.5 kg per person in the Netherlands and 16.8 kg in the UK. The Netherlands total fat consumption was 48.5 kg per person or about 133 g per day, which is very close to UK consumption at 136 g per day.

Composition

Chemically, oils and fats consist of glycerol esterified with three fatty acids: a triglyceride. There are more than 50 different fatty acids, for instance linoleic acid, oleic acid, palmitic acid, stearic acid and butyric acid. The fatty acids can vary structurally in:

— chain length, with 2 to 24 carbon atoms. Fats are easier to digest the more fatty acids with short chains are present.
— the number of double bonds between the carbon atoms. If there are no double bonds they are termed saturated fatty acids. Where there are one or more double bonds they are termed mono- or poly- unsaturated respectively.

A high percentage of unsaturated fatty acids in triglycerides results in a more liquid consistency (oil) at room temperature; a high percentage of saturated fatty acids in triglycerides gives a more solid consistency (fat) at room temperature.

In vegetable oils the double bonds present in the unsaturated fatty acids are called cis double bonds. They can, by different means — for instance in the preparation of margarine — be changed into trans double bonds. Cis and trans fatty acids are geometric isomers: they have the same chemical formula but different characteristics. Trans fatty acids are said to increase the serum cholesterol and so contribute to the causes of heart and circulation diseases as a consequence of raised serum cholesterol values. To fight heart and circulation diseases, the use of unsaturated fatty acids with cis double bonds, especially *cis-cis*-linoleic acid, is preferred to the consumption of animal fats, in which trans fatty acids can naturally occur.

Natural fats and oils contain as well as glycerides a number of other components in very small quantities, such as the chemically related waxes, phospholipids and hydrolysis products (di- and mono-glycerides and fatty acids) and chemically non-related compounds such as sterols, pigments, (carotenes, chlorophyll), vitamins A, D, E and K and odour and taste compounds.

Possible health hazards
Erucic acid occurs in glycerol esters in such products as cabbage- and rape-seed oil and, through these, in margarine, low fat margarine and mayonnaise. Excessive use of foodstuffs prepared with such oils can led to toxic symptoms (heart complaints). The law lays down maximum permitted contents for erucic acid in this context.

Cabbage and rape seed yield oils which are very similar in composition. Rape oil from rape seed contained in former days around 50 per cent erucic acid; such oil would now be used only as lamp oil. More recently, only a variety of cabbage seed which yields an oil low in erucuc acid is used in the food industry in northern Europe.

Detrimental compounds can occur in animal fats as a result of production (veterinary products) or from the environment (pesticides and environmental contaminants) via the animals' feed, and can build up in the fat tissue.

In both vegetable and animal oils and fats, residues can occur from pesticides which are used in their production.

13.2 PRODUCTION OF OILS AND FATS

For the extraction of oils and fats three methods are in use.

— *Rendering*: mostly used for fat tissue of slaughtered animals.
— *Pressing*: of oil-containing seeds and fruits of oil-containing plants.

— *Extraction*: from fat-containing material by means of organic solvents.

The **rendering** of animal fat is done in autoclaves at temperatures between 60°C and 150°C. The fat comes out of the fat cells, leaving a solid residue of tissue (greaves) behind. The molten, rendered fat is then poured off as an oil layer and filtered.

Depending on the desired quantity and quality of oil, a hydraulic press or screw press is used in **pressing**.

Cold pressing is done with a hydraulic press and gives in comparison a low yield of good quality oil (few contaminants). The seeds or fruits are cleaned, coarsely ground and sometimes heated with steam to a maximum temperature of 35°C. A higher temperature is undesirable because vitamin E would be inactivated (natural antioxidant), double bonds would be broken, through which free fatty acids would form and loss of odour would occur. The oil is pressed out between presscloths and pressplates. The 'cake', which still contains 4–5 per cent fat, is used as cattle feed.

Warm pressing is done by roller or screw presses. This gives a high oil yield, but a higher level of contaminants in the oil. The seeds or fruits are cleaned, ground into flour and heated by steam to around 80°C. In a continuous process the flour is transported via an Archimedes screw in a somewhat cone-shaped housing with slits, through which the oil can drain. The residues serve as cattle feed.

In **extraction**, the fat-containing material is treated with organic solvents such as hexane. As a rule, the solvent is brought into contact with the ground oil-containing material in a counter-current flow. The oil solution is evaporated by distillation and the last traces of solvent are driven off by steam. The solvent is also removed from the extracted material. This residue ('cake') contains less than one per cent fat, but considerable protein and fibre, and can serve as cattle feed.

The crude oils and fats are mostly refined for consumption with the aim of removing compounds which have a detrimental influence on colour, taste and keeping qualities. Cold-pressed oil does not need to be refined because it contains few or no contaminants.

Depending on the type of contaminant, the refining process consists of a number of treatments.

— **De-sliming** The oil is mixed with hot water, to which phosphoric acid or citric acid is added. The phospholipids are transformed into an insoluble form and separated by centrifugation.
— **De-acidification** Free fatty acids are bound (saponified) by means of an alkaline solution (lye). The insoluble soap can be separated. The free fatty acids can also be distilled under defined conditions of temperature and vacuum (physical method).
— The de-acidified, neutral oil can then be **bleached** by a treatment at 90°C with Fuller's earth and then **filtered**.
— The almost neutral and colourless oil is finally cleared of undesirable odour and taste compounds with the aid of hot steam (**deodorized**).

Only a small percentage of the edible oils and fats are used in households directly as such. The main part of the refined oils and fats is a raw material for the production of margarines, halvarines, mayonnaises, and fat for frying, baking and roasting.

For technical reasons, the industry gives a number of treatments to oils and fats (hydrogenation, fractionation and esterification) with the aim of changing the consistency of the fats or oils.

— **Hydrogenation** (hardening). By this means, a firmer consistency is given to oils. During the hardening process hydrogen is added to the double bonds of the fatty acid molcules under the influence of a catalyst — usually nickel — at a temperature of approx. 150°C. The saturation of the double bonds results in a rise in the melting point of the oil; so it acquires a firmer consistency. A disadvantage of the hardening is that cis double bonds are totally or partly transformed into trans double bonds.
— **Fractionation** is used to separate fats into fractions with different melting points. During the fractionation the fats are heated until everything is liquid. Then cooling follows, in such a way that the fat fractions (fatty acids) crystallize out separately one after another, after which, on each occasion, the solidified fat can be filtered off. An example is the removal in this way of stearic acid from beef fat. Beef fat without stearic acid is less hard and less brittle than untreated beef fat and is even suitable for heating to a very high temperature (deep frying). The treated beef fat solidifies less quickly on cooling and smells less penetratingly than beef fat.
— **Esterification** is used to give a suitable firmness and spreadability to fats. The esterification takes place at 90°C in the presence of a catalyst. The fatty acids of the triglycerides are mutually interchanged; they alter their position. This results in a change of consistency.

13.3 FAT AND OIL TYPES

Animal fats and oils are liberated from the fatty tissue of slaughter animals and fishes (among others the herring family) by rendering.

Fatty tissue occurs in slaughter animals mainly: in the abdominal cavity, around the kidneys and the peritoneum, under the skin and in and around the muscle tissue. The domestic use of the rendered fats is only of minor importance. They are mainly the raw material for the fat-processing industries.

In Table 13.1 the most commonly occurring animal fats and oils are listed with their physical characteristics and an indication of the flavour intensity.

Vegetable oils and fats are obtained from seeds or fruits rich in oils. They have mostly a soft yellow to light green colour and a slight smell, and taste of the seed or fruit from which they are pressed or extracted. Some kinds are suitable, after extraction, as table oils; others, especially those that are cheap or have a less popular taste, are used in the preparation of such products as margarine. Groundnut, olive, corn, sesame, soy, sunflower and safflower oil are used as table oils. Industry uses, amongst others, cabbage seed, rape, soy, cottonseed and palm oil and palm-kernel fat.

Types of vegetable oils and fats

Groundnut oil from peanuts (groundnuts) is mainly imported from America, Africa and Brazil.

Table 13.1 — Kinds of animal fats and oils

Name	Colour	Consistency	Flavour intensity
Beef fat	Light yellow	Hard	Strongly pronounced
Calves' fat	White	Fairly soft	Less pronounced
Pork fat (lard)	White	Soft	Slightly pronounced
Sheep fat	White	Hard and crumbly	Strong
Horse fat	Yellow to orange	Soft	Strong
Fish oil	Yellow	Liquid	Pronounced fish taste

Babassu oil is from the fruit kernels of the babassu palm; the fat has the same characteristics as palm-kernel fat. The palm grows mostly in Brazil.

Coconut oil from the coconut has in our climate mostly a solid consistency (coconut fat) and is mainly imported from Indonesia.

Corn oil is from the maize seeds (cornseed oil/maize oil), and comes mostly from North America.

Palm oil from the flesh of the palm fruit, has a red colour because of the high β-carotene content (0.1 per cent).

Palm-kernel fat is obtained from the kernels of the palm fruit. Palm oil and fat are imported from South American and African tropical areas.

Olive oil comes from the olive; the flesh of the fruit contains 30–50 per cent oil and the pips five per cent. This oil is especially used as a table oil in the countries around the Mediterranean.

Rape oil from rape seed and *cabbage seed oil* from cabbage seed are both used as cheap oil for margarine manufacture. They come from the Northern Countries of the EC.

Safflower oil comes from the seed of a blue kind of thistle, which is grown in the United States of America.

Sesame oil from sesame seed is often imported from China, Mexico, India and Venezuela.

Soya oil from the soya bean is especially grown in the United States, China and Brazil.

Sunflower oil comes from the seeds of the sunflower. This oil is imported from Russia and Mexico.

13.4 PRODUCTS WHICH CONTAIN OILS OR FATS

Salad oil is one description often used for table oil. This consists of one or more vegetable oils.

Cooking/frying oil consists of one or more vegetable oils. Oils with a very pronounced taste like olive oil and safflower oil are less suitable; more suitable are groundnut, corn, soya and sunflower oil. Soya oil is the least stable oil; this means it polymerizes quickly on heating. To overcome the disadvantage of this instability, soya (frying) oil is lightly hardened, which improves the heat stability a great deal.

Cooking/frying fat consists of 100 per cent animal and/or vegetable fats (mostly hardened oils). The fat is suitable for heating to a high temperature (175°C) for a longer time.

Baking and frying products consist of about 96 per cent of fat; the remainder consists of (among other things) milk solids, water, vitamins A and D and salt. The kind of fat used varies, both (hardened) oils and/or animal fats are suitable. These products are especially suitable for the frying or roasting of meat. They are not suitable for deep-frying or for spreading on bread instead of margarine or butter.

Mayonnaise is an emulsion of oils (minimum 80 per cent) in water and vinegar. Egg yolk is used as the emulsifier.

Salad cream is an emulsion, which contains 25–80 per cent oil and also egg yolk as emulsifier. In some countries another name for salad cream is chip sauce. In practice, salad cream and chip sauce differ in taste and consistency. Salad cream contains a bit more sugar and vinegar, while chip sauce is slightly thicker.

'**Low fat mayonnaise**', because of its composition, also comes into the category of salad cream. This mostly contains only 40 per cent oil, but the use of thickeners gives it a mayonnaise-like consistency.

13.5 QUALITY DETERIORATION, SPOILAGE AND STORAGE

Oils and fats deteriorate in quality when they are stored. This deterioration progresses to spoilage after long storage.

Changes which can occur are:

— tainting: absorption of odours from the surroundings
— auto-oxidation
— oxidation of (or development of rancidity in) fats
— hydrolysis of fats.

Lowering the temperature results in a slowing-down of these reactions, and the addition of anti-oxidants slows chemical reactions down, but light increases them. Oils and fats should therefore preferably be stored in the dark, cool and well sealed from the air.

By heating fats and oils for long periods at temperatures of about 180°C (which happens in deep frying) increased hydrolysis occurs under the influence of water, sometimes followed by polymerization and the formation of toxic compounds (acrolein). For this reason, frying fats and oils should be replaced when they become brown and viscous.

Some storage times of pure fats and oils, and of products prepared with oil or fat, adequately packaged and stored cool and in the dark, are noted in Table 13.2.

13.6 LEGISLATION

Regulations for **edible oils and fats** set standards for the soundness of oils and fats, the labelling and the presence of additives such as anti-oxidants, dyes, anti-foam and odour and taste compounds.

Table 13.2 — Storage times of oils, fats and products

Product	Storage time
Table oil and salad oil	6 months–1 year
Frying fat	Approx. 1 year
Mayonnaise and salad cream	2–3 months

Mayonnaise and salad cream regulations lay down standards concerning the raw materials, the presence of erucic acid and pathogenic organisms.

Labelling regulations lay down standards concerning the way in which the raw materials used have to be described.

14

Margarine

14.1 INTRODUCTION

Margarine is similar to butter in appearance and composition, but differs in that the fat does not, or only a very small part does, originate from milk.

The product 'margarine' was developed in the year 1867 by the French chemist Mège Mouriès. He made a replacement for butter by order of Napoleon III. The army of this emperor lived on a ration of bread and butter, but this was scarce and perishable. The new product had to be cheaper and had to be easier to store than butter.

The product made by Mouriès was different in composition from the present margarine. His basic compound was molten beef fat. After cooling to 25°C, the liquid phase (oleomargarine) was separated from the solid phase. The oleomargarine was mixed with (acidified) whey and chopped udder. After thorough churning, a product developed which look a bit like butter. The margarine product was taken up around 1880 by, among others, the Dutchmen Jurgens and van den Bergh in Oss. The animal fats were then already partly replaced by vegetable oils, to obtain the right consistency. Later a process was developed to solidify liquid oil (hardening), so that the right consistency was more easily obtained.

Consumption
The high price of butter is one of the reasons for more margarine being used. Besides this, the present margarine industry is able to make products with a good taste, smell and consistency and of a different composition (low fat margarine). Margarine use in the Netherlands was in 1985 about 11.7 kg per person and for low fat margarine this was 2.6 kg. Total UK margarine consumption amounted to 7.3 kg per person.

Composition
Margarine is an emulsion of water or skimmed milk (water phase) in an oil–fat mixture (fat phase).

The **fat phase** is composed of refined vegetable and/or animal oils and fats (minimum 80 percent). The mixture has such a composition that the margarine is spreadable at room temperature in the winter as well as in the summer. The most used vegetable oils are soy, sunflower, rape, palm, coconut and babassu. Fish oil, beef fat and lard are used from among the animal oils and fats.

The **water phase** consists of water and/or pasteurized skimmed milk (maximum 16 per cent). This is sometimes mixed with a pure culture of lactic acid bacteria. This acidification is meant to develop a butter aroma in the margarine (diacetyl formation).

Finally, emulsifiers (mostly lecithin), colour, smell and taste compounds, milk sugar, food acid, vitamins (A and D) and sometimes salt and a preservative are added.

The desired melting or solidifying point can be given to the oils and fats by hardening, fractionation and/or esterification (see section 13.2).

14.2 MANUFACTURE

During the manufacture of margarine the basic compounds come into a mixing vessel (votator), after which a number of treatments follow each other in a continuous process.

The emulsifying takes place at about 40°C. The not-very-stable emulsion becomes more stable during the cooling down and crystallizes. The solidified margarine is kneaded to obtain the right spreadability and smoothness and is then packaged.

14.3 KINDS OF MARGARINE

All margarines contain a minimum of 80 per cent fat.

Margarine is made with animal and/or vegetable fats and oils. The consistency is fairly hard. It is mostly packaged in block form (small packs). The description 'table margarine' on a small tub refers to the good spreadability, even if the margarine is stored in the refrigerator. Margarine with the labelling '(pure)vegetable' or 'vegetable margarine' contains in the fat phase only vegetable oils and fats. The oils are sometimes hardened, depending on the desired firmness of the product.

Camping margarine is margarine based on oils and fats which keeps a good consistency unrelated to the storage circumstances. It is packaged in cans and because of this it can be kept longer in closed packaging than the ordinary margarine.

Crust margarine is suitable for use in the bakery for the preparation of puff pastry. The fat choice makes this margarine firm and tough. It is only for sale in large quantities to bakeries, but not for the consumer.

Polyunsaturated margarine differs from normal margarine in a number of points. Sunflower oil is mainly used. This oil is fractionated to obtain a good spreadable margarine; the high melting oil fraction is filtered out and some palm and palm kernel oil is added. The margarine manufacture takes place without the addition of salt, and the choice of the other ingredients is (as much as possible) in agreement with wholefood principles. The linoleic acid content in such margarine is around 40 per cent.

Diet margarine
On the market there are products for:

— *Linoleic-acid-enriched diet* The fat phase consists of vegetable oils and fats rich in linoleic acid. These margarines contain mostly 60–65 per cent linoleic acid in the fat phase.
— *Salt-restricted diet* These kinds of margarine are not allowed to contain more than 40 mg sodium per 100 g product. They are suitable for people who follow a sodium-restricted diet and are permitted to be labelled 'unsalted'.

Low fat margarine
All kinds of **low fat margarine** contain 40 per cent fat, mostly of vegetable origin. The high moisture content (around 50 per cent) needs extra emulsifiers and thickeners and makes low fat margarine unsuitable for baking and frying. The manufacture is near enough identical to that of margarine. Low fat margarine also exists for linoleic-acid-enriched diet, for salt-restricted diet, or unsalted.

14.4 QUALITY DETERIORATION, SPOILAGE AND STORAGE

Margarine and low fat margarine have a limited storage life. These products are sensitive to light and oxygen during storage; these increase oxidation of the fats and oils from which rancidity occurs. Also heat or changes in temperature can lead to changes in taste and disturbance of the emulsion. The last is visible as a separation of oil and water.

Low fat margarine is also sensitive to mould growth.

Margarine and low fat margarine can best be kept in the refrigerator for 6–10 weeks.

14.5 LEGISLATION

Margarine Regulations regulate the composition of margarine and halvarine and describe the labelling for margarine products in general as well as those for salt-restricted diet.

15

Cereals

15.1 INTRODUCTION

Cereals are the seeds of cultivated grasses (*Gramineae*), that is: wheat, rye, barley, oats, rice, corn and millet. Buckwheat is also counted among the cereals because of its use; however, botanically it belongs to the family of the knotweeds. Rice includes the cultivated sort and the wild version (black rice) which is also harvested and sold.

The use of cereals has been known since the earliest times. From excavations in such places as Iraq (previously Mesopotamia) it has been determined that man has used wild cereals for about 9000 years. Mostly mixtures of cereals were flattened and softened, after which they were put in the sun on hot stones or baked in hot ash. Some wild kinds of wheat still occur in the Middle East, including einkorn (*Triticum monococcum*) and emmer or twocorn (*Triticum dicoccum*). The harvesting of the grains of these wild kinds was a difficult problem because the grains fell out of the ear soon after ripening.

Because of these reasons men began, during the cultivation in the past 5000 years or so, at the same time a selection in the choice of their plants. Even now the improvement process is being continued, for instance to select strains for the precise resistance against diseases or to increase the yield per hectare.

In every part of the world cereal products form an important part of the daily food. Some reasons for this are the high yield per hectare compared with the harvest of other crops, the good keeping qualities and easy transport, the high nutritional value and the neutral taste.

Consumption

Cereals can be eaten in various ways: as a starch source in a hot meal, or processed into products such as bread, rusks, pastry, thickeners for soups or sauces, dough products, breakfast cereals and even alcoholic drinks such as beer and gin. It is not easy to give a consumption figure because of these multiple uses, but figures are available in some countries for supplies moving into consumption. For example in the UK in 1985, 68.1 kg per person of cereal products was supplied to the population.

Composition

The division of protein, fat, carbohydrates and water in the different kinds of cereals is given in Table 15.1.

Table 15.1 — Summary of the nutritional composition of cereals, expressed as percentage of the whole

Nutritional substance	Wheat (whole grain)	Rye (whole grain)	Barley (hulled grain)	Oats (hulled grain)	Rice (non-polished)	Corn (whole grain)	Millet (hulled grain)	Buckwheat (hulled broken grain)
Water	13.2	13.7	11.7	13	13.1	12.5	12.1	14
Protein	11.7	11.6	10.6	12.6	7.4	9.2	10.6	10
Fat	2	1.7	2.1	7.09	2.2	3.8	3.9	2
Carbohydrates	69.3	69	71.8	62.9	75.4	71	70.7	72

Source: Franke, W., *Nutzpflanzenkunde*.

The biological value of cereal protein is not as high as animal protein. For most of the cereals, lysine is the missing amino acid; for maize it is tryptophan. To increase the biological value, cereal protein has to be supplemented (for instance with milk products, egg, meat or peanuts). Some cereals like wheat, rye, oats and barley contain gluten proteins. Because of this, these cereals are suitable for the baking of bread. More than half the fatty acids in the fat are polyunsaturated. Vitamin E is also present which can act as a natural anti-oxidant. Starch is found in the cereal kernel in the form of grains. This consists of two kinds of polysaccharides: the amylose inside and the amylopectin on the outside of the starch grain. The bran which is found on the outside of the cereal kernel consists of fibre compounds (among them cellulose and lignin).

The major minerals found in cereals are iron, calcium and phosphorus. Cereals contain generous quantities of vitamins of the B-complex (B_1, B_2 and nicotinic acid).

Possible health hazards

Cereals can contain compounds either by nature or as a result of their production which can have detrimental effects on consumers.

Phytic acid in the bran hinders the absorption of calcium and iron because insoluble phytates are formed from the acid and the minerals.

Gluten protein can cause an over-sensitivity reaction (coeliac disease). Wheat products especially, which contain much gluten, have then to be avoided in the food, especially by young children. Special gluten-free products are available for patients with coeliac disease.

Ergotine, a very poisonous mycotoxin, can occur in rye. The toxin is produced by a mould. The presence of the mould can be recognized by the blue-purple grains (sclerotium) in the head.

By the use of pesticides and soil sterilization compounds and fungicides, chemical pollution can appear in and on the cereals. Cereals can also be contaminated with weed seeds and mould spores.

15.2 STRUCTURE OF CEREAL FRUIT

Cereal fruits belong to the one-seed dry fruits, in which the fruit wall and the seed coat have fused together.

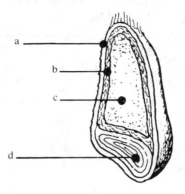

Fig. 15.1 — Diagram of a cereal grain. (a) Bran. (b) Aleurone layer. (c) Starch body. (d) Embryo.

Cereal fruits or grains grow in ears (wheat, barley, rye), in panicles (rice, millet, oats) or in cobs (maize). Buckwheat is a herb-like plant with a three-cornered pod fruit, which contains one seed.

In all cereals, the grains are surrounded by the chaff. These are the withered or dried sepals and crown petals from the cereal flowers. In some cereals, the chaff has partly fused onto the fruit wall. These are called 'covered seed' or 'covered' grains (oats, rice and barley). When the chaff has not fused with the grain, the cereal is called a 'naked seed' or 'naked' (wheat, rye, corn, millet and buckwheat). Covered cereals need an extra treatment after threshing to separate the chaff, which is indigestible for humans, from the corn. Fig. 15.1 shows the structure of a cereal fruit. It is divided into starch body, aleurone layer, bran and embryo.

The **starch body** or endosperm, which makes up about 80 per cent of the corn, is rich in starch and sometimes also contains the gluten proteins. It also contains vitamins of the B-complex, iron, phosphorus, fat and water.

The **aleurone layer** is a thin wall rich in protein, which separates the starch body from the bran. This protein is scarcely digestible by humans. In the aleurone layer are also B-vitamins and minerals.

The **bran** together with the aleurone layer constitute 15–18 per cent of the grain. The bran (the fused fruit wall and seed coat) is called beeswing. Bran contains mainly cellulose, minerals and phytic acid.

The **embryo** forms circa 2 per cent of the grain. The embryo is rich in a type of fat with a high content of polyunsaturated fatty acids, and also contains, amongst other things, vitamin E, vitamins from the B-complex and minerals. Enzymes also occur in the embryo, which after damage to the grain can cause decomposition of the fats present in the embryo.

15.3 PRODUCTION AND DISTRIBUTION

Cereals are grown all over the world; each kind needs a certain climate and a certain soil. Rice grows mainly in tropical and sub-tropical areas; wheat grows best in a temperate or continental climate on clay.

In Northern Europe nearly all kinds of cereal are grown with the exception of millet and rice. Wheat, rye and barley are grown as winter and summer cereals. This means that the sowing can be done in the autumn or in the spring. The harvest is in the early or late summer. Oats, corn and buckwheat are sown once a year in the spring and harvested in the latter part of the summer.

In recent years much attention has been given in current agriculture to seed improvement and selection of cereal crops. The aim of this is to obtain a high grain yield per culm and little straw. With wheat the aim is to approximate the quality of hard wheat (see section 15.4). Also about 200 different pesticides are being used against, amongst other things, mildew, yellow rust and several diseases that occur during ripening. The soil is regularly treated with fertilizers to prevent exhaustion. In alternative agriculture much attention is given to crop rotation, a correct choice of variety and good soil structure. By ploughing in the straw as a byproduct of the cereal, the soil structure is improved. This can also be done with the aid of green manures, for instance kinds of clover and other legumes, which are alternated with the cereal crops. Now, in traditional agriculture there is also a tendency to combine biological and chemical control and good soil cultivation. The ripe grain is taken from the ears with the help of combines. At the same time the chaff is loosened from the naked grain. The straw which stays behind is pressed into bales, chopped, ploughed under or is sent to the straw-processing industry. The cereal grains are then pre-cleaned, that is, a little chaff and other contaminants are removed, and dried. The chaff stays on the corn of the covered cerals after threshing. By an extra treatment (shelling or hulling) this is removed.

The cereal is now ready to be traded. Some countries import much bread wheat and cereal for concentrates for cattlefeed from Russia and America amongst others. the (imported) cereal is stored in silos where it is protected against heating and attack by insects and microorganisms by blowing cool air through and, if necessary, poisonous gases. The moisture content of the corn in the silo is about 14 per cent.

Treatments of the whole grain
Several treatments can be applied with the aim of obtaining products which for instance are easier to digest, more quickly cooked or easier to use.

Huling or shelling
This is done by rubbing the grains intensely over each other (barley, rice) or by shooting them against a type of washboard (oats). By this treatment the indigestible chaff which is fused with the fruit wall is separated from the corn.

Polishing
By this the bran and the embryo are removed from the grain. The external appearance of the grain is also improved (polished rice and pearl barley).

Roasting
The grains are dry-heated, which improves the digestibility, shortens the cooking time, develops aromatic compounds (for instance in the preparation of cornflakes and roasted muesli) and increases the keeping quality, because the product dries further.

Popping
This can be done in three ways, with the main aim of obtaining an airy and crunchy end product.

— *Oven popping*: the cereal is first pre-cooked, partly dried and then puffed up in an oven by sudden heating.
— *Gun popping*: the corn is, after the previously mentioned pre-treatments, exposed to a high temperature and sudden pressure decrease, through which it is puffed up.
— *Popping in hot oil* (popcorn): after popping, the products are sometimes roasted with honey or sugar.

Flattening
The corns are soaked, perhaps more or less steam-cooked and flattened between rollers into flakes or rolled grains. For instant flakes, which do not have to be cooked, heated rollers are used. The flakes are dried after flattening and sometimes roasted.

Milling
There are two main milling methods: flat milling and high milling. **Flat milling** still takes place in the water- or wind-mill. The grain is ground fine between two horizontal cylindrical stones with the lower one in a fixed position and the top one turning round. Grooves have been made in the stones which become smaller and less deep from the centre to the outside. The cereal lands first in the grooves after it has been led on to the stones from the middle. By the turning movement of the top stone the grains are cut and ground. The product which leaves the stones on the outside is **wholemeal flour**. A slightly lighter-coloured flour can be obtained by sieving.

High milling or **roller milling** is only done in the factory and is differentiated from flat milling, because the breaking of the corn, the grinding into **flour** and the separation of grain fractions happens in several stages.

The milling process takes place in a series of rollers in a machine in which the rollers have been installed in groups of two. The rollers turn with different speeds against each other. Depending on the aim for which they are used, the surfaces are either corrugated or smooth. Overall the milling process can be divided into two phases.

In the first phase, *breaking*, the grains are cracked open with the help of corrugated rollers. With flat sieves (sometimes 10 to 12 on top of each other) the different parts of the grain are sorted. The top sieve is very coarse, the lowest has a very fine mesh. One can differentiate:

— *loaded bran*: the coarsest parts of the endosperm and bran;
— *semolina*: coarser and smaller parts of endosperm mixed with fine bran particles;
— *flour*: the finer ground endosperm, nearly without bran particles.

In the second phase, the *grinding*, with the help of smooth rollers the sieved parts are finally ground into flour in 10 to 15 separate grinding treatments. After each milling process, a sieving process follows. In this manner, a number of kinds of flour are obtained with different contents of protein, fat, ash and fibre. In the first part of the milling process, whole white flour is left with little bran and low ash content; flour with a higher ash content and a greyer colour falls from the last milling rollers.

Because several fractions of the corn are left over in high milling as a result of the milling process, these can again be added to the flour, so that any desired end product can be produced (for instance by adding bran, wheatmeal is obtained).

The percentage of the grain which is used in the making of the flour or meal is called the *extraction rate*.

Malting
The cereal is soaked for 48 hours at a temperature of around 25°C, by which the moisture content increases to 45%. The grains sprout during this process. The aim of malting is to obtain a product with a high content of enzymes (catalase and amylase), which are needed to change the starch to di- and monosaccharides. Also, by malting, the starch is for a large part changed into a water-soluble form. The malting of barley is used for beer production; the malting of wheat is used in bread production.

15.4 KINDS OF CEREALS AND THEIR PRODUCTS
Wheat
Wheat is the most cultivated cereal in the world and is a very old cultivated grain. It is grown over the whole world. Sowing and harvesting follow the progress of the sun through the seasons. Because of that, wheat is harvested somewhere in the world throughout the whole year (see Fig. 15.3).

Wheat likes rich, moist soil, preferably chalk containing clay. The plant has also demands regarding light and warmth. If because of little sunshine there is a shortage of light, the harvest will be less good. Winter wheat will start to freeze at a temperature of about −20°C or lower. Most of the wheat is grown in Russia, USA, China, India and France. In the Netherlands it is only grown in small quantities.

The wheat plant is an annual. The stalks stand firmly upright with thick full ears with golden-yellow grains, which are situated loosely in the head chaff. The ears rarely have any awns. The kernel is large, stout and round. There is a groove on the back. The fruit wall has fused with the seed coat.

From olden times three kinds of wheat are known: **einkorn, emmer** and **spelt**, which are all covered seeds. From the oldest culture form, the einkorn, emmer has been developed. Spelt is still grown in a very limited way. The other kinds do not play

Fig. 15.2 — Wheat.

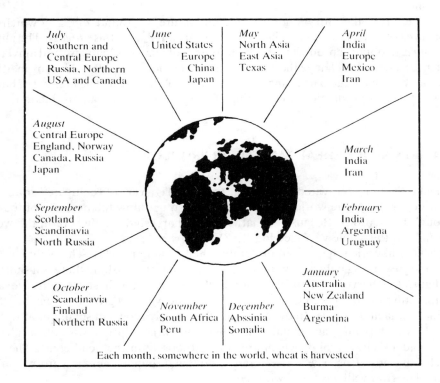

Fig. 15.3 — Survey of wheat harvests.

a role of importance any more. But from these the present varieties have been developed: durum wheat and bread wheat.

Durum wheat has been developed from emmer. Durum wheat grows especially in hot dry areas (including Canada, Egypt, Australia and Southern Europe). This type of wheat is too hard for bread (too much tough gluten), but is much used for the production of dough products (pasta) and semolina.

For the baking of bread, **bread wheat** (*Triticum aestivum*) from spelt, is grown. This kind grows in temperate and colder zones, including in the Netherlands. Bread wheats can be divided into hard and soft wheat, depending on the baking quality, which is determined by the gluten in the corn. Hard wheat is more suitable for bread-making, and grows in areas with a continental climate. Soft wheat is used for the production of baked goods, for instance cookies and biscuits, and is grown in areas with a maritime climate.

Products of wheat

Wheat grains are produced which can be prepared in the same way as rice. The cooking time is about two hours.

Strong flour is obtained by the milling of wheat kernels. The wheat germ and/or husks are completely or partly removed.

If all the naturally occurring compounds of the specific cereal occur in their natural ratio, it is then called **wholemeal.**

Flour is the flour of grains in which the embryo and the husks are not visible to the eye. The extraction rate is about 75 per cent. For **patent flour**, which is white in colour, the extraction rate is 50 per cent. **Quick flour** is patent flour which has been treated (granulated) to make stirring easier.

For the preparation of bread, bakers' products and products such as pancakes, **powder, flour** or **mix** are for sale, such as cake flour or pancake mix. Several additives may be present, for instance emulsifiers, thickeners, smell and taste compounds, a raising agent and flour improvers. **Self-raising flour** may consist only of flour and a raising agent. The labelling 'ready to use' is only allowed if by the addition of water and then baking an immediately edible product is obtained.

Semolina is obtained by grinding the grain with or without bran and embryo and is available in several coarseness.

Embryos (wheatgerm) remain as a byproduct of the flour milling. Oil can be obtained (wheatgerm oil). The embryos are also sometimes used in muesli and other breakfast cereals or sold as wheatgerm.

Bran is also a byproduct of flour milling, and consists mainly of cellulose indigestible to humans. Because of the benefit to the digestion it is added to food (such as muesli, yoghurt, porridge).

Gluten is a byproduct from the wheat starch industry and is used as an aid in bread production.

Flakes are whole or broken grains which are soaked before flattening. They are then dried and often roasted or partly steam-cooked ('instant').

Puffed wheat is obtained by popping whole or broken wheat. Sometimes it is then roasted with honey or sugar. Puffed wheat is used as a breakfast cereal and as a snack.

Dough products (pasta) is the combined name of products such as macaroni, vermicelli, spaghetti and noodles. From wheat semolina, flour and/or meal and water, perhaps mixed with egg, lecithin and salt, a tough dough is kneaded from which various shapes are pressed. Then the shapes, perhaps after a process of pre-steaming, are dried and packaged. If in the production only semolina or strong wheat is used then this is mentioned on the packaging. Sometimes the pasta is sold fresh (not dried). Yellow dye can be added to vermicelli. Green- and red-coloured dough products are imported from Italy. These are coloured with spinach and tomato puree respectively. Macaroni and spaghetti are also available in the wholemeal version. The basic flour in this case is wholemeal flour.

Couscous is in fact the name of a dish: steamed wheat semolina with a (lamb) meat sauce, especially eaten in Morocco. From flour (hard wheat) and water a dough is formed that after drying is crumbled into coarse semolina granules. These are steamed for 30 minutes.

Bulgur, bulghur or boulgour, consists of whole wheatcorns, which are brought to germination, pre-cooked or steamed, and then broken (cracked). Often the cracked grains are also roasted. Bulgur is eaten in Turkey. The cooking time is about 30 minutes.

Seitan is a product which is used as a protein source in macro-biotic nutrition. It consists of wheat gluten, which is obtained by the repeated rinsing out of a dough of wheatmeal and water. Seitan contains 10 per cent protein. (Compare: meat 20 per cent protein, cheese 23 per cent protein.)

Spelt is a kind of wheat which is still grown on a very limited scale in the Swiss and French Jura. It is hard to grow and the harvest varies, and is often low. Spelt cannot cope very well with fertilizer — the grain then grows very tall and is easily blown down by changing weather. Spelt also belongs to the covered seeds. After the harvest the corns have to be de-husked. The grains have a round shape and are light reddish in colour. The baking qualities are excellent. It is very suitable for bread-making because of the high gluten content. In France, spelt is harvested in the green unripe state and later roasted; this results in a tasty product. A large part of green spelt is bought by the soup factories: flour from green spelt gives a delicious aroma to soup.

Maize

The maize plant originates from Central America (Mexico). Several varieties are known, of which some possess a high starch content and others a high sugar content; for instance sweet corn, which harvested unripe is eaten as a vegetable. The annual plant needs a great deal of sun and warmth and copes badly in wet conditions. Some varieties also grow in temperate zones. Maize is grown a lot in the USA, China, Russia, the Balkan countries, Italy and southern Germany. The harvest is from August until October; it takes place when the grains are not quite ripe, because otherwise they cannot be cooked.

In some European countries so-called cut corn (whole maize) is grown. This is especially used as cattle food. Starch (cornflour) and sweeteners (glucose and fructose) are also obtained from it.

The maize grains grow and ripen in the pollinated flower, surrounded by large bracts. They are solid, trapezium-shaped seeds with rounded-off sides. The colour is mostly golden yellow, but all sorts of shades occur: dark red, blue, violet and nearly black. The colours can also vary from cob to cob.

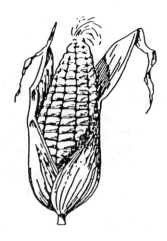

Fig. 15.4 — Maize.

Characteristic for corn varieties is their size. Dwarf kinds exist which are scarcely one metre high, while giant kinds can reach a height of five metres with stems of about 5 cm diameter. The larger the corn plant, the larger the cobs are. The length of cobs varies from 5 to 40 cm.

Products of maize
Grains are sold in dried form. Some kinds are suitable for the starch industry; others can be used for popping.

Flour consists of dried ground grains. In Mexico, ground corn is used for the prepartion of tortillas (little pancakes).

Semolina contains coarse and fine corn particles which stay behind after the grinding of the grains. From this, polenta (a solid, tasty porridge) is made in Italy.

Embryos are used for the production of corn germ oil, which is rich in linoleic acid and β-carotene.

Starch or **cornflour** is obtained from the starch body. Cornflour mixed with aroma, colour and taste compounds is sold as custard powder.

Flakes are make by flattening, and perhaps roasting, the grains.

Popcorn consists of dried popped corns, which are afterwards often roasted and sweetened with honey or sugar, or are lightly salted and sometimes even supplied with a colour.

Corn flakes are produced by cooking broken maize grains with water, salt, sugar and malt extract. After this the porridge is dried and flattened between cooled rollers into flakes and then roasted, which makes them crunchy.

Oats
Oats are grown in Northern Europe, USA, Canada and Russia, especially in moist, moderate areas. Oats do not form an ear, but have a plume-shaped head. The seeds of this annual are covered; after threshing, oats have to be shelled. The oat grain is

long and small in shape, grooved and bleached-grey in colour. The embryo contains more fat than those of other kinds of cereals.

Products of oats
Grains, after shelling, are prepared like rice (cooking time about 40 min). To inactivate the enzyme lipase the de-husked grains are treated with hot steam (lipase can cause fat spoilage which causes a rancid bitter taste). The cooking time will decrease by treatment with steam.

Flour is used as a binding compound in soups.

Flakes or **oats**, are broken or unbroken flattened and perhaps roasted grains. The flakes can also be more or less steam-cooked, which produces cooked (instant) or extra-quick cooking oats. Oats are eaten with milk as porridge or as part of mixed breakfast products. Flakes can also be used as a binding compound in soup or vegetables. Oat flakes are used in muesli, with or without other cereal flakes, dried fruit and nuts.

Rye
Rye was originally grown to the east of the Caspian Sea; now it is a widely grown agronomic crop in Russia, Poland, West Germany and the Netherlands. Rye grows at low temperatures, can develop well on poor soil and does not suffer many plant diseases. The plant can be annual or bi-annual.

Rye has a long thin stalk with a greyish small ear with short equal awns. The grain is longish and grey-blue to yellowish in colour.

Products from rye
Grains of rye need a cooking time of 3 hours.

Meal, obtained by grinding the grain, is used for the production of ryebread, and *knackebrod*.

Flour from rye has about the same potential as ryemeal.

Flakes, which are flattened and roasted grains, are sometimes used in mixed breakfast products and are recognizable by the green-blue colour.

Barley
Barley is, next to wheat, the oldest cereal. It is grown all over the world, in tropical areas as well as in Scandinavia, and even in the mountains, for instance in the Alps, in the Caucasus and in Tibet. Barley has, of all the cereals, the shortest growing period. To illustrate: the average vegetative period for barley is 100 days, for wheat 320, for maize 120, for rye 310. Barley makes low demands on the soil; it grows on dry, cold soil as well as on acid soil. But it has a preference for long light days through which the vegetative period can be shortened (even to 70 days).

Important barley-growing countries are Russia, China, Canada and France, among others.

Depending on the time of sowing, barley may be termed winter or summer barley.

Characteristic for barley are the long awns. The grains are oval and gleaming gold in colour. The chaff and seed coat are firmly fused together.

Fig. 15.5 — Oats.

Fig. 15.6 — Rye.

Products of barley
Grains can be prepared in the same way as rice; the cooking time is about one hour.

Groats is the name for shelled, polished grains. By polishing, several forms can develop. Examples are Alkmaar groats, which is oblong in shape, and pearl or round barley. By pre-steaming groats, the cooking time can be shortened (quick-cooking groats).

Grits are corns shelled and cut in pieces. This is done with a grit cutter. Grits have a shorter cooking time in comparison with the grain.

Fig. 15.7 — Barley.

Meal is obtained by grinding the whole shelled corn.
Flour is sieved meal, obtained after grinding and sieving of the shelled grain.
Flakes are often steamed and roasted after the flattening of the grain. Barley flakes are also called groat malt.
Popped groats (also called manna) consists of blown up and coloured grains.

Rice
The origin of the rice culture lies in India and South-East Asia. From there it penetrated into China and Japan. Around 400 BC Alexander the Great brought rice to Europe. From Europe, rice-planting was transferred (around 1700 AD) to America (Virginia).

At the present time, 90 per cent of the world production takes place in eastern Asia (China, India, Indonesia and Bangladesh). However, these countries export little: the main part of their production is for home use. The main export of rice to western Europe comes from the USA. Europe itself also grows rice on a limited scale (Spain and Italy).

Rice culture is possible between the latitudes 45°N (Italy) and 40°S (the line Argentina–Australia). To grow rice a temperature of 25–30°C and a generous water supply are required. There are two methods of rice cultivation:

— **the wet method:** this takes place in lower lying areas especially in deltas (wet rice) and on higher land on terraces, the sawahs (sawah rice). The plants stand for a large part in water, for good irrigation canals are dug on the fields and around the

Fig. 15.8 — Rice.

rice fields. Little dikes are built to keep the water on the land. On hilly terrain, terraces are built to let the water flow from the top down as along steps. The rice so grown is harvested three to four times per year.
— **the dry method:** this is used in higher sites, even up to 2000 m high (upland or ladang rice). There are only two harvests per year. This rice has no importance for export.

The rice plant is an annual and can grow up to 1.5 metres tall. Depending on the variety and soil, the vegetative period is three to nine months. In Asia the harvest is still done by hand and sickle; in the USA it is done by machine. The rice grains grow in a panicle or halm; the grains are covered seeds. Unmilled rice is called paddy.

Cultivated rice is of the genus *Oryza*, with a large number of different varieties. Rice can be divided according to shape (long or round). The shape also determines the cooking characteristics. On this basis, rice can be divided into:

— **dry cooking rice;** also called patna rice; for instance Siam, Surinam and American rice with long glassy hard grains which after cooking stay dry and separate.
— **wet cooking rice;** also called pudding rice; for instance Japanese and Italian rice. The grains are short and oval or round in shape and often show white spots. The centre is soft and chalky. After boiling, the whole becomes somewhat porridge-like because more starch is made free in the liquid; as a result the grains stick slightly together. This is enough to stop them falling apart. Wet cooking rice is used for the preparation of rice pudding and porridge.
— **sticky rice;** also called ketan or waxy rice. This is a rice variety of which nearly all the starch consists of amylopectin. After cooking, the grains lose their shape; a sticky mass results.

Fig. 15.9 — Rice cultivation.

In Africa, a red *Oryza* variety is grown. In eastern Asia and North America the wild rice (*Zinzania aquatica*) with brown-black slim grains occurs. This plant does not really belong to the genus *Oryza*, but shows similarities. The plant grows especially along the banks of lakes and rivers. The grains were gathered as food by the Indians; on a limited scale this black rice is still gathered in the state of Minnesota. It is appreciated because of its aromatic taste.

Products of rice
The undamaged hulled **grain** is also called full, brown, red, or silverskin rice. If a small part of the outer layer (bran) is cut from the corn then it is called 'half-full' rice. The cooking time of full and half-full rice is 45–60 minutes.

If all the bran is polished away and only the white flour body stays behind, **'white' rice** is obtained. The cooking time of white rice is about 20 minutes. To beautify and to prevent spoilage, white rice is often glossed (with potato flour). Before cooking, the grains have to be well rinsed to prevent sticking after preparation.

Quick-cooking rice has a shorter cooking time, because the rice has been pre-steamed under pressuire. Through this the rice has been partly cooked and has also absorbed some water. Before packaging, the rice is again dried. White as well as brown rice is available in quick-cooking form.

Par-boiled rice is soaked in the unhusked state, steamed under pressure, then dried, husked and polished. Because of the absorbtion of water and the heating, nutritional compounds migrate, including vitamin B_1 from the germ and the bran to the starch body. The corn looks yellowy after polishing and is slightly swollen. The cooking time is about the same as that of white rice.

Cracked rice or broken rice is available in several coarsnesses, as a byproduct of the rice husking. The cooking time is considerably shorter. Broken rice is used for the production of rice flour among other things.

Grits consists of ground rice grains, with and without bran and embryo, ground into smaller or larger pieces.

Flour is the powdery product without bran or embryo. Rice flour is also called rice meal. Rice flour is suitable as a binding additive for porridges; it is also used in gluten-free diets.

Flakes are steamed, flattended rice grains, which are often roasted afterwards. They are used in mixed breakfast products among other things.

Popped rice consists of pre-steamed, puffed up corns. They come on the market roasted with honey or sugar as a breakfast cereal or they are used in rice wafers.

Rice crackers are made of cooked rice which is kneaded into a dough, shaped into a form, baked, and finally dipped in a flavouring liquid.

Mihoen or Chinese vermicelli is made of rice flour. Mihoen is white in colour and an extremely fine thread. It is prepared in the same way as bahmi and also serves as a binding compound in soups.

Millet

Millet is a collective name of several cultivated plants which belong to the family of the *Gramineae*. Millet grains grow in panicles as well as in cobs. The grains are mainly golden yellow with a light sweet wheat taste, but they can also be variegated (black, white, red, yellow). Commonly available are:

— **sorghum**, also nigger corn or kaffir corn, a panicle millet variety which originates from equatorial Africa. Presently this is grown a lot in the south of the United

Fig. 15.10 — Sorghum.

States, India and China. The plant can develop stems 5 metres tall with panicles of 10–60 cm. The ears supply large golden yellow grains.
— **millet,** a plume-millet type from Central Asia, and also called German millet, because it was formerly grown in Germany. The height of the plant is 120 cm.

Millet needs warmth during growth and is sensitive to frost. The plants need much light, but it can cope with dry periods and makes low demands on the soil.

Fig. 15.11 — Millet.

Millet is grown in sub-tropical areas from South America to Russia, from India to Africa (Nigeria).

In America the greater part of the millet production is used as cattlefeed; about 10 per cent goes to the starch-processing industry. Also, small quantities of oil and glucose are obtained from millet. In some Third World countries, millet is an important staple food. It is used for porridges, pancakes, flat unrisen loaves (often mixed with cassava and wheat). In Africa, millet is sometimes malted and processed into a sweet porridge, or an alcoholic drink is brewed from it (Bantu).

The vegetative period for millet is 100 days. The harvest takes place when the halm is still partly green, as ripe grains fall from the chaff and are difficult to retrieve. The corns are shelled before use, though millet belongs to the naked cereals. The pod is not digested by humans.

Products of millet
Grains after shelling are prepared and eaten as rice. The cooking time is about 15 minutes.

Flakes are the shelled, pre-steamed, flattened and, perhaps, roasted grains.

Buckwheat
The origin of buckwheat lies in China. The triangular shape of the buckwheat grain with its red-brown shell is characteristic, reminiscent of a beech nut. The name buckwheat is derived from the word beech-wheat. Before use, the shell has to be removed. The fruit is used as cereal; it is not suitable for bread making. The quick-growing annual herb can grow up to 60 cm tall. First, trusses of aromatic flowers develop on the red stem, from which fruit develop within three months.

Buckwheat demands little from the soil.

The harvest is very variable. Reasons for this are the plant's sensitivity to the weather and the uneven ripening of the grains. They have to dry on the field for a number of weeks after mowing. Because of these reasons, buckwheat is at the moment only little grown as a cereal.

Buckwheat is sometimes planted for the production of the very aromatic buckwheat honey, because of its aromatic flowers.

Fig. 15.12 — Buckwheat.

Products of buckwheat

Grains after shelling are prepared and eaten as for rice. The cooking time is 15–20 minutes.

Kasha is a product of shelled and roasted buckwheat grains. The cooking time is about five minutes; kasha can also be eaten like a little 'nut'.

Grits are coarsely broken shelled grains (corner pieces) and can be used as binding agents in porridge and soup.

Flakes are the shelled, pre-steamed, flattened and perhaps roasted grains, which are used as binding agents in porridge and soup.

Buckwheat flour is obtained by grinding and sieving the fruit of buckwheat. It is used as a binding agent in soup and porridge. Buckwheat flour is sometimes a constituent of mixed flours, for instance pancake flour.

15.5 QUALITY DETERIORATION, SPOILAGE AND STORAGE

Cereal products can spoil and/or can be attacked by the following.

— **Lactic acid bacteria,** which make moist flour and macaroni sour.
— **Decomposition of fats,** whereby wholewheat flour products especially can become rancid.
— **Moulds,** which grow especially in moist and warm storage. Several harmless moulds occur on and in cereals. Exceptions are the poisonous mould in rye (ergot) and fire-blast mould in millet.
— **Insects,** including the flour mite, flour moth, corn weevil and beetles which can also attack cereals and cause spoilage. The *flour mite* is a small reddish spider

(1–2 mm) which occurs in cereals. A temperature of 25°C and moist circumstances are the optimal conditions for the bug. Attacked products can be recognized by a sweetish smell and taste. Flour mites can be demonstrated in flour by the appearance of unevenesses on the surface after it has been made smooth. The *flour moth* is a little grey butterfly (1 cm). The development from egg to caterpillar and moth takes place in storage above 13°C. The caterpillars spin sticky threads which cause little clumps in flour and meal, and are especially noticeable in the corners of packaging. The *corn weevil* is a dark brown beetle (4–5 mm). Females bore holes in the grain with their snout and lay their eggs. From this, white larvae develop which feed on the grain contents. Through this the cereal becomes unusable. *Beetles* and their larvae can be noticed in a range of cereal products.
— **Mice and rats**, which are especially a problem in the storage of corn. The presence of these rodents can be recognized by their droppings and by the visibly attacked corn.

Storage

Small as well as large stocks of cereals and cereal products can best be stored cool, dry and airy. Good ventilation, especially in large stocks, prevents over-heating and the development of insects. In cereal silos, gases are used, nitrogen among others, so that insects cannot develop. The storage time for cereals and cereal products is up to one year or even longer. Wholewheat products spoil sooner because of the high fat content in the embryo. The storage time for these is three to six months.

15.6 LEGISLATION

The **Flour Regulations** set down rules for the composition, labelling and use of additives in the production of cereal products.

Some countries prohibit potassium bromate in flour for the production of bread. The **Pesticides Laws** set rules about the use of pesticides during production.

16
Thickeners

16.1 INTRODUCTION

Thickeners or binding compounds are materials which alter the viscosity and/or structure of liquids or liquid products. In the kitchen, products such as potato starch, cornflour and gelatine have been used for years. These compounds are often less suitable for use in industry. The high temperatures during the sterilization process or the long storage in the deep-frozen state can influence the binding capability badly. The food industry has developed techniques to alter existing binding compounds in such a way that they beocme more suitable for various industrial applications (the modified starches).

Besides binding compounds, the industry also uses compounds to maintain existing structures in products or to reinforce them, for instance emulsifiers and stabilizers. Some compounds have a double function, and can function in the product both as a binding compound and as a stabilizer. Included under thickeners are products based on carbohydrates, products based on fats, products based on proteins and pudding powders.

16.2 PRODUCTS BASED ON CARBOHYDRATES

Green plants produce glucose by the process of photosynthesis. The largest part of the glucose is stored in the form of starch grains as a carbohydrate reserve in seeds, root parts and bulbs. Another part of the glucose is changed into cellulose and used for building the cell wall. Some plants do not store their reserve as starch, but as other carbohydrates such as inulin, carrageen or agar-agar.

Starch
Starch is made up of two polysaccharides: amylose and amylopectin. The separate starch types differ from each other by their contents of amylose (14–30 per cent) and amylopectin (86–70 per cent).

Amylose consists of long straight chains without branching. Through this a large surface area is available for contact with other amylose chains. Between these separate chains exist great attraction forces from the formation of hydrogen bonds. Between amylose and water molecules hydrogen bonds are also formed which enables amylose to bind a lot of water in the formation of a firm network (gel).

Amylopectin consists of long chains with branches. Because of the globular or ball-shaped structure, through which little contact surface exists, there is no great mutual attraction between the amylopectin molecules. The water-binding ability of amylopectin is large but the gelling ability is indeed small, because few bonds are possible between the different amylopectin molecules.

Starch is difficult to dissolve in cold water. During boiling in water swelling and stiffening occurs. The viscosity of the liquid increases. A gel forms after quickly cooling down. On very slow cooling under certain circumstances, such as the cooling or bread, retrogradation occurs. This is the result of the binding of the long chains of the amylose. The longer the chains the greater the chance for binding, through which water, which was bound before, is expelled. To lessen the chance of retrogradation, the starch is changed in such a way, that it is like amylose, but with shorter chains. This process is called **the modification of starch** and can be done by physical, chemical and enzymic means.

Two methods for the modification of starch are as follows.

— Starch, treated with a little water, is allowed to stiffen and dry on hot rollers. By doing this the starch becomes soluble and will swell in cold liquids (instant pudding powder).
— Starch molecules are esterified, by which the network in the gel becomes firmer. The result is that the gel can cope better with the sterilization process, can cope with storage in the deep-freeze or copes better with the effect of acids in viscous products (mayonnaise).

Kinds
Potato starch is obtained from potatoes, which are grown on peat and have low protein content. Such potatoes are not used for consumption. The potato is grated finely and rinsed several times. The starch settles and is separated from the water by centrifuging. It is then dried and packaged in powdered form. Potato starch is blue-white in colour; it glistens and creaks when rubbed in between the fingers. It is used to thicken transparent liquids, for instance fruit juices or soups. It becomes somewhat glue-like if it is used in large quantities.

Maizena or **cornflour** is obtained from corn. The extraction process follows the same path as that of potato starch. Maizena is white in colour, does not glisten and creaks less than potato flour. It binds semi-transparent liquids. It is used for the thickening of milk dishes, soups and sauces. Custard powder is obtained when a yellow dye and a synthetically prepared vanilla compound are added to cornflour.

Wheat starch is made from wheat flour. This is obtained by rinsing wheat flour repeatedly with cold water, which separates the starch out. At the same time the gluten protein is also freed. This is separated from the wheat starch and goes in dried form to the bread factories. The wheat starch is dried and packaged as powder. The white wheat starch has the same usage as potato starch. Wheat starch is for sale to the consumer for the baking of gluten-free bread. The food industry uses it in modified form as a thickener, for example in salad creams. It is also used as a glossy coating on paper to make it less moisture absorbing.

Sago means 'bread' in the language of the Papuans. The sago palm grows in the Moluccas among other places. The tree flowers only once after about 15 years. For this, reserve food is stored in the stem for three years. This is harvested as sago. The local people use it for bread-making. The sago is rinsed several times in cold water so that the starch forms a sediment on the bottom. The wet starch porridge is pressed through sieves with different pore sizes and gathered on hot moving plates. Through this the starch already partly stiffens on the outside, and small and large grains develop (pearl or Siam sago). The starch porridge is also sometimes dried and processed into a powder. About 200 kg sago can be obtained from one palm. The sago grains are somewhat yellow-brown in colour as a result of the heating process. The powder is white in colour, and odourless and tasteless. The grains swell in water and keep their shape after cooking. Sago powder thickens transparently and stickily. It is used in milk puddings, soups and fruit juices, among other products.

Tapioca is obtained from the root of manioc or cassava, which is rich in starch. This grows in South America, Africa and Asia. The tuber can reach a length of 90 cm, with a diameter of 10 cm and a weight of 5 kg. The tubers are shredded and the starch is rinsed out in cold water. The moist starch is heated in rotating drums. From this process, irregular, half-transparent flakes are obtained. The starch is also processed into grains of different sizes, in the same way as with sago (pearl tapioca). The starch porridge is also dried and processed into a powder. Tapioca flakes and grains are coloured yellow-brown by the drying process. Tapioca powder is white, it has no smell and no taste. During cooking, the flakes and the pearls do not lose their shape through the partial thickening. Tapioca thickens transparently and is used in porridges and soups.

Arrowroot is produced from the root stock of the marantha or West Indian arrowroot, which grows especially in the West Indies and the Bermuda Islands. The starch is obtained by grinding and rinsing, and is then dried in the sun and powdered. Arrowroot is available as a white powder, and has no taste or smell. It thickens transparently and is used for clear soups (turtle soup) and fruit juices.

All these thickeners do not spoil in practice. However, they have to be kept dry, otherwise lumps form. Moist starches go mouldy or become sour through the action of lactic acid bacteria.

Other carbohydrates
A number of these thickeners are unavailable or not readily available to the householder consumer. The food industry uses these thickeners for technical reasons, in combination with starch-containing thickeners. The thickeners discussed below are mentioned with an E-number (EC-context) in the Emulsifiers Regulations which apply for most European countries.

Kinds
Alginic acid and alginates (E400 to E405) are obtained from several brown seaweeds or algae. The seaweeds grow along the coasts of the United States and Great Britain

among other places. Alginic acid is obtained from algae (cut into small pieces) by treatment with dilute alkali. Alginic acid consists mainly of mannuronic acid and guluronic acid. The acid is soluble. The salts of alginic acid, alginates (including sodium and calcium alginate), are insoluble. They are used in milk products and lemonade syrups. They are also used as stabilizers in the foam of beer, whipped cream and ice-cream. Alginates are not digested, and because of this are used as fillers in slimming products.

Agar-agar (E406) is obtained from red seaweeds which grow in the tropical waters of Japan, Mexico, California and Australia. The seaweed is cooked and then the thick liquid agar is separated from the seaweed. After drying and bleaching, it is processed into a powder, flakes or bars. Agar-agar is composed of the carbohydrate galactose. Agar-agar has no smell, taste or colour. It forms a transparent jelly on cooling below 40°C. Because of the high melting point, it is used in the tropics to prepare jelly pudding. It is used in the vegetarian kitchen instead of gelatine. In the Jewish kitchen, agar-agar is especially used for the thickening of milk puddings.

Carrageen (E407) occurs in red seaweed which grows along the Irish (Irish moss) and Scottish coast. The weed is treated with sodium hydroxide to free the carrageen. This is dreid and ground into a powder. Carrageen is, like agar-agar, a galactose polymer. It is used in puddings and sauces. It is also used as a stabilizer in whipped cream, ice-cream, and beer foam.

Carob seed flour (E410) is the ground starch body of the seeds of the locust bean or carob tree, which grows along the Mediterranean. The seeds grow in brown-black pods which are about 20 cm long and 2 cm wide. The seeds are removed from the wilted pods and ground. The flour is composed of the carbohydrates galactose and mannose. Carob seed flour is white-yellow in colour and has no smell or taste. It is used in ice-cream and, because of the fact that the carbohydrates are difficult to digest, it is also used as a filler in diet products. For the consumer, the flour is for sale in wholefood shops.

Guar seed flour (E412) is obtained from the seeds of the pods from the guar plant, which grows in India and South America. The processing is done in the same way as for the carob seed flour. Guar flour is composed of galactose and mannose. A white powder, with no smell and taste, it is used for the same purpose as the carob seed flour.

Tragacanth (E413) gum tragacanth is the name of a gum which comes from a shrub which grows in Asia. Exposed to the air the liquid gum hardens to yellow, horny pieces. Those pieces may be ground into a flour. Tragancanth is composed of several carbohydrates, including xylose, arabinose and galactose. The yellow powder and the yellow pieces have no smell or taste, and are used in sweets and ice-cream.

Gum arabic (E414) is the dried gummy exudate from the stems and branches of a shrub which belongs to the Aracia family. This Acacia grows mainly in Africa, India and Australia. The gum runs out of damaged branches and stems and hardens in the air into white, yellow or reddish pieces. These are ground into powder. The composition of gum arabic is similar to that of tragacanth. The gum is especially used in sweets and ice-cream.

Gum xanthan (E415) is produced from several kinds of carbohydrates by means of a fermentation process. Gum xanthan contains glucose and mannose. The gum is dried after fermentation and ground into a powder (cream colour) and industrially processed as a clear thickener, for various applications.

Pectin (E440) is a carbohydrate which is naturally present in all plants (0.2–1 per cent). Pectin is responsible for the firmness of the cells in the plant tissue. Unripe fruit particularly contains much pectin, for example quinces, apples, ctirus fruits and blackcurrants. As the fruit ripens, so the pectin breaks down under the influence of enzymes, and reduces the firmness of the fruit. Pectin is produced in industry by boiling apple residues and beet pulp. The pectin dissolves in the liquid and separates out under acid conditions. Pectin is sold as a powder or as a liquid. It is also for sale combined with sugar as preserving sugar. Pectin is used in the production of jam, marmalade and jelly.

Sodium carboxymethylcellulose (E466) is a white-to-yellow-grey powder, which is obtained from plant fibre. It is used as a thickener in meat products and ice-cream.

16.3 PRODUCTS BASED ON FATS

Fatty thickeners act only as emulsifiers. The most used kinds are:

— **egg yolk** with lecithin and phospholipids as the emulsifying components, especially used in mayonnaise and salad cream;
— **natural lecithin**, obtained from soya, used in margarine;
— **esters of mono and diglycerides** (E471, E472), used for example in jams.

16.4 PRODUCTS BASED ON PROTEINS

Gelatine is produced from the protein collagen from the bones of slaughtered animals. After cleaning and breaking the bones, the fat is removed with petroleum spirit and the calcium salts are extracted with hydrochloric acid. The bones are heated in water in autoclaves so that the insoluble collagen is converted into the soluble gelatine. The gelatine solution is reduced to a syrupy mass and bleached with sulphur dioxide. The mass is poured into moulds if the gelatine is being processed into leaves. After cooling, the solidified mass is cut into thin slices, which are dried on grids. Powdered gelatine is obtained if the liquid gelatine mass is spray-dried. Leaf gelatine is available in colourless and red-coloured forms. A leaf weighs 1.7 g. Powdered gelatine is white in colour. Gelatine is used for aspic, jellies and puddings. It can also be processed into jelly powder.

Milk proteins (casein and whey proteins) are used in small quantities (less than three per cent) in the food industry. Casein is hard to dissolve and is neutralized into caseinate, which is soluble. Caseinates are used as emulsifiers and/or stabilizers in cooked kinds of sausages†, processed cheese, ice-cream and coffee whiteners. The various whey proteins can be separated from each other and can be used as such or after modification. They are used for their emulsifying activity and also for their foaming ability in bakery products, toppings and acidic sauces.

† The meat product industry also uses phosphates besides milk protein as stabilizers.

16.5 PUDDING POWDERS

The kinds of starch mentioned can be processed into **pudding powders**. These powders are made of both modified and unmodified starches and all sorts of other ingredients and additives (including sugar, coffee extract, cocoa, crystallized fruit, salt, dyes, emulsifiers and stabilizers). Custard is one type of pudding powder.

The pudding powder is called **instant pudding powder** if it is suitable for preparation with cold milk.

Besides pudding powder based on starch, **jelly powder** and **mousse powder** are also for sale. **Jelly powder** is based on food gelatine and/or modified starch and is used for the preparation of a jelly pudding. **Mousse powder** provides an airy and foamy dish and is based on starch and/or egg. Both powders are also available in instant form. They can contain the same additives and taste compounds as the pudding powder based on starch.

The keeping quality of pudding, jelly and mousse powders is two to three months depending on the compounds present.

16.6 LEGISLATION

Starch Regulations may contain rules concerning the naming and composition of starches, whether or not they have been modified.

Emulsifier Regulations may deal with the purity requirements of thickeners.

17
Bread

17.1 INTRODUCTION

Bread is made from dough which is prepared on the basis of meal or flour from wheat and/or rye, with water and/or milk, and, as leavening agent, baker's yeast or sourdough.

Gradually bread production has grown from a purely hand-made work which was associated with domestic tasks into a fully automated mechanical industry. Through the ages, the preparation, composition and quality of bread have been subjected to large alterations. Bread was already being baked some 6000 years ago. However this did not look like the bread we know now. In the beginning, the corn was smashed between stones. From the meal, with the addition of moisture, a flat cake was shaped, which was baked on heated stones or in hot ash. Barley, as well as oats, rye, wheat, buckwheat and maize were used, dependent on what was available in the area. Sometimes a spontaneous rising of the bread mixture took place through the working of microorganisms which are now known under the name 'sourdough'. By the use of yeast (already known around 1300 BC) wheat loaves could be given more form and an airier structure. Beer yeast, among other agents, was used. Since the end of the nineteenth century baker's or compressed yeast has been used instead.

Consumption

In the Netherlands bread consumption has decreased by 42 per cent in the last 25 years. However there are indications in the direction of a stabilization of bread consumption. There was even an increase in 1984. New views on nutrition and new baking techniques and also the influences of other cultures have resulted in a larger assortment in loaves, rolls, stick bread and fruit breads. The alternative culture has rediscovered wholemeal sourdough bread as a natural product.

Table 17.1 gives the annual consumption figures for bread in kg per person for the Netherlands and UK for a number of years.

The preference for certain kinds of bread is a personal one. Availability and

Table 17.1 — Consumption figures for bread

Year	Bread consumption (kg per person)	
	Netherlands	UK
1959	86.8	
1964	75.9	
1970	64.6	
1975	62.1	49.7
1980	57.6	46.0
1981	56.8	46.1
1982	56.3	45.8
1983	56.7	45.4
1984	58.6	45.1

Source: Dutch Bakery Institute and Annual Abstracts of Statistics.

keeping quality also play a part in the choice. The fact that bread can be eaten without further preparation, has a neutral taste, and is relatively cheap ensures that bread is the main component of at least one meal each day for very many people.

Table 17.2 shows that in the last ten years there has been a move towards the

Table 17.2 — Percentage consumption of different bread types

Bread type	% total bread consumption	
	1960	1980
White	61	34
Brown (including wholemeal)	30	57
Other	9	9

Source: Dutch Bakery Institute.

consumption of more brown loaves in the Netherlands (a trend which is present now in many other countries).

Composition

Bread in most parts of the EC is mostly baked from wheat flour or meal and/or rye flour. Besides these ground cereals, other basic compounds and additives are also used. By the addition of bread salt (iodine-containing baker's salt), a contribution is made to the consumers' daily requirements of iodine. Table 17.3 shows a diagram of the nutritional components which are present in 100 gram (about 3 slices of bread) of four frequently bought kinds of bread.

Possible health hazards
Phytic acid occurs especially in the germ (or embryo) and the outer layers of the cereal grain. This means that wholemeal flour contains more phytic acid than white flour.

Phytic acid combines especially with iron and calcium in the digestive tract of humans to form phytates, which are very hard to dissolve and cannot be absorbed.

During the proving process, the enzyme phytase is active and breaks down the phytic acid; so the harmful activity of phytic acid is reduced. The longer the proving time is, the more complete the breakdown by the phytase is. This is also the case for bread baked with sourdough as leaven.

17.2 PRODUCTION AND DISTRIBUTION

Bread production requires:

— basic materials (flour/meal, leaven, salt and water);
— additives (bread improvers) and ingredients to get variations in taste and appearance.

Basic materials
For bread production, per 100 parts in weight of flour or meal, 1.75–2 per cent yeast, 2–2.2 per cent salt and 50–60 per cent water are used.

Wheat or rye meal or flour
The choice of the flour or kind of meal determines the kind of bread. Wheat and/or rye meal or flour is used. Most of the bread is made from wheat, because of the good baking qualities of this kind of cereal. The baking quality of wheatmeal or wheat flour is influenced by climatic circumstances and inherited factors. The quantity and the composition of the protein complexes (gluten) is especially of importance.

Gluten consists of about 40 per cent of gliadin and about 60 per cent of glutenin. Gliadin dissolves in water to give a sticky mass; glutenin swells in luke-warm water. Glutenin is responsible for the viscosity of the dough. With soft wheat (seaboard climate), a relatively large number of starch grains are free or at least not enclosed in the gluten matrix. In hard wheat (continental climate), the starch grains are in comparison more enclosed. The gluten also varies in firmness or ability to stretch. The difference in baking properties between kinds of wheat is connected with this. During the kneading, a gluten network develops in which the air and the gas formed by the yeast are trapped. Through this, the wheat dough from wheat flour and meal can become light. If no gluten is present or is of bad quality there is little or no rising of such dough.

In the production of rye bread, fine or coarser ryemeal and even whole corns are

Table 17.3 — Weight of the various nutritional components in 100 g bread

Nutritional components	Weight per 100 g bread			
	Brown	White	Wholemeal	Rye (light)
Water	40 g	40 g	40 g	40 g
Carbohydrates	43	46	40	40
Protein	8	8	7	7
Fat	2.5	2.5	2	1
Minerals:				
Ca	20 mg	10 mg	20 mg	25 mg
Fe	1.5	1	2.5	2.5
K	200	100	250	300
Vitamins:				
B_1	0.12	0.05	0.18	0.15
B_2	0.08	0.04	0.07	0.10

Source: *Nederlandse Voedingsmiddelentabel* (34), 1983.

used. In rye, the starch grains lie loose. The proteins are different from those of wheat; they are also surrounded by mucus compounds. As a result of this, a cohesive mass cannot be formed. Rye bread is more solid, more compact and heavier than wheat bread because of this.

Water

Through the addition of water to flour, a partly plastic, partly elastic mass develops during mixing. At about 20°C the starch grains swell somewhat (15 per cent); the gluten protein, glutenin, increases greatly in volume and absorbs about 200 per cent water. The capacity to absorb water (hydration capacity) is dependent mainly on the cereal variety. Hard wheat flour with strong gluten absorbs moisture at a quicker rate and has a larger absorption capacity than flour from soft wheat.

    The hydration capacity is increased by fine grinding. A finer meal absorbs more water, more quickly. The presence of salts, however, lowers the hydration capacity. The optimal quantity of water varies with the meal or the flour which is used, and lies between 50 and 60 per cent calculated on the quantity of flour or meal. If too much water is used, the elasticity and firmness of the gluten decreases; if too little moisture, the starch and proteins cannot absorb sufficient moisture and the yeast cannot absorb sufficient nutrients.

Salt

Usually 1.5–2 per cent bread salt or cooking salt is added.
    Salt is necessary for a good structure. It firms the gluten and promotes a good bread taste. Salt slows the yeasting down, through which this fermentation process

takes place more slowly and gradually. In doughs without salt, proving takes place more quickly, but the bread is flabby and has a more crumbly sructure than bread prepared with salt.

Leavening agents
As a leaven for bread-making, baker's yeast or sourdough can be used.

Baker's yeast (Saccharomyces cerevisiae)
Until the nineteenth century, beer yeast was mainly used. Yeast is now specially grown for bread production and delivered to the baker as compressed yeast. The quantity of added yeast for white bread is 1.75–2 per cent, calculated on the flour. If fat is also added then the percentage of yeast has to be higher because the fat forms a layer around the yeast cells as well as the starch grains.

Single sugars are broken down to carbon dioxide and alcohol by the yeast cells. These sugars may come from the flour itself, or be added. The diagram below shows how starch is broken down into single, fermentable sugars by enzyme action in the yeast-containing bread dough.

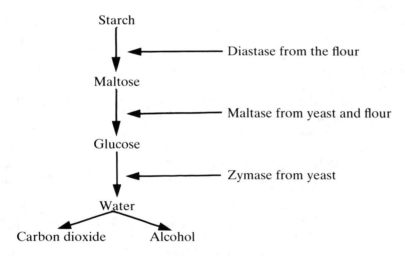

With a higher yeast percentage, the gas development and so the proving of the dough will take place faster; the yeast taste will also be more pronounced. The optimal temperature for good yeast activity is around 30°C.

Sourdough
The use of sourdough for bread production is very old. A spontaneous souring and fermentation takes place by leaving a quantity of flour mixed with water for a few days with occasional stirring. Enzymes from the flour, microorganisms (including lactic acid bacteria) and yeasts from the air become active. First the mixture becomes sour. Then the yeasting phase starts. To encourage a good fermentation, fresh flour and water has to be added. The total fermentation time is five to six days. This sourdough is added in small parts to bread dough. The quantity varies from one-sixth

to a half of the quantity flour. The more sourdough is used, the sourer the bread will be. If bread is produced with sourdough, the labelling 'sourdough' may be used. The proving time of sourdough bread varies from two to five hours. If a small piece from the fully proven dough is kept behind, this can be used the next day for new bread.

Bread in which a special sourdough is used as leaven is valued in anthroposophical circles. This special sourdough is for sale in the Netherlands as (Sekowa) baking ferment and is produced from ground wheat and rye, fermented honey and some pea flour. It consists of dry granules, and if stored cool can be kept for at least one year. By the addition of water and flour, a so-called 'basic dough' is obtained comparable with sourdough, in which honey yeast cells are responsible for the fermentation and the proving. Also because of the honey this bread tastes less sour than ordinary sourdough bread.

Additives (bread improvers)
These compounds may be added during the production of bread doughs to alter the baking and/or bread properties. Milk compounds, egg compounds, gluten, (defatted) soyflour or flour from other pulses, malt or malt extract, sugars and/or honey, among other substances, are used to improve the baking properties.

The appearance, structure and keeping quality of bread can be influenced by the addition of emusifiers, edible oils and fats, food and acetic acid, flour improvers ($L(+)$cysteine and L-ascorbic acid), dyes in, for instance, raisin bread and preservatives in pre-packed ryebread. The action and the effect of a number of additives is described below in more detail.

Milk or milk compounds (milk powder) makes bread softer and more supple. The crust is less crisp, and also the nutritional value is increased.

Egg compounds give a (white) loaf more smell and colour. The crust also colours more and faster during baking.

Gluten is obtained from the production of wheat starch as a side product. Adding a few per cent to bread flour or meal can result in a noticeable improvement in the bread quality.

Soyflour, de-fatted or not, contains a high percentage of proteins. The addition of a few per cent has an influence on the absorption and retention of water and on the final nutritional value.

Malt or malt extract is added because of the sugar-forming ability of this compound. From this, more nutritional compounds will be released for the yeast cells and so the proving will be improved. Also the taste of the bread and the colour of the crust will be somewhat different.

Sugars play a part in the crust colouring and the taste of the bread. They can also serve as nutrition for the yeast cells and enlarge the water-binding ability. Sugars may be added in the form of saccharose, maltose, galactose, lactose, glucose or glucose syrup, fructose and/or honey.

Fat enlarges the bread volume. Tenderness and crumbliness (factors which determine the bread consistency), alter when fat is added. Fat spreads itself over the starch grains and the gluten strings. This fat film makes the gluten more supple. As a result of this, the carbon dioxide formed can puff the dough up more easily. The pores in the gluten network are less permeable because of the fat film; the gas is held longer. After baking and cooling, the fat will 'close' the bread, so that the water as well will be retained longer. Bread stays fresher longer as a result of this.

Emulsifiers reinforce the action of fat. Their purpose is to disperse the fat better through the dough and by doing so increase the useful effect of it. So they, too, influence the structure, the volume and the storage characteristics. A much-used emulsifier is lecithin.

The **flour improvers** L-ascorbic acid and L(+)cysteine influence the mobility of the protein molecules. L(+)cysteine weakens the gluten; the dough becomes supple and easier to knead. L-ascorbic acid makes the dough firmer again. During baking, the latter is converted into water and carbon dioxide. This is important in the baking of crusty bread such as stick bread and crisp rolls.

17.3 THE BREAD PRODUCTION PROCESS BASED ON YEAST DOUGH

The conventional white bread production consists of a number of steps, which all need a certain time. The total production process takes three to four hours. By the use of new technological developments (fast and continuous kneaders), it is possible to shorten the bread production time to 1.5–2.5 hours. This is shown summarized in Table 17.4.

Table 17.4 — Treatments and the time taken in the different bread production methods

Treatment	Time in minutes		
	Conventional	Fast kneader	Continuous kneader
Weighing of ingredients, bringing moisture to 30–35°C, mixing and kneading	210–40	15	
First proving	30–40	35	
Knocking down (second proving)	15–25	35	45
Division into dough balls, ball-proving	15–25		
Shaping and marking, putting in tins, tin-proving (after-proving)	approx. 60	approx. 30	approx. 30
Baking	30	35	35

In former days **kneading** was done by hands or feet. Later it was done with a one- or two-armed mechanical dough kneader or with a quick kneader. The quick kneader kneads very intensively so that the kneading time is shortened and the number of treatments can be somewhat limited. The use of continuous kneaders has shortened bread production time to 1.5–2 hours. These kneaders work so quickly and intensively, that dough can be shaped and marked without further intermediate treatment. It can then, after proving, be baked in the tin. To reach an acceptable end result, the use of bread improvers is desirable or necessary.

The **proving process** follows in four stages: first proving, second proving, ball-proving and tin- or after-proving. For the part yeast plays in this, see section 17.2.

Knocking down (several kneadings) ensures that superfluous carbon dioxide is pressed out. The dough can then maintain a finer structure. The yeast cells get new food through this treatment.

Division: The dough is divided into pieces which produce the desired weight after baking.

Rounding and final moulding: A round shape is the most suitable form for moulding. After the ball-proving the ball is rolled out into a flat shape which is then rolled into a cylinder shape. This cylindrically shaped dough roll is the same length as the baking tin. The superfluous gas has been pressed out by the rolling during the moulding. This produces a straight- or square-baked 'tinloaf', in which the gas bubbles, pulled more or less oblong, run parallel to the long side of the loaf. For 'cross-baked' bread the dough roll is divided into four. The short rolls are turned a quarter and placed next to each other in the tin. The direction of the gas bubbles is now crosswise. This bread gives less resistance on cutting and seems softer.

'Platebread' and 'floorbread' are made in the same way as straight-baked tinbread. Platebread is baked on plates. The sides of the loaves are sometimes baked together, and then the loaves have white sides after separation from each other because of this. Floorbread is baked on the oven floor.

Baking: The gas in the dough expands at oven temperatures of 230–250°C. The water vapour pressure increases. These factors ensure an enlarging of the dough volume. The dough structure is fixed at 60°C. The starch grains gelatinize, through which they retain water. The proteins coagulate, and the gluten network stabilizes. Alcohol, carbon dioxide and water vapour escape. The empty cavities fill with air. The high temperature dries the outside completely. The temperature can increase up to 150°C, whereby a number of chemical reactions occur through which a crisp, brown-coloured crust is formed. Among other reactions, the starches break down to dextrins, the sugars caramelize and the proteins and carbohydrates combine in the Maillard reactions.

After baking, but before cooling down, the crust is washed. The moisture dissolves the dextrins. Because of the still-high temperature the moisture evaporates and a glossy dextrin layer remains.

With present-day methods, there is the possibility to interrupt the proving process. The advantage is that the dough is kneaded a day before baking. The baking can be started early the next day, if wished, and perhaps at another place. After a short second proving the dough is stored at 0°C. The proving process is then as good as stopped. If the time between the dough-making and the baking has to be extra-long then the proven dough is frozen at −15°C. This is especially applicable for dough for rolls. This method makes it possible to make dough in a central place and to deliver it to branches which possess a baking oven. These can supply their customers with freshly baked rolls and bread from their own bakery at any desired time.

17.4 KINDS OF BREAD

Wholemeal bread is produced from the whole grain of wheat or rye or from mixtures of these cereals. Wholemeal wheat and wholemeal rye loaves are made using yeast or sourdough for lightness.

Kinds of bread

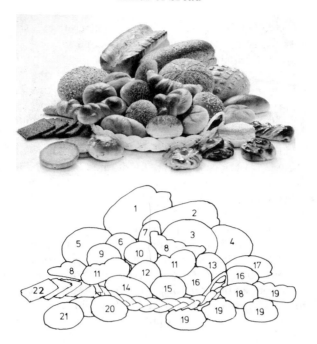

1 White split tinloaf
2 Wholemeal loaf
3 Sesame seed round
4 Squared loaf
5 Wheat round
6 Floor roll
7 Pan roll
8 Croissant
9 Roll
10 Tiger roll
11 Luxury plait
12 Sesame roll
13 Wheatmeal roll
14 Crusty long roll
15 Poppy seed roll
16 Emperor loaf
17 Soft long roll
18 Cream bun
19 Coffee pastry
20 Currant bun
21 Dutch rusks (beschuit)
22 Pumpernickel

Fig. 17.1 — Some kinds of bread.

Brown bread or wheatmeal bread is produced from meal and flour with an extraction rate of 80–90 per cent. The bran in it is easily visible with the naked eye. There are several shapes and finishes possible.

Malt bread is brown bread to which malt has been added.

White bread is produced with wheat flour, of which the extraction rate is 70–75 per cent. The bran in it is not easily seen with the naked eye. White bread is available in many different shapes, finishes and with very many kinds of additions. The bread usually has a special name then such as:

— milkbread, if a certain quantity of milk constituents are present
— cream bread or butter bread, if milk fat is added.

Tiger bread is a plate- or floorbread that before the baking has been brushed with a mixture of riceflour, sugar, oil, yeast and water. Through this it obtains a spotted crust-effect.

Casino bread is baked in a smooth or ribbed tin with a lid. It is suitable for special purposes because of the square or round shape of the slices.

Split tin is a tinloaf, in which the dough is cut with special scissors before baking. Some find this crust tastier.

Sliced bread: To the basic bread recipe, additives, such as fats, are added to get a better keeping quality. This bread is sold sliced and pre-packed.

Stick bread is a crisp-baked water bread, long and thin in shape. It is French in origin.

Currant bread, raisin bread and fruit bread: These kinds of loaves have to contain at least 30 per cent of the constituent mentioned in their name. The term fruit may be replaced by 'ginger' or 'nut' depending on the added constituents.

Christmas bread is a fine kind of current or fruit bread, which is baked on the plate and contains a layer of marzipan.

Rolls: Several kinds of dough, specially fine milk dough, but also water, wholemeal, and brown bread doughs are processed into rolls and sold with a hard or soft crust and depending on the shape and finish as tiger rolls, poppy seed rolls, sesame seed rolls, long rolls, soft rolls, plaited luxury rolls, currant rolls and mini-loaves. These last are baked in small tins, the others on the plate.

Croissants are crescent-shaped rolls of French origin, which are made of a thin white bread dough through which extra fat or butter is rolled. It is extra light and flaky after baking.

Rye bread is made of whole or broken rye grains (dark rye bread) or of rye flour (less dark). Dark rye bread was originally a speciality of the northern provinces (Friesland). No yeast is used and the dough requires a very long baking time. It is a heavy bread without a crisp crust. Of the lighter coloured kinds Limburgs, Drents and Brabants rye bread are especially the most well known. To these doughs based on rye flour, yeast as well as sourdough is added to obtain a certain lightness. They are somewhat lighter in colour as well as in digestibility.

Crisp bread is made of wholemeal. The light kinds are made of wheat, the dark ones of rye flour. Flour, water, salt or skimmed milk powder and yeast or baking powder are mixed into a dough. This is rolled out and baked in a tunnel oven (for about seven minutes) after proving. The moisture content is very low. The bread is very crisp. Sometimes sesame or poppy seeds are added.

Pre-baked bread is baked in such a way that the crust has not or not yet sufficiently coloured. The bread baking has to be finished off.

As a result of the growing interest in natural foodstuffs, the demand for bread which is baked without additives and improvers increases. Some of those available are ecological bread, Demeter bread, Manna bread, soy bread and Allinson's bread.

Ecological bread has the following characteristics:

— the cereal used is organically grown;
— no unnecessary additives are used;
— the corn is ground in windmills;
— the flour has an extraction rate of 100 per cent;
— it is baked by craftsmen.

Ecologically naturally leavened bread (sourdough) also exists, as does ecological yeast bread.

Manna bread is used by people who eat macrobiotically. It can be compared with ecologically naturally leavened bread. However instead of baker's salt it contains sea salt. It is always floorbread.

Demeter bread is baked with flour of bio-dynamic origin or with Lemaire-Boucher flour. These wholemeal loaves are mostly baked with baking ferment, but also yeast doughs are used. Demeter loaves are especially used in anthroposophical circles.

Allison's bread is a wholemeal wheat tinloaf baked with yeast and water. The production process differs, because more moisture is used and the proving and baking times are longer. The bread was developed in England around the beginning of the nineteenth century, and is named after a London doctor, Allison, who was an advocate of eating wholemeal bread produced as naturally as possible.

Soy wholemeal bread is baked from a mixture of wholemeal and soy products (soy meal or grits). Not more then 50 per cent of the weight of the flour may be soy products. This bread has a distinctive structure, colour and smell and a higher protein content.

Gluten-free kinds of bread are available for gluten-free diets.

Low-salt bread is for sale for low-sodium diets. This bread may also be labelled as 'unsalted'.

17.5 BAKERS' PRODUCTS

In baker shops, next to their own baked bread, a more or less extensive assortment of bakers' products is also for sale. Some bakers produce these additional products in their own bakery; others sell factory products. Bakers' products greatly in demand include Dutch rusks (beschuit), melba toast, breadcrumbs, biscuits, crackers and cookies and koek.

Beschuit/rusks are made from rolls baked from fine bread dough in flat round beschuit/rusk tins. The rolls are cut in half. The half rusk rolls are baked in the oven until they are toasted light brown. Depending on the type of dough used and the shape of the tins the following are produced:

— large and small rusks;
— wholemeal rusks;
— mixed corn rusks;
— cinnamon rusks;
— low-salt rusks.

Because of the toasting process in the oven, rusks are easier to digest than rusk rolls. Rusk rolls are also for sale as such.

Melba toast is baked from fine bread dough in different tin shapes. They are then cut into thick or extremely thin slices. They are dry, crisp and cruncy after toasting in a tunnel oven. By long toasting of very thinly cut bread the moisture absorbing ability is decreased making the toast suitable for special purposes.

Breadcrumbs are made from white bread which is dried then ground.

Crackers, tea crackers or **matzos** are made of flour, water and perhaps salt. The dough is rolled out and baked in a tunnel oven. They are called cream crackers if fat is added.

Biscuits and cookies are shaped in between engraved rollers from special doughs. The baking is done mostly in tunnel ovens. It is piped into the desired shape if a batter is the basis. Cookies and biscuits are sometimes filled or covered with a fat-containing icing or dried fruit. Very little has been arranged in the regulations for this assorted group although in 1979 the Dutch Marketing Board formulated rules for *speculaas*, which is a traditional Dutch spiced pastry. These regulations controlled the quantity and the kind of spices which can be used.

Cut koek is a bakers' product composed of starch-containing components of which at least 50 per cent is ryeflour and sweetening components. Sugar (white, brown), inverted sugar, candy sugar, glucose syrup, molasses and honey are among the sweet components used. Cut koek exists under many names. There are breakfast koek, pepper koek, old wives' koek, honey koek (at least 25 per cent honey), ginger koek (at least 15 per cent ginger), candy koek (at least 6.5 per cent candy), Groninger koek, Friesian koek, and fruit koek. All kinds have to have the characteristics of cut koek, described in the regulations.

17.6 QUALITY DETERIORATION, SPOILAGE AND STORAGE

The question of quality is closely connected with the ideal which is envisaged. Bread can differ a great deal. This is dependent upon such factors as:

— the type of cereal used (wheat, rye);
— the leaven used (yeast, sourdough);
— the additives used;
— the production process;
— the way of baking.

The typical smell of bread is at its most noticeable as the bread is just coming out of the oven. The smell decreases as the bread cools down. Every loaf comes out of the oven with a crispy crust, but this only stays for a short time. The smell of bread (bread aroma) consists of two components: the smell of the crumb, in which the yeast is clearly recognizable; and the smell of the crust, where the browning reactions determine the smell. On storage, changes occur in the smell of the crumb as well as the crust. Old bread smells differently from fresh bread.

Staleness
Bread is fresh as long as the crumb is not completely cold. As soon as the bread has become cold, the process of becoming stale begins. In this staleness process, which takes place more or less continuously, four aspects can be distinguished:

— deterioration of the softness;
— reduction of smell and taste;
— loss of crispy crust;
— drying out.

Deterioration of the softness

This means that the crumb becomes tougher. The adhesion becomes less, so that it crumbles sooner. This is partly the result of drying out, but especially of **retrogradation**. This process occurs at temperatures between 60 and $-20°C$, but most markedly between $10°C$ and $-7°C$. Gelatinized starches in bread (amylopectin and amylose) retrograde. This means, they release moisture which was retained by gelatinization. A part of the amylopectin and amylose will again crystallize. These crystals are visible and felt in the tougher bread crumb. The released moisture moves to the crust. For amylopectin this reaction is reversible by reheating the bread above $60°C$ ('pep up'). Bread then appears for a short time softer. For amylose, a 'pep up' has no effect. These crystals stay visible in the somewhat tougher bread crumb.

Reduction of smell and taste

Loss of bread aroma is a result of oxidation processes. During the cooling of bread, air moves from the crumb to the outside. Later, cold outside air moves inside. The oxygen combines with smell and taste components from the crust and the crumb. In the crumb, compounds such as aldehydes, ketones, esters and diacetyl are present which are formed during the yeasting process and determine the smell of fresh bread. When the fresh air supply is limited or stopped by adequate packaging, the oxidation processes will be slowed down. These processes can be stopped by storing well-wrapped bread at deep-freeze temperature.

Loss of the crispy crust

This is the result of the movement of moisture from the more moist crumb to the drier crust. By storing the bread in a bin or in packaging at $10–15°C$, this moisture movement can be somewhat slowed down. By deep-freezing, this moisture movement is slowed down even more, although a slight amount is still possible. It becomes visible in a white ring in the crumb and in a loosening top crust.

Drying out

This always occurs during storage of bread. Drying out can be slowed down somewhat by storage in a bin or by packaging. Plastic packaging is the most effective.

Stale bread is for most consumers less attractive, but is not harmful to health.

Spoilage

There are two forms of spoilage which can occur in bread and can have harmful results, namely *mould formation* and *rope*.

Mould formation

This occurs when especially the more moist kinds of bread such as rye and Allinson's bread are kept for a long time in warm, moist surroundings. Some kinds of moulds form mycotoxins. Mouldy bread is better not consumed.

Rope

This is caused by spore-forming bacteria. On cutting or breaking the bread, thin threads are visible. The bread smells unpleasant and shows sticky patches. Rope occurs only sporadically in warm summers and then only in brown and rye bread.

17.7 LEGISLATION

The legislation for bread and bakers' products is put down in a number of regulations. In the **Bread Regulations**, the minimum rules are formulated which concern the quantity of dry matter in bread. In this also are found the aspects concerning the naming, the permitted additives and the production of different kinds of bread.

18
Pulses

18.1 INTRODUCTION
Pulses is the collective name for crops which belong to the legume family. People throughout the ages have eaten pulses. The Bible relates the story about the sale of Esau's birthright to Jacob for a bowl of lentil soup. In the old Chinese Empire (about 2000 BC) the soya plant was looked upon as one of the five holy grains. (The others were rice, wheat, barley and milllet.)

If seeds are eaten in the dry and ripe state, for instance green peas and brown beans, they too are classified in nutritional information as pulses.

Consumption
The consumption of pulses per person in the Netherlands is about 1.9 kg per year whereas it is about 0.4 kg per year in the UK. The consumption includes home-grown products as well as imported pulses such as soya beans, lentils and white beans from Asiatic countries.

The most-eaten are peas, including split peas and brown beans.

Table 18.1 — Composition of pulses, expressed as a percentage of the whole

Type of pulse	Water	Protein	Fat	Carbohydrates
Brown and white beans	12	20–21	1.5	43
Peas				
Lentils				
Soya beans	9	34	16	36

Sources: *Nederlandse voedingsmiddelen tabel* (34), 1983; Souci, 1973.

Composition

Table 18.1 shows the most important nutritional compounds which occur in pulses. The fat in the soya beans contains about 55 per cent linoleic acid. Pulses contain many minerals, including iron and phosphorus and the vitamin B-complex.

Possible health hazards

Raw or insufficiently heated pulses contain **fasine**, a protein-containing compound, which is poisonous to humans. Fasine can cause clotting of the red blood cells.

In raw soya beans there are poisonous **lectins**, also protein-containing compounds, which can cause clotting of the red blood cells. Lectins also slow the enzyme trypsin down.

Goitrogens ('goitre inducers') can also occur in raw soya beans.

The toxic compounds lose their detrimental action when the pulses are heated for a long time (e.g. *boiled* for 10 min), germinated or fermented.

18.2 PRODUCTION AND DISTRIBUTION

A large proportion of the available pulses in Northern Europe are grown locally, for example beans, white as well as brown, and peas. Soya beans are imported from Brazil and China. Lentils are still imported from Egypt and Asia minor.

The harvest of peas in Europe takes place in July and August; that of beans in September and October. The peas are mown by machine when the pods are yellow and the plant has as good as died down, beans are mown when the leaf has completely dropped.

The harvested plants are dried and are then threshed, which frees the seeds. The seeds are mechanically sorted for size, colour and soundness, and dried and packaged. The new harvest of peas is for sale in the last half of August, the new harvest of beans in October and November.

18.3 KINDS OF PULSES

Beans, peas and lentils belong to the pulses group, and each can be subdivided again into a number of varieties.

Beans

There are many kinds of beans. Some, such as the white and brown bean, also come from Northern Europe. However the majority of beans are imported.

The taste of **brown beans** is more appreciated than that of white beans. They are more easily processed in soups and dishes because of the soft structure. Some varieties are: the small round bullet bean, which does not disintegrate during cooking and is suitable for dishes in which whole beans are wanted; and the large oblong bean, which disintegrates during cooking and is especialy used for the preparation of bean soup.

White beans are firmer and crumblier in structure than brown beans. Some varieties are: the large kidney-shaped beans ('bent beaks'), which stay whole during cooking; the small round beans (Walcherse beans); and the large flat beans, which are slightly sweet in taste.

Kinds of pulses

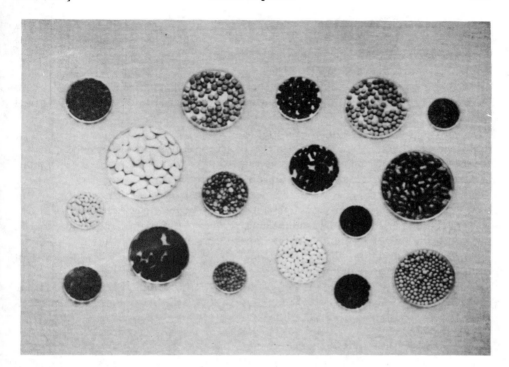

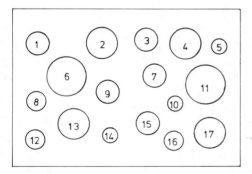

1 Brown beans	10 Mung beans
2 Chick peas	11 Borlotti beans
3 Kapucijners	12 Grey peas
4 Green peas	13 Red kidney beans
5 Red lentils	14 Lentils
6 Butter beans	15 White beans
7 Black kidney beans	16 Adzuki beans
8 Flageolets	17 Soya beans
9 Split peas	

Fig. 18.1 — Several kinds of pulses.

Borlotti beans are similar in shape to brown beans, but they are beige and purple-red speckled. During cooking, the colour of the skin becomes an even light brown. Borlotti beans are used in industry for the production of brown bean soup.

Flageolets are small, flat, oblong, light green little beans. They are usually only available canned.

Aduki or **adzuki beans** are small dark red beans originating from Japan. They taste sweet after cooking. They are also sprouted (see Mung beans).

Black beans or black kidney beans are small and deep black in colour and are greatly used in the Surinam kitchen; they are, for instance, served with white rice.

Lima beans or butter beans are flat white beans the size of fresh broad beans. They are imported from South America.

Red kidney beans are kidney-shaped and red-brown in colour. They have a sweet taste after cooking. They are imported from South America and processed in mixed bean dishes.

Soya beans are known in many varieties. They vary among themselves in shape (oval, kidney-shaped), size (0.5 to 1 cm) and colour (light yellow to brown or black). In Northern Europe, mainly the light yellow, round soya bean is available, originating from South America and China. The main difference between soya beans and the beans already mentioned is that the soya beans are harvested fully grown but not completely ripe. The reason for this is that ripe soya beans are difficult to cook. Soya beans have a specific taste and smell, which differs from that of other beans. The bitter taste is caused by the decomposition of phospholipids (fat-like compounds) during the preparation. The typical smell is caused by the decomposition of compounds which are related to lignin.

Mung beans are small olive-green round beans from China, also called *Katjang idjoe*. Usually the bean itself is not used, but the soft, white sprout of the germinated bean (beansprouts).

Peas

The pea family is less extensive than the bean family. Most kinds are grown in Northern Europe.

Green peas are mostly used for the production of soup. Some varieties are: the round green pea and the marrow fat pea (angular green pea). Split peas are round green peas, cleaned of their skin, in which process they split in two.

Yellow peas are available as whole round peas and in the split form. They are less used in the Netherlands than the green peas.

Kapucijners (Capuchin maple peas) are peas which originate from Turkey. They are square in form. The colour varies from green to dark brown.

Maple peas look in shape like kapucijners, but are more brown–red in colour with a marbled appearance and a crinkled seed coat.

Chick peas look more like shelled hazelnuts. They are round and yellowish in colour with an embryo point sticking out. In the Asiatic and southern European kitchen this pea is often used in soups and stews. It is also used in the Jewish kitchen.

Lentils

Lentils have a flat, round, discus-like shape. They vary in size and colour (green, yellow, red, orange, brown and black) and can be traded whole or split. In the Indian, Turkish and Moroccan kitchen, lentils are often used.

18.4 PULSE PRODUCTS

The preparation of dried pulses requires a lot of time. The cooking time varies from 45 minutes for split peas to 3 hours for soya beans, but it can be shortened by soaking the pulses. The pulses are often pre-treated in preservation processes to shorten the preparation time for the consumer, and then they scarcely need any preparation time at all.

Some examples of pre-treated pulses are:

— pulse soups, in dried form, and canned or deep-frozen in liquid form, perhaps processed into a meal soup;
— pulses, soaked and sterilized in cans or bottles, with and without a sauce (beans in tomato sauce);
— ready-to-eat meals in which the pulses have been processed (brown bean platter).

Among pulses, the soya bean has a place apart, not only because of its different composition but also because of the different processing possibilities. The soya bean is little eaten in the Netherlands; however interest in products from the soya bean is increasing. Available among other products are soy flour, tempeh, tofu, soy milk, miso, toatjo, textured protein meat-substitute, soy oil and soy sauce.

Soy flour is made from the skinned, flattened and roasted soya beans which are ground into flour. Besides full-fat flour (20 per cent fat) de-fatted flour (1 per cent) fat is also available. By means of extraction, soy oil is taken out of the flour. Soy flour is yellow–white in colour and has a nutty taste. It is used in bakery products, confectionary, diet foods (gluten-free diets), meat products, soups and sauces.

Tempeh is a fermented soyproduct inoculated with kinds of moulds, which can be used as a meat replacement in a hot meal. The mouldcake, in which the beans are still visible, can be marinated, stewed, baked or cooked.

Tofu, soy cheese or **tahoe** is renneted soy milk which looks like fresh cheese. It is sold packaged in blocks, moist, or dried. Tofu can be used in the same way as tempeh.

Soy milk is produced from cooked soya beans, which are pureed. The white liquid, which is separated by means of sieves or a centrifuge from the soymass, is soy milk. Soy milk is available in liquid and in powdered forms. It is used, among other things, by unweaned babies with a cows' milk allergy.

Miso is a brown paste made from soya beans inoculated with types of mould and fermented perhaps in combination with cereals (rice, barley, wheat or rye) and salt. Miso is used in soups and sauces to improve the taste.

Toatjo looks like miso and is also a thick paste of soya beans inoculated with kinds of mould and fermented, in which salt and flour are processed. It is sometimes mixed with sambal. It is used in the Indonesian kitchen in combination with tempeh.

Soy meat, also called textured soy protein (TSP), is similar in appearance, taste, consistency and nutritional value to meat, but is in general cheaper than meat. Soy meat is for sale in cans (sausages and slices) and in dried form (pieces and granules). It is also used in ready-made meals and soup powders. Soy meat can be made tastier by marinating, coating with crumbs, frying, deep-frying or processing into stews and ragouts. Dried soy meat has to be soaked before use. Other products based on soya are: soya oil (see section 13.3) and soy sauce (see section 5.11).

Some other products based on pulses are: imitation marzipan, in which white beans are processed, together with or in place of almonds, to imitate the much dearer marzipan; and pulse flour, for instance from peas, which is used as a thickener in soups.

18.5 QUALITY DETERIORATION, SPOILAGE AND STORAGE

Pulses have to be sound, neither mouldy nor musty. Pulses of good quality have a soft, thin skin. They swell nicely in water and cook quickly. The longer the pulses are stored the harder the skin becomes and the longer the cooking time becomes.

Beans can start to germinate under the influence of light, moisture and warmth. Pulses lose colour if stored in the light. There is a chance of mould forming through the absorbtion of moisture. Good packaging prevents attacks by insects or rodents.

Dried pulses can best be stored cool, dry, dark and well packaged. The storage time is a maximum of one year. Pulses over one year old are hard to cook. Soya beans must be stored for a shorter period because of the high fat content, through which the chance of rancidity is increased. Soaked pulses in cans or bottles can be stored for one year or longer. Bottled preserves have to be stored in the dark because of discolouration. The storage times and places of soy products vary, depending on the composition and processing. In Table 18.2, storage times and storage conditions are given.

Table 18.2 — Storage times and storage conditions for some soy products

Product	Storage time	Storage condition
Soy flour (fat)	3 months	Cool, dry
Soy flour (de-fatted)	9 months	Cool, dry
Tempeh	3 days	Refrigerator
	1 month	Deep-freeze
Tofu	1 week	Refrigerator
	1 month	Deep-freeze
Soy milk (liquid)	3–6 months	Cool, dry
Soy milk (powder)	6 months	Cool, dry
Miso	1 year	Refrigerator
Toatjo	1 year	Refrigerator
Soy meat (dried) / Soy meat (canned)	1 year	Cool, dry

18.6 LEGISLATION

Pulses Regulations set out, among other stipulations, the rules for the exact labelling and visual characteristics of pulses. There may also be **Soaked Pulse Regulations**, which contain rules concerning additives and labelling, such as drained weights.

19
Potatoes

19.1 INTRODUCTION

The potato is a stem tuber which grows in the soil. The wild potato originates from Peru (South America).

In the sixteenth century the Spanish conquerors who followed in Columbus's footsteps to America acquainted Europe with the potato. Potatoes as we know them now, do not look at all like the product that the Spanish 'conquistadores' found, namely small watery little tubers, mostly greenish in colour.

Although the potato was introduced in the sixteenth century, it took until the eighteenth century before it was eaten on a large scale. Strange tubers were distrusted. It was also known that the foliage of the plant, which belongs to the family of the deadly nightshade (as does the tomato and the sweet pepper), is very harmful. It was believed that the tuber from the potato plant was also poisonous.

The Frenchman Parmentier, a pharmacist who, imprisoned in Prussia, had got to know the potato as prison food, made the potato popular. Back in France he introduced the potato to the French court with great success during the time of Louis XVI. The name Parmentier is immortalized in a number of potato dishes.

Ireland was one of the first European countries where the cultivation of potatoes was done on a large scale. The main reason for this was the favourable climate for potato cultivation. In Ireland this extraordinary expansion of potato cultivation was fatal for a large number of people. In 1846, through blight, a plant disease which is caused by moulds, the whole harvest was ruined. As a result of this, half a million people starved to death and many impoverished Irish emigrated to America.

Around 1850 the potato was a favourite national food in many parts of Northern Europe, because of the low price. The poor people used to eat potatoes two to three times a day, often with vinegar and melted fat.

In the Netherlands, much attention has been given to variety improvement and method of cultivation. These are some of the reasons why Dutch potatoes are known over the whole world.

Consumption

In the Netherlands, potatoes for consumption are eaten either freshly cooked at home or processed into products such as chips or crisps.

A small part of these potatoes are selected for turning into seed potatoes in specialized seed potato firms.

In the starch industry, potatoes for industrial use are processed into glue, glucose syrup and potato starch among other products.

For the cattle feed industry, special potatoes are grown which are more suitable as cattle feed.

The consumption of fresh potatoes has decreased sharply in the last 100 years as a result of changed eating habits. The consumption of potato products, however, has greatly increased over recent years. For a summary of the consumption figures see Table 19.1.

Table 19.1 — Consumption of potatoes and potato products per head of population for the Netherlands.

Calendar year	Fresh potatoes kg	Potato products kg	Total kg
1850	250	—	250
1960	98	2	100
1970	75	12	87
1985	62	23*	85

* The assorted potato products consisted of: pre-fried products (about 16 kg), snacks (crisps etc.) (5 kg), dried products (1.5 kg).

Source: Potato Marketing Board.

In the UK, the consumption of potatoes in the home has decreased from 60.1 kg to 55.6 kg per person, over the period 1977 to 1987. The consumption of potato products was about 7.10 kg per person in 1987.

Composition

The potato contains roughly: water 77%, carbohydrates 19%, protein 2%, vitamins B-complex and C†, and minerals including potassium and phosphorus. (Source: *Dutch Foodstuff Table* (34) 1983.)

19.2 POTATO CULTIVATION

The potato is a stem tuber, which is a thickening of the stem or rhizome in the soil. The buds or eyes of this thickened stem, full of reserve food, will sprout under suitable circumstances and form a new plant. This will form new stems, which in their turn can grow into tubers. Seed potatoes are selected from the previous year's

---

† New potatoes contain in comparison a lot of vitamin C (25 mg per 100 g potatoes). During storage the vitamin C content decreases.

harvest. These seed potatoes are stored during the winter. They are then planted, either chitted (sprouted) or not, in the spring.

If they are chitted before they are planted in the spring, the growing process starts quicker. About two to three weeks after the potatoes have been planted, little plants develop above the soil, and can reach a height of 70 cm.

The potatoes are ready to be harvested when after about five months the halm turns yellow-brown and dies down.

The average harvest from one seed potato is 15 new potatoes.

Potato varieties

A number of characteristics of potato varieties are important for the consumer, including the following.

— **Taste**: this is determined by the variety and the type of soil. Potatoes grown on clay have a more appreciated taste than potatoes grown on sand.
— **Colour** of the skin, the number of eyes and their depth.
— **How they cook**: whether or not potatoes cook 'floury' depends on:
  — *the ratio of protein to starch*: more protein gives a firmer potato and more starch a more 'floury' one;
  — *the kind of soil*: a potato grown on clay cooks more floury than a potato grown on sand;
  — *the ripening time*: in a short ripening time not so much starch has been formed in comparison, as a result of which the potato is firmer and cooks less floury.

Depending on the method of cooking, eating potatoes are divided into four types (A, B, C and D).

— **Type A** has a firm consistency after cooking. The outside is smooth and shiny, often wet and somewhat transparent. This potato is used in salads, among other things.
— **Type B** is a fairly firm potato, which cooks a little floury. This one is suitable to be eaten boiled, but also for frying and to make into mashed potatoes or use in stews.
— **Type C** is more mealy. The consistency is less firm. Potatoes of this type are not so suitable for frying, but good for boiling for floury potatoes, for baking in their skin or for mashing.
— **Type D** boils very mealy and often breaks. This potato is for preference used for the preparation of mashed potatoes. In this type are also included the 'off-cookers': that is, potatoes which break up on the outside while the inside is not yet cooked.

The potato grower divides the potatoes in three groups depending on the time taken to mature, and consequently the harvest time.

— **Early or new potatoes** with a growing time of three months. They are harvested in June when they are not completely full-grown. Early potatoes contain less starch than full-grown ones and do not boil floury. The potatoes are easily freed from the skin; the skin has not yet corkified and can be removed by firm rubbing. Around the end of December new potatoes arrive on the market under the name Alpha or Malta from warmer countries including from Italy and Egypt. These

foreign potatoes are mostly grown from Dutch seed potatoes. They can be planted and harvested sooner in a sub-tropical climate.
— **Medium early potatoes** have a growing time of 4–5 months and are harvested in August and Sepember.
— **Late potatoes** are harvested in September and October after a growth of about six months. These potatoes have a high starch content and can be stored for a longer period.

N.B. New potatoes must not be thought of as 'Kriel' potatoes. These are sorted, very small potatoes of different varieties. New potatoes do not have to be small.

In Table 19.2, visual characteristics, availability, and how they cook are given from a number of potato varieties.

Table 19.2 — Summary of some potatoes for eating

Variety	Visual characteristics	Availability	How they cook
Alpha/ Malta	Light yellow peel; round oval; shallow eyes; light yellow flesh.	March to August	B–C floury
Ersteling	Yellow peel; long oval; shallow eyes; loose peel; light yellow to yellow flesh.	May to September	A firm not floury
Dore	Brown-yellow peel; round oval; shallow eyes; yellow flesh.	June to October	B firm to floury
Lekkerlander	Brown-yellow peel; round; shallow eyes; light yellow flesh.	June to October	B–C floury
Eigenheimer	Light yellow peel; oval and irregular in shape; deep eyes; yellow flesh.	June to May	C–D very floury to 'off-cooking'
Bintje	Light yellow peel; round oval; eyes; light yellow flesh.	August to June	B firm to floury
Irene	Red peel; round; fairly shallow eyes; yellow flesh.	September to June	C–D very floury to 'off-cooking
Nicola	Yellow peel; long oval; shallow eyes; yellow flesh.	October to June	A firm, not floury

19.3 POTATOES AND POTATO PRODUCTS

Potatoes can be bought loose from the farm, at the market or at the greengrocers. They are mostly sold washed and packaged in bags of 2.5 and 5 kg in self-service shops and supermarkets. The bags have air holes to let the potato continue to breathe.

Potatoes can also be bought peeled. If these peeled potatoes come into contact with oxygen a discolouration occurs as a result of enzymic action. This results from

Potatoes and potato products

Rode Eersteling

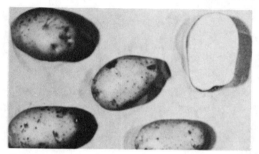

Eersteling

Irene

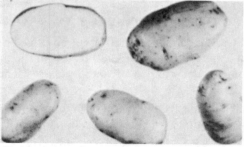

Bintje

Fig. 19.1 — Several potato varieties.

Fig. 19.1 (*cont.*) — Several potato varieties.

the amino acid tyrosine changing into melanin, a brown-red dye. The discolouration is avoided by storing the potatoes under water, packaging under vacuum, or by using sulphurous acid.

Potatoes can be prepared or processed in several ways. Included in this assortment of products are the following.

— Pre-fried and frozen products (chips, potato croquettes, rosti; low-fat or oven chips which are first pre-fried in the factory so that they can be baked in the oven without the use of fat.)
— Pre-fried and cooled products (chips, which are used in restaurants etc. and are also available to the general public).
— Fried potato products, packaged with or without salt and/or taste compounds (crisps etc).
— Dried potato products (mashed potato powder, rosti. In the potato processing industry chip-powder is used, a half-processed product. After mixing with water, a doughy mass results, from which chip-shaped bars are pressed).
— Sterilized potatoes (canned pommes parisiennes).

19.4 QUALITY DETERIORATION, SPOILAGE AND STORAGE

In the grading of potatoes for eating, special attention is given to the number of discoloured, glassy and damaged potatoes. On the basis of the number of defective specimens found the potatoes are grouped into quality classes. The quality characteristics are:

— **Blue sensitivity**, which means that the potato, by bumping or falling, especially during transport, is susceptible to a blue colouration in the tuber tissue. The blue sensitivity can be decreased by warming the storage rooms, where the potatoes are kept at 4–7°C, up to about 12°C before transportation.
— **Discolouration** under the skin as a result of harvest and transport-damage.
— **Glassiness**, as a result of particular conditions during the growth of the plant. In

very dry and warm weather the growth of the young potatoes is arrested. When the weather changes, the young plants start to sprout again and form new potatoes. This is called 'through growth'. The first formed tuber can be glassy if the starch has been extracted by the second tuber. Glassy potatoes stay hard after cooking and are not suitable for consumption.
— **The dry matter content** of the potato, also described as the underwater weight, is an indication for the starch content. In glassy potatoes the dry matter content is very low.

Changes occur during the storage of potatoes, which can cause spoilage.
— The starch of potatoes, as result of ongoing respiration, is changed by enzymes into fructose and glucose. The glucose is oxidized into carbon dioxide and water. Heat is given out. If the potatoes are stored between 0 and 4°C then the change from starch into glucose still takes place, but the oxidation of glucose slows down. These potatoes taste sweetish on consumption. This taste sometimes disappears again if the potatoes are stored at a higher temperature, so that the oxidation starts again. Reducing sugars (glucose and fructose) cause unwanted discolouration (Maillard reaction) during baking or frying. In the chips and crisps industry, potatoes with a low content of reducing sugars are preferred for processing. By means of blanching, before frying, part of the sugars is removed. It is also possible to keep the content of reducing sugars low with the aid of enzymes.
— If stored too warm and in the light the potato can start to sprout. This uses up the reserve compounds and the vitamin C content. The potato becomes shrivelled.
— Moisture loss occurs with too warm and/or too dry storage. The potato becomes soft and shrivelled.
— Mould formation and rotting occurs with moist storage. Potatoes stored in a closed plastic bag will eventually go mouldy and rot, as a result of the water which is liberated during respiration. For this reason potatoes should be removed from plastic packaging.
— During storage in the light, solanine is formed, visible as green areas. Solanine is a poisonous compound, high doses of which give symptoms such as nausea, headache and stomach ache. Green areas should always be cut away. Blue areas, resulting from bumping or falling, should also be cut away.
— Storage below 0°C can lead to freezing of the potato, through which the cells are damaged. After thawing, the potatoes are wet and can quickly go rotten and mouldy. Glassy potatoes also rot faster.

The best way in which to store potatoes is in a well-ventilated, dark room at a temperature of 5–8°C.

Potatoes can be irradiated or treated with an anti-sprouting compound to prevent the formation of sprouts.

Only potatoes which are stored for a long period, the so-called storage varieties, are treated with an anti-sprouting spray or powder. Potatoes may not be treated if sold before December because of the possible detrimental side-effects of the compound used.

In the home, potatoes can be kept for 2–6 weeks. For storage on a large scale, the storage time resulting from very favourable storage circumstances (good storage temperature, ventilation and air composition) is nearly one year.

Peeled potatoes, vacuum-packed or treated with sulphurous acid, can be stored for 5–7 days in the refrigerator.

Deep-frozen potato products can be kept for about three months.

Crisps and other snacks, either gas-packed or not, can be stored for about three months.

Mashed potato powder can be stored for at least one year in the original packaging in a cool, dry place.

Canned or bottled potatoes can be stored for one year or longer.

19.5   LEGISLATION

Grading Regulations may set the minimum quality standards and describe gradings procedures for potatoes.

Preserved Potato Regulations may limit, among others, the use of sulphurous acid with peeled potatoes.

20
Vegetables

20.1 INTRODUCTION

In horticulture, vegetables is the collective name for herbaceous plants, parts of which, cooked or raw, are eaten by man. In practice, fruits such as cucumbers, tomatoes and sweet peppers are also included.

In the Netherlands and other European countries professional horticulture has developed from the cultivation of monastic and castle gardens around 1500. Exports began in the second half of the nineteenth century. At the moment, exports of vegetables are of great importance, and tomatoes, cucumbers, lettuces and onions are the most important Dutch products.

Consumption

For a number of years, on average about 80 kg per person per year, or 220 g per day, have been used in the Netherlands. A bit more than half is consumed fresh, the rest as canned, deep-frozen or dried vegetables. The use of fresh vegetables is higher in the spring and summer than in the autumn and winter. In order of importance, according to the number of kilograms sold, the most vegetables eaten fresh in the Netherlands are: cauliflowers, cucumbers, tomatoes, carrots, endive, onions, chicory and lettuces. It is estimated that about one-fifth of the quantity of vegetables consumed is grown in private gardens. In the UK, the total consumption of vegetables is about 55 kg per person, of which about 60 per cent is consumed fresh. The most popular fresh vegetables are, in order of importance, carrots, onions, tomatoes, cabbage and cauliflower.

Composition

Vegetables can make an important contribution to the supply of vitamins C and B, β-carotene, minerals and cellulose. The composition varies for each kind. The seeds have in comparison a high protein and carbohydrate content. 100 g edible vegetable contains roughly (source: *Nederlandse Voedingsmiddelentabel* (34), 1983):

water		90–95 g
protein		0.5–5 g
fat		0.5 g
carbohydrates		0.5–15 g
minerals	Ca	10–200 mg
	P	12–125 mg
	Fe	0.5–5 mg
	Na	5–200 mg
	K	200–800 mg
vitamins	β-carotene	6 mg
	C	5–100 mg
	B_6	0.07–0.22 mg

Possible health hazards

Some vegetables contain natural detrimental compounds. They can also accumulate harmful compounds from the surroundings. Among others can be found the following.

— **Oxalic acid** in, for example, spinach, rhubarb and purslane. It forms an insoluble compound with calcium (calcium oxalate).
— **Fasine** in pulses (see section 18.1).
— **Nitrate** in all vegetables (see section 4.4). The compound is a nutrient for the plant. Under unsuitable hygenic circumstances nitrate can be changed by bacteria into nitrite, which attaches itself to the red blood pigment. The bacteria are most active at a temperature of 10–50°C. That is why little nitrite formation occurs under cooled storage. Heightened nitrate contents are often found through (excessive) use of fertilizers for instance in radishes, purslane, spinach and lettuce and also in greenhouse-grown vegetables.
— **Goitrogens** (cause goitres) in different kinds of cabbage. From these compounds an antagonist to the thyroid hormone can be formed. The enzymatic formation of these compounds with their antagonistic action can be prevented by cooking.
— **Herbicides and pesticides**. These can be used during the growing period. The residues are found in the harvested vegetables. The harm caused depends, of course, on the kind of compounds and the quantity of the residues.
— **Heavy metals and polycyclic compounds**. These occur in vegetables because of industrial pollution. They are detrimental through their accumulative characteristics.

20.2 PRODUCTION AND DISTRIBUTION

Production

Vegetable cultivation in Northern Europe takes place both in the open and in greenhouses.

The most important products from the open cultivation are cauliflowers, carrots, leaf vegetables, onions, chicory, Brussels sprouts, beans and asparagus.

The most important greenhouse products are tomatoes, lettuces and cucumbers.

Open field cultivation

Locally grown vegetables come in their season in large quantities from the open fields and are usually not expensive then. The vegetable preserving industry works almost solely with produce grown in the open, particularly because of the lower cost price. They have agreements with contract growers, who do not deliver through the vegetable auctions, but direct to the industry. The quality depends on the weather, the condition of the soil, pests and diseases.

Ecologically, biologically and bio-dynamically grown vegetables also appear on the market (see section 1.2).

Research has shown that there are differences between the end products of the traditional production methods and the alternative methods. So, for instance, the moisture content of leaf vegetables is higher when fertilizer and/or chemical pesticides are used, and the nitrate content in leaf vegetables lower if no fertilizers have been used.

Greenhouse cultivation

Plants grow faster when cultivated under glass. They can also be presented on the market for a greater part of the year and plants from tropical or subtropical areas can be grown. Cultivation under glass can be subdivided into production in **cold frames**, for which only the warmth of the sun is used, and production in **greenhouses**, which can be centrally heated. With the climate computer, any desired climate can be created in the greenhouse. The climate control comprises the regulation of the moisture content, the quantity of sun and light, the temperature and a calculated quantity of nutritional compounds, so that the plants can grow to their optimal capacity. It is also possible to obtain special effects such as blanching† and/or forced‡ vegetables.

A new development is **substrate cultivation**. The plants stand in blocks of stone wool on a layer of plastic. This reflects the light, as a result of which the plants grow better and faster. A nutrient solution diluted with (rain) water is added several times a day to the plants (in drop form). The stone wool can be re-used; to destroy possible disease germs it can be steamed. The costs are only a fraction of the sterilization costs of the soil.

The advantages of substrate cultivation are a higher yield and less heating costs. The disadvantages are water pollution through the excessive fertilizers and the pollution of the environment by the large quantities of plastic refuse and stone wool. Horticulturists use substrate cultivation mostly for mono-cultures, which means they

---

† BLANCHING or etiolation has the aim of producing a blanched (not green) vegetable. Plants in poor light will form no or very little chlorophyll. The stalks and/or leaves stay whitish. For instance: white chicory, asparagus and celery.

‡ FORCING is used for vegetables to produce them for the market at a time which is completely different from their normal season. In our climate with changing seasons periods of growth and rest follow each other. The growth of plants is especially influenced by the number of sun-hours, a certain temperature and moisture. In climatized greenhouses seasons can be imitated, so that for instance, forced rhubarb is available in the heart of winter and blanched chicory in the summer.

concentrate on the cultivation of one product, for instance, cucumbers or sweet peppers or tomatoes. In 1984, in the Netherlands the number of hectares of standing glass for substrate cultivation was already 2000.

Distribution
The growers may be members of a cooperative auction association. If so, they are obliged to deliver products to the auction of the association. From there the vegetables are sold to exporters, preserving industries and wholesalers. The price at the auction is determined on the basis of the supply on that day.

The vegetables are here also separated according to quality, as regulated in the **Ordinance Quality Regulations for Vegetables and Fruit** of the Marketing Board for Vegetables and Fruit.

In the Netherlands the auctions are associated to the Central Bureau of Horticultural Auctions in the Netherlands (CBT). Similar practices exist in other parts of Europe.

For almost all products minimum prices are set by the CBT and no products are allowed to be auctioned off below these prices. The aim of the minimum price regulation is the support of the auction price level to avoid total market disruption.

If during the auction there is no interest in a product offered, the price on the auction clock will drop below the set minimum price: it 'turns through'. For 'turned through' products, the grower obtains compensation, which has been set beforehand by the CBT and is lower than the minimum price. This compensation comes from the minimum-prices fund. To provide this fund, a levy is imposed per quantity of each separate product brought to the auction. From this fund the advertising for each product is also financed. The 'turned through' products may not be sold afterwards for fresh consumption. They are offered to the preserving industries for a low price, processed into cattle feed or destroyed.

The alternative-grown vegetables are distributed through different distribution centres. The price is not dependent on the quantity offered, but on the cost price (see section 1.2).

20.3 KINDS: DIVISION ACCORDING TO EDIBLE PART

In Western Europe, we use mainly cultivated vegetables from the western world, but also exotic and wild-growing kinds. Vegetables can be divided into categories according to the part which is eaten. For a summary and division of vegetables and vegetable crops on this basis, see Table 20.2.

Root, bulb and tuber plants
Among these are very well-known kinds such as the swede and the beetroot, but there are also less well-known ones such as the Jerusalem artichoke and parsnip. The onion kinds are not so much used as a vegetable in the Dutch kitchen, but are more added to dishes to give flavour.

The **Jerusalem artichoke** is a winter-hardy root which looks like a knobbly potato.

Sec. 20.3]   **Kinds: division according to edible part**   243

Table 20.1 — Supply chart of fresh vegetables (Largest supply ▬▬, smallest supply ——)

Product	Jan.	Feb.	Mar.	Apr.	May	June	July	Aug.	Sep.	Oct.	Nov.	Dec.
Asparagus					▬	▬						
Aubergines						▬	▬	▬	▬	—		
Beans												
broad						▬	▬	▬				
French						—	▬	▬	▬	▬		
runner						—	▬	▬	▬	▬		
scarlet runner							▬	▬	▬			
Beetroot	▬	▬	▬	—					—	▬	▬	▬
Brussels sprouts	▬	▬	▬	—					—	▬	▬	▬
Cabbage												
Chinese									▬	▬	▬	▬
green	▬	▬	—			—	▬	▬	▬	▬	▬	▬
pointed					—	▬	▬	—				
red	▬	▬	▬	▬					▬	▬	▬	▬
white	▬	▬	▬	▬					▬	▬	▬	▬
yellow	▬	▬	▬	▬					▬	▬	▬	▬
Carrots												
washed				▬	▬	▬	▬	—	—	▬	▬	▬
winter	▬	▬	▬	▬	—			—	▬	▬	▬	▬
young					—	▬	▬	▬	—			
Cauliflower				—	—	▬	▬	▬	▬	▬	▬	—
Celeriac	▬	▬	—						▬	▬	▬	▬
Celery												
blanched								—	▬	▬	▬	▬
leaf	▬	▬	▬	▬	▬	▬	▬	▬	▬	▬	▬	▬
Chicory	▬	▬	▬	▬	—					—	▬	▬
Cucumbers				—	▬	▬	▬	▬	▬	▬	—	
Endive				—	▬	▬	▬	▬	▬	▬	—	
Gherkins							—	▬	▬	—		
Kale	▬	▬	—									▬
Leeks	▬	▬	▬	▬	—			—	▬	▬	▬	▬
Lettuce	—	—	▬	▬	▬	▬	▬	▬	▬	▬	—	—
Mangetout					—	▬	▬	—				
Mushrooms	—	—	—	—	—	—	—	—	—	—	—	—
Onions	▬	▬	▬	▬	—				▬	▬	▬	▬
Parsley	▬	▬	▬	▬	▬	▬	▬	▬	▬	▬	▬	▬
Peas						▬	▬	—				
Purslane					—	▬	▬	—				
Rhubarb			—	▬	▬	—						
Radishes			—	▬	▬	▬	—					
Scorzonera	▬	▬	▬	—						—	▬	▬
Spinach		▬	▬	▬	▬	—	—	—	▬	▬	—	
Sweet peppers						—	▬	▬	▬	—		
Swede	▬	▬	▬	▬	—					—	▬	▬
Tomatoes				▬	▬	▬	▬	▬	▬	▬	▬	
Turnip tops	—	▬	▬	▬	▬	—						

Source: Central Bureau of Horticulture Auctions.

These are eaten as a vegetable, but also as a substitute for potatoes. The root appears to contain compounds with an appetite-restricting action. Extracts of the Jerusalem artichoke can be found in slimming preparations.

Table 20.2 — Division of vegetable crops and vegetables

Root, bulb and tuber crops	Leaf vegetables	Fruit and flower crops	Stem vegetables	Legumes
Beetroot	Am choi	Artichoke	Asparagus	*Unripe seeds:*
Carrot	Brussels	Aubergine	Bamboo	Broad beans
Celeriac	sprouts	Broccoli	shoots	Kapucijners
Florence	Chicory,	Cauliflower	Cardoon	Peas
fennel	blanched	Courgette	Celery	
Garlic	Chicory,	Cucumber	Leaf beet	*Unripe pods:*
Kohlrabi	green	Gherkin	or chard	Beans,
Jerusalem	Cabbage,	Mushroom	Rhubarb	French
artichoke	Chinese	Okra		Beans,
Onion	Cabbage,	Pumpkin		runner
Parsnip	pointed	Sweet corn		Beans,
Radish	Cabbage,	Sweet pepper		scarlet runner
Radish,	red	Tomato		Beans,
black	Cabbage,			yard-long
Spanish	white			Mangetout
Radish	Corn salad or			
Japanese	lambs lettuce			
Scorzonera	Cress			
Shallot	Endive			
Swede	Kale			
Turnip	Leaf beet			
	or chard			
	Leek			
	Lettuce			
	Pak choi			
	Pursalene			
	Sea			
	vegetables			
	Seaweed			
	Sorrel			
	Spinach			
	Stinging			
	nettle			
	Turnip tops			
	Water cress			

Parsnip is a white root and one of the oldest native vegetable crops. It has now been supplanted in the Netherlands by the orange-coloured carrot.

Kinds of carrots: There are **summer carrots**, which are sold with the green still on or as washed carrots, and the larger **winter carrot**, which has a more pronounced taste and a deep orange colour. Because of its good storage quality this kind of root is available practically the whole year round.

Turnip (light green-white to violet-coloured) is braised in butter especially in restaurant kitchens and served as an exclusive vegetable.

Swede is a specific winter vegetable. The fairly large root is yellowy-brown coloured. It is usually cheap.

Kohlrabi is a ball-shaped, thickened stem. The ball (white-green or violet-red) grows above the soil and out of it grow stalks with leaves. It can also be eaten raw.

Florence fennel originates from the Mediterranean area. The thickened leaf

Fig. 20.1 — Carrots.

stalks form a light-green bulb. The taste is like aniseed. It is eaten raw as well as cooked.

Celeriac is a ball-shaped root and related to other kinds of celery. It is eaten especially cooked in the winter, but also raw, and is used as a herb instead of leaf celery.

Beetroot is a purple-red-coloured kind of beet. The summer beet are sometimes sold in bunches with the green still attached and are more tender then the winter or storage beetroot. The winter beetroot are mostly sold pre-cooked or steamed.

Radishes, black Spanish radish and **Japanese radish** are in so far as taste is concerned much alike. They all three contain mustard oil which causes the hot, bitter taste. Common radishes, round as well as oblong and red, white or red-white, are mostly sold in bunches with the green still attached. They are mostly eaten raw.

Black Spanish radish has a black peel with white flesh. The taste is sharper than that of the common radish.

Japanese radish is a hot-tasting white root. Black Spanish as well as Japanese radish are especially eaten raw grated or used in stews.

Scorzonera is a dark brown, long, thin root. It is also called 'winter asparagus', because of its white colour, mild taste and consistency after cooking. It is available fresh paricularly in the months of October and November.

Onion in the Dutch kitchen is especially used as a flavouring ingredient. It is related to leek, garlic and chives. There are many varieties.

Fig. 20.2 — Beetroot.

Fig. 20.3 — Radishes.

Fig. 20.4 — Scorzonera.

Silverskin onions are small, white onions, which are especially used for pickling. They are not available fresh.

Spring onions, salad onions or **summer onions** are harvested in bunches with the green leaves and the scarcely developed bulb. They are used fresh in salads, the bulb as well as the green part.

Storage onions are harvested when the bulb has developed well and the green dies down. The bulb and the green dry somewhat further after the harvest. There are a number of kinds of storage onions, which differ in colour, smell, taste and price. Yellow, red and white onions are for sale.

Shallots look like little yellow onions, but they grow in trusses around a basic root. They have a strong taste and are only used in dishes as a flavouring ingredient.

Garlic is the most piquant member of the onion families. Cooking reduces the taste. It is especially imported from France, fresh as well as dried.

Leaf vegetables

Leaf vegetables in general should be used or prepared as soon as possible after harvest. They quickly lose moisture and then become limp and less attractive.

Endive is available nearly all the year (greenhouse and open ground). Curled endive with a deep-cut curling leaf is of limited availability, is more expensive, but is more suitable for the preparation of salads. The ordinary endive is also eaten cooked and has often a somewhat bitter taste.

Turnip tops are usually eaten raw and are available for a limited time.

Spinach from the open soil is only for sale for a limited time, but there is also greenhouse spinach. The preserving industry (deep-freeze) works particularly with open soil spinach on a contract basis.

Sorrel looks like spinach, but is gathered from the countryside or harvested by amateur gardeners. This also applies to the **stinging nettle**.

Purslane contains much oxalic acid, has a very specific taste and is mainly eaten cooked.

Corn salad or **lambs lettuce** has oblong, dark green leaves, which are set into a rosette shape. They are only eaten as a salad.

Watercress has round green leaves and is grown in running water. It is sold in bunches and eaten raw (mostly without any additions) or is used as a garnish. It is in taste similar to **garden cress**. Garden cress has very small star-shaped leaves on a thin white stalk. It is mostly grown and sold in small cardboard boxes.

Chicory is the name for heads of the chicory (endive family) blanched under the soil or in other ways. The taste is somewhat bitter.

Green chicory and **red chicory** (radicchio, also called red lettuce) are both of the chicory family. The taste can be very bitter. It is especially cultivated by amateur gardeners. Red chicory is used for the decoration of dishes especially in restaurants.

Leaf beet or **chard** are varieties of the red beetroot. However, the dark green young leaves or the white fleshly leaf veins and part of the full grown plant are eaten. Either way they are mostly eaten cooked.

Kale is a typical Dutch winter vegetable. The strongly curled leaves only are cooked and processed into a stew.

Savoy cabbage (green and yellow) has crinkly leaves and a somewhat loose head. Both kinds are appreciated for the mild (cabbage) taste and are mostly eaten cooked.

White and **red cabbage** have firm heads and can be stored longer than the Savoy because of this. They are available all through the year, at least the keeping kinds. White cabbage is processed into sauerkraut, but also eaten raw and cooked. Red cabbage is eaten cooked sweet-and-sour, and raw as a salad.

Pointed cabbage is a white summer cabbage with a pointed head. It has a not very pronounced cabbage taste and has a shorter preparation time in comparison with white cabbage.

Brussels sprouts are small, firm, light-green-coloured side buds from the sprouting cabbage plant. They are real winter vegetables. They are available fresh, perhaps cleaned, from October until March. They are also deep-frozen and canned.

Chinese cabbage, pak choi and **am choi** belong to the same family. They are oblong in shape, have wide, white veins and large green leaves. Chinese cabbage is more popular with the Dutch and is also more readily available than pak choi and am choi. These last two kinds are often used in the Surinam and Antillies kitchen.

Lettuce varieties are mainly eaten raw and used to garnish dishes. There are many varieties which differ from each other in colour, firmness and head formation. The kinds of lettuce vary (sometimes considerably) in price depending on the place and the method of production. **Cabbage lettuce**, **Iceberg lettuce** and **oakleaf lettuce** are available. Sometimes very exclusive kinds of lettuce are available, such as crispy lettuce (crisp lettuce with a strongly cut-in somewhat curling leaf, which looks like curly endive) or red lettuce (see red chicory).

Fig. 20.5 — Endive.

Fig. 20.6 — Corn salad.

Sea vegetables are for sale locally. They grow and are gathered at places which are regularly flooded by sea water. They have a salty taste and a fleshy leaf. Among others, sea spinach, samphire, starwort and sea kale are gathered.

Seaweed is only imported dried in the Netherlands, particularly from Japan, Seaweed is used daily in small quantities in the preparation of macrobiotic food.

Leeks are especially valued in the autumn and winter as a vegetable. The white part as well as a part of the green leaves are used. Leeks are especially grown in open ground. They can be stored reasonably well for a few months. Leeks are usually also an ingredient of vegetable soup mixtures.

Fig. 20.7 — Chicory being forced on water.

Fruit and flower plants

For some kinds, the whole fruit is eaten with or without seeds, as in the tomato, cucumber and aubergine. Sometimes only a certain part of the fruit is eaten, as in artichokes and sweet corn.

Of the **artichoke**, a green flower bud of a kind of thistle, the lower thickening of the flower petals and the flower base is eaten after cooking.

The violet **aubergine** is not so well known in the Netherlands, even though they are grown in Dutch greenhouses. It is always eaten cooked, often stuffed.

Gherkins are especially cultivated on a contract basis for the preserving industry. They are only pickled in vinegar and are rarely available fresh.

Cucumber is available the whole year through and is mostly grown as a greenhouse vegetable. Cucumber is mainly eaten raw, but also processed into soup or braised in butter.

Courgette looks very much like cucumber and is also related. The peel is however green speckled and somewhat rough. A very young courgette can be eaten raw. But they are mostly eaten prepared, for instance boiled or fried in butter.

Pumpkins belong, like the courgette, to the cucumber family. They exist in many varieties, small and large kinds, which again differ in colour and shape. The summer kinds have a very short cooking time. The storage or winter kinds have a more fibrous substance. They are often cooked in butter or sweet-and-sour and processed into soups and sauces.

Some variety names are: the giant pumpkin (marrow pumpkin) which weighs 15–20 kg and is green or orange in colour; the smaller pear-shaped green chayote; and the flattened, white patty pan with a scalloped edge.

Sweet peppers (also called green peppers) are related to the chilli peppers (see Chapter 23). They are often cultivated and eaten in the Balkan countries. The somewhat pear-shaped fruits can be red, green, yellow, white, flecked and violet-black. They are used raw as well as cooked as a vegetable.

Cauliflower is a flower plant like **broccoli**. The white cauliflower is cultivated in the open ground as well as in greenhouses, as is the green broccoli, which originally was cultivated in Italy, China and the United States. There is also a violet-red broccoli variety and a soft-green cauliflower variety.

Fig. 20.8 — Cauliflower.

Okra is a small, green, pointed fruit (about 15 cm). The skin feels stubbly-rough. It is mainly used in the Creole-Surinam kitchen and in the south of the United States. During cooking, a mucilaginous, starch-like liquid separates from the fruit. Because of this it also serves as a thickener in stews and soups.

Mushrooms: see Chapter 22.

Sweet corn is a maize variety of which the unripe grains are eaten cooked. They are imported from Israel, southern France and the United States. Very small cobs, which are scarcely developed and are about 5–10 cm long, are sold as sweet-and-sour preserves.

Tomatoes can be eaten raw or cooked, or are added to soups, sauces and stews. They are available all through the year, because they are imported or grown in greenhouses.

A special variety is the **beef tomato**, which is larger and squarer in shape and less juicy and because of this is more suitable for use in salads. Besides this, there is available **mini** or **cherry tomatoes** (small and round) and Italian **plum tomatoes** (more suitable for use in soups or sauces).

Stem vegetables
From some plants the stems are eaten and not any leaves possibly present.

   Asparagus (white) is only available fresh in the period from April until the end of June. Asparagus is cultivated in special asparagus beds; it is grown in the dark to keep it blanched. The vegetable has to be peeled and cooked.

   Green asparagus, particularly cultivated in France and Italy, grows above the soil. It is less tender and is less appreciated in some countries.

   Celery is available during the whole year. It is imported from Israel, France and the United States. The stems are eaten cooked as well as eaten raw. The outer stems are often fibrous. The plant is covered to obtain a white colour, so that no or very little leaf green can be formed.

Fig. 20.9 — Celery.

   Rhubarb has a pink colour and is eaten cooked with suger. The best qualities are harvested in the early spring. Rhubarb contains a lot of oxalic acid.

   Bamboo shoots are the young shoots from bamboo (a kind of grass) which grows in tropical countries. In most European countries young bamboo shoots are available only in tins. They are used in Indonesian and Chinese dishes.

Cardoon is an artichoke-like thistle plant, of which the stalks are eaten. It is cleaned and prepared as celery, and has recently become available within Northern Europe.

Legumes

The pods of some kinds are eaten raw: such as French beans, runner beans, scarlet runner beans and mangetout. Sometimes only the unripe seeds are eaten: as in peas, broad beans and kapucijners. The cultivation of bean plants takes place in market gardens, especially under open ground cultivation. The largest supply is in the months of July and August.

French beans are available under different names, for instance princess bean, sugarbean or salad bean. They are available throughout the year and are then imported from countries such as Egypt and Israel. Very young harvested French beans are sold as '*haricots verts*'. These very small pods have scarcely developed seed. They are available fresh almost the whole year, but are very high in price. They are regarded as luxury vegetables and especially used in restaurants. In general they are cheaper canned.

Yard-long beans are long, thin, shoe-lace-like beans (14–25 cm) of tropical origin. They are similar in taste and colour to French beans and are used in the Surinam kitchen.

Fig. 20.10 — Yard-long bean.

Mangetout are very young fruits, which have no membrane on the inside of the pods. The seeds are scarcely developed. Pod and peas together are eaten, as the name 'mangetout' implies. Fresh they are always high in price.

Runner beans are available fresh in the summer and have a large, flat pod. They have a long cooking time and because of this are cut into thin slices before cooking. They are also preserved in salt and then after about three months are suitable to eat and for sale as 'salted runner beans'.

Peas are harvested unripe. The green-coloured, small young peas are very tender and sweet in taste. The later harvested larger peas are harder and mealier, because the largest part of the sugars is changed into starch. They are only available fresh in the summer months for a limited time. They are often canned or deep-frozen after harvesting.

Kapucijners, when they are harvested young, are one of the more delicious vegetables. The pods are violet or green in colour.

Broad beans are a delicacy if harvested young. The skin which surrounds the young bean is somewhat bitter in taste. The older the bean becomes the more leathery that skin becomes and the more bitter the taste. They are sold fresh for only a short time. They are available canned or deep-frozen.

20.4 VEGETABLES AND VEGETABLE PRODUCTS

Because of the convenience aspect, the presentation of ready-prepared or pre-processed vegetables will stimulate the consumer to use more vegetables with meals.

Pre-processed vegetables can be divided in three groups: pan-ready fresh vegetables, vegetable preserves and products based on vegetables.

Pan-ready fresh vegetables

The vegetable has been cleaned, perhaps cut up, washed, weighed and packaged. By such pre-processing the chance of spoilage becomes larger and the storage time is limited. The sale of pan-ready vegetables has not yet been regulated by law. Marketing Boards think that storage at 1–3°C is desirable.

Vegetable preserves

By preservation the storage time is lengthened and changes occur in the structure, the colour, the smell and the nutritional value. Sometimes a totally different end product is obtained because the plant material changes considerably — it 'dies'. Methods for improving the storage time, not counting cooling and irradiation, are (in descending order of importance): sterilization, deep-feezing, drying, pickling and salting. The changes which occur or become visible as a result of the chosen preserving method are sometimes wanted, as is the case with the making of sauerkraut and salting and pickling; sometimes unwanted, for instance the loss of turgor of deep-frozen vegetables.

Changes in structure

A living plant cell has a certain pressure (turgor) which makes the plant tissue firm. The tugor disappears a certain time after the harvest and the tissue becomes limp. By heat treatment or deep-freezing the internal structure is disturbed and the turgor disappears completely (by heat) or partly (by deep-freezing). Pectin in the cell wall dissolves on cooking in water, especially in the presence of acid and sugar. Through this, vegetables become softer in consistency. Also pectolytic enzymes break down the pectin, as also happens in vegetables in acid or salt.

Changes in colour

The colour in fresh green vegetables is especially determined by the fat-soluble chlorophyll. In yellow-, orange- and red-coloured vegetables the fat-soluble β-carotenes or the water-soluble flavonoids dominate.

The stability of **chlorophyll** is linked to the magnesium atom. By increase in temperature and/or acid conditions chlorophyll is changed into phaeophytin. This has a brown-green colour. This reaction progresses less quickly and less intensively in more alkaline conditions and at a lower temperature.

The fat-soluble **carotenoids**, including β-carotene, bixin and lycopenes, vary in colour from yellow to red. They are fairly heat-resistant and oxidation-sensitive under the influence of light.

Among the water-soluble flavonoid pigments are the red **anthocyanins**, which are very pH sensitive. Metal ions (iron, tin, aluminium), oxygen and ascorbic acid have a negative influence on the stability, as have some enzymes. Vegetables with anthocyanins in them have to be preserved and stored in glass or lacquered cans.

The yellow-coloured **flavonoids** quickly colour brown under the influence of enzyme systems (tyrosinase) with air as catalyst, while sulphurous acid, L-ascorbic acid and/or a salt solution slow this down.

Changes in smell

Vegetables have in general a characteristic smell, which is mild in undamaged tissues. The smell is caused by volatile compounds such as esters, ketones, alcohols and aldehydes and sometimes sulphurous acid. In bitter-tasting vegetables, the smell of the flavonoids, which are soluble in water, dominates. Organic acids, turpentine-like compounds and sugars also have an influence on the smell and the taste. On heating, the very volatile compounds in particular escape. Industry collects the volatile oils and may add them again to the product. If cell tissue is damaged, such as happens through cutting, enzymes are liberated and all sorts of changes take place, including marked change in smell.

Changes in nutritional value

When vegetables are heated in water, losses of about 50 per cent of the water-soluble vitamins occur. This is also the case in preparation in the home.

The nutritional value in fermented vegetables changes, sometimes considerably. For instance: in the preparation of sauerkraut the sugars are changed into lactic acid.

In a number of preserves the changes are very noticeable. In the following section, acidified and salted products are fully described and the use of chemical preservations is discussed.

Acidified and salted products

Acidification of vegetables includes pickling in an acetic acid solution and the formation of lactic acid by lactic acid bacteria.

Pickling in an acetic acid solution is mostly done with firm vegetables such as gerkins, onions or cucumbers. The vegetables are lightly salted to extract water from them and for the taste. They are then pickled in an acetic acid solution, after which they receive a heat treatment (pasteurization) to improve the keeping quality.

Sauerkraut is sliced white cabbage, to which sometimes some Savoy cabbage and about 1.5% per cent salt is added. It is then fermented by means of lactic acid bacteria. The cabbage has to be put under pressure and covered to seal the whole from the air. All this improves the liberation of cell moisture, and a very speedy conversion of the sugars into lactic acid occurs under aerobic conditions. The lactic acid content in the end is 1.4–1.8 per cent (pH about 3.6).

By the **addition of 33 per cent salt** to, for instance, runner beans, a product is obtained with a limited keeping quality. The sliced vegetable is covered with salt, layer by layer. The whole is covered and put under pressure. The runner beans stand after some time in a brine solution, in which nutritional elements (vitamins and minerals) are dissolved. The salt must be rinsed out as much as possible in the preparation to make the product edible, as a result of which the nutritional value is lowered even more.

Sometimes **chemical preservatives** are added in combination with another storage-improving method. Permitted preservatives include benzoic acid in gerkins and onion, and sulphurous acid in dried blanched vegetables; also anti-oxidants including EDTA (ethylenediaminetetra acetate) in asparagus, onions, mushrooms and cauliflowers.

Products based on vegetables
The usefulness and the price are strongly dependent on the type of product and the additives used. Available among others are vegetable purees, juices and spreads, soup vegetables, mixed vegetables and kinds of ketchup.

Vegetables can be processed into puree, for instance tomato and carrot puree, or into juice, for instance beetroot, carrot and tomato juice. The taste of the product may be adjusted with salt and/or herbs and spices.

A special group are the **homogenized vegetables**, suitable as baby, toddler or invalid foods. Vegetable preserves and products based on vegetables are also available for low-sodium diets.

Vegetable juices can be produced in different ways, in which fermentation with the help of lactic acid bacteria and enzymic maceration with the help of pectinase are more and more applied (see Fig. 20.11).

The purpose of macerating or softening the vegetable material is to enable the enzymes to work. In the production of vegetable juices, pectinase will break down the pectins in the cell walls, causing the cells to become more permeable; the juice will then be more easily extractable. The pectins will have also lost their gel-forming characteristics, and thus the vegetable juices will also stay fluid after cooled storage.

20.5 QUALITY DETERIORATION, SPOILAGE AND STORAGE

Harvested vegetables remain a living product. Metabolic processes, growth of microorganisms which are present on vegetables, and attacks by caterpillars and aphids continue.

These processes can be slowed down by correct storage conditions and temperature. A deterioration in quality of the fresh product is, however, unavoidable. The changes which in the end spoil vegetables during storage are:

— *Moisture loss through evaporation*. The vegetable becomes limp or dries out. This is particularly quickly seen in leaf vegetables.
— *A continuing ripening*, through which the fruit-vegetables become over-ripe and change in colour. They also usually lose their firmness.
— *A renewed growth*, which especially takes place in biennial plants, for instance in onions and carrots.

Sec. 20.5] Quality deterioration, spoilage and storage

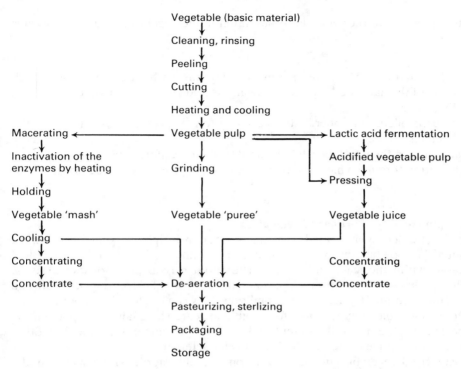

Fig. 20.11 — Diagrammatic summary of the production of vegetable juices. (Source: *Voedings-middelentechnologie* 12 april 1984.)

— *Enzymatic and/or oxidation processes* which cause a brown colouring, for instance brown edges in endive.
— *A yellow colouring of green vegetables* which are stored in the dark. The unstable chlorophyll is broken down and is not replenished because of the absence of light.
— *Mould growth on moist vegetables.* Mould is visible as black or brown spots.
— *Storage at too low a temperature disturbs the metabolism* in such a way, that the (detrimental) metabolic products are no longer further broken down. The product is 'poisoned' by these break-down compounds. Discolouration also frequently occurs and the product loses the firm consistency, so that the spoilage becomes visible. Tomatoes and sweet peppers which are stored too long too cool sometimes show this form of spoilage.
— *Rotting* as a result of the autolytic and enzymatic processes, through which the cells lose their turgor. Moisture is liberated and all sorts of, sometimes detrimental, break-down compounds are formed.

Storage advice for fresh vegetables

Vegetables should if possible be bought fresh each day for use in the kitchen. If one wants to stock up, some fresh vegetables are more suitable for storage than others; for instance winter carrots, storage kinds of cabbage and storage onions.

Storage and care receives a lot of attention in the vegetable trade. Techniques used are:

— Cool and dark storage in sheds or in cooled rooms, perhaps preceded by a quick cooling of the harvested product by means of cold air, water, or water combined with a vacuum.
— CA (controlled atmosphere) storage suited to the specific storage needs of the vegetables.
— Packaging of the vegetables to be stored to avoid evaporation.
— Irradiation through which ripening and/or mould growth is slowed down.

The keeping quality of preserved vegetables
Canned or bottled (heat-treated) vegetables can, in principle, be kept for an almost unlimited period, but the product changes in quality. It is advisable to use leaf vegetables within the year; other kinds can be stored in a cool place for several years.

The quality of deep-frozen vegetables depends on the freezing method and storage temperature. Under optimal conditions they can be stored for about a year.

Dried vegetables can be stored for about one year. The quality is dependent on the packaging, the moisture content and the storage temperature amongst other factors. Long storage decreases the cooking characteristics.

Pasteurized vegetables in vinegar are commercially sterile with a shelf-life of 2 years. They have a limited keeping quality after opening the packaging and should then be stored refrigerated.

20.6 LEGISLATION

The legal regulations are usually divided into regulations for fresh vegetables and regulations for preserved vegetables. Fresh vegetables are often allocated in quality classes (extra-I-II-III). The Marketing Boards for Vegetables and Fruit have worked these rules out closely for each product and have described them in quality regulations. A number of visible characteristics such as colour, shape, size and soundness are taken into consideration. Imported vegetables are also given a class-labelling next to a name label, with the country of origin indicated.

Pesticide and **Herbicide Regulations** may set down rules for the type of pesticide or herbicide, the maximum permitted quantity of residues which are allowed on freshly harvested vegetables and the time at which they can be harvested after the treatment. Some regulations regulate additives to vegetable preserves.

There are regulations or codes of practice describing, among other requirements, the labelling on the packaging of preserved vegetables. For instance 'extra fine' on a can of peas depends upon the diameter of the pea, and, in the case of vegetables canned in water, the drained weight has to be mentioned next to the net weight.

21

Fruit

21.1 INTRODUCTION

Fruit is the collective name for plant fruits which in general are eaten raw and have a pleasant sweet or sour taste. Botanically, a fruit is the edible part of the flower or what develops from it after pollination.

Some fruits, such as tomato, sweet pepper, aubergine, cucumber, are counted as vegetables.

Seeds and nuts develop on or in the fruit. They differ somewhat in taste, use and nutritional value from fruits.

Consumption

For years the consumption of fruit has been nearly the same in the Netherlands, about 82 kg per person per year. This is a bit more than 200 grams per person per day. In the UK, fruit consumption has increased from 35.6 to 43.5 kg per person per year over the period 1972 to 1987.

Much of the fruit bought in Northern Europe is imported. The demand for locally grown fruit has increased in the last few years, especially from the ecological viewpoint of the consumers. However there is also more demand for exotic kinds such as kiwi, persimmon and pineapple.

Composition

Fruit makes an important contribution to the supply of moisture, vitamin C, minerals and fibre. The quantities vary for each kind. Fruits contain a fair amount of organic acids, among others citric, tartaric and malic acid. The acids give rise to a fresh acid taste and stimulate the flow of digestive juices. The carbohydrates in fruit are present in the form of pectins, sugars (di- and mono-saccharides) and cellulose. In bananas they occur in the form of starch. Fruit in general supplies little energy.

On average, 100 g fruit contains (source: *Nederlandse Voedingsmiddelentabel* (34), 1983):

water	80–90 g
protein	1–1.5 g
fat	—
carbohydrates	3–22 g
vitamins	
C	5–150 mg
β-carotene	0.08–2.5 mg
B_6	0.03–0.35 mg
minerals	
Ca	10–60 mg
Fe	0.3–1 mg
K	75–350 mg

Possible health hazards

By nature edible fruits contain few detrimental compounds.

Serotine can occur in bananas, gooseberries, blackcurrants and pineapples. This causes hardening of the heart muscle. However, more than one kilogram of these fruits per day has to be eaten for a considerable time to produce any detrimental effects.

Detrimental compounds which may be found in and on fruit are the result of the use of certain production aids or originate from the environment. The following may be found:

— **Herbicides** and **pesticides**. Often used are organophosphorus compounds, which can cause acute poisoning, and organochloro compounds, which are difficult to break down and therefore accumulate in the food chain.
— **Diphenyl**. Citrus fruits, because they are very mould-sensitive, are sprayed after the harvest with diphenyl, a mould-preventing compound. Apples are also sometimes packaged in paper impregnated with diphenyl.
— **Heavy metals** and **polychlorobiphenyls** (PCBs) from the environment (air, water).

21.2 PRODUCTION AND DISTRIBUTION

Production

The growing of fruit requires much specialized knowledge and specific care. The cultivation of apples or pears (trees) is of a different kind from the cultivation of currants (shrubs) or strawberries (herb-like plant)

Traditionally the cultivation of fruit is an open ground culture. Glass is used to bring the harvest date forward, for instance with strawberries, and also to permit the harvesting of fruit which cannot be cultivated under normal circumstances, for instance grapes.

The Northern European climate is very suitable for the cultivation of apples, pears, cherries, currants and strawberries. However, sometimes late frosts in the blossom time and hail storms during the fruit development can cause much damage. The cultivation of fruit is very labour-intensive. The grower needs varieties which are easy to harvest. This has already led to the disappearance of tall standard apple and pear varieties. Plant breeders are trying to create varieties with higher yields, a greater resistance to diseases and longer keeping quality after harvesting.

In fruit cultivation, use is made of products during production which help to obtain a more attractive end product. Among other things **growth regulators** are used. These influence the hormone metabolism of the plant. The regulator ethylene stimulates flowering, ripening of fruit, colour formation and fall of fruit and leaves. Ethylene made chemically is added, but it is also formed naturally by the fruit.

A number of growers work in contract cultivation for the preserving industry; others work along the lines of biological and bio-dynamic agricultural and horticultural methods.

Distribution

The sale of fruit is mainly done through auctions. The grower, in the Netherlands usually a member of a cooperative auction, rarely sells his own product himself, but offers his selected fruit to the auction. Fruit auctions work in the same way as vegetable auctions (see Section 20.2).

21.3 KINDS OF FRUIT

The number of kinds of fruit is very large, certainly when viewed world-wide. In Northern Europe many kinds and varieties are available. The kinds of fruit can be divided according to botanical charcteristics. In Table 21.2, division into groups of pip, citrus, currant and stone fruits is based on practical similarities which they have (storage requirements, transport, keeping quality and usefulness) and not on the division into families, such as is usual in botany. The heading 'more common' in this case means kinds of fruit of which the fresh and/or processed forms are easily and generally available. The less common or exotic kinds are not generally available fresh or preserved, or are usually very expensive. Because of this, these products will only be bought by consumers for special occasions.

Fruit when bought has to be fresh and should have the right degree of ripeness. Ripe fruit has mostly a somewhat soft consistency, is aromatic and sweet and has the correct green-yellow and/or orange-red to violet-black colour.

The number of fruits mentioned in Table 21.2 is so great that mention of varieties, characteristic qualities and harvest times would fill a copious book, and so is outside the scope of the present work. However, an exception is made for apples, pears and citrus fruits, as these kinds of fruits are the most consumed and processed by the food industry in Northern Europe and hold an important place in those countries' food-eating habits. A number of different kinds are described in short, and some kinds are described in detail so that they can be recognized.

Table 21.1 — Availability of fresh fruit

	1	2	3	4	5	6	7	8	9	10	11	12
Strawberries					□	■	■	■				
Apples	■	■	■	■	□	□		□	■	■	■	■
Red currants						□	■	□				
Black currants							■	□				
Blackberries								□	■	□		
Grapes		□	□			□	□	□	■	■	■	□
Raspberries							■	□				
Cherries						□	■	□				
Gooseberries						□	■	□				
Melons						□	□	■	□	□		
Pears		□	□	□				■	■	■	■	□
Peaches						□	■	■	□			
Plums						□	□	■	■	□		

Key:
1=January; 2=February; 3=March; 4=April; 5=May; 6=June; 7=July; 8=August; 9=September; 10=October; 11=November; 12=December.
■ Greatest supply
□ Smallest supply
Source: Centraal Bureau van de Tuinbouwveilingen.

Apples

A large proportion of the apples consumed in Northern Europe are grown where they are eaten. But apples are also imported from other EC member states and from Florida, California and New Zealand.

The estimated 2000 or so apple varieties grow especially in the northern hemisphere in a temperate climate. More and more apple varieties appear through breeding. The apple gets its popularity from a number of charcteristics favourable both to the producer (such as a high yield per hectare and good keeping qualities), and to the consumer (who appreciates the smell, taste and consistency).

Apple varieties vary in shape (round, oblong or flattened), colour (green-yellow to blood red), taste (sweet to acid) and consistency (crisp to mealy).

The consumer distinguishes, on the basis of taste and quality, between **eaters,** which are suitable to be eaten as they are (Golden Delicious, Lombarts Calville, James Grieve), and **cookers,** which are more suitable for making into puree (Golden Reinette, Jonagold) or for making into tarts and pies (Cox's Orange Pippin, Golden Delicious). The fruit grower differentiates on the basis of harvest time and keeping qualities:

— **Early apples**, which are harvested from the end of July and are available until the end of September (Yellow Transparent, Benoni, James Grieve);
— **Autumn apples,** which are harvested from mid-September and are available until the end of March (Cox's Orange Pippin, Laxton's Superb);

Table 21.2 — Kinds of fruit

Fruit group	More common kinds	Less common and/or exotic kinds
Pip fruits	Apples Pears	Persimmon Quinces Medlars Figs Indian figs
Citrus fruits	Lemons Grapefruits Mandarins Oranges	Seville oranges Kumquats Limes Limquats Pomelos Ugli fruits
Currants or berry fruits	Red currants Strawberries Bananas Blueberries Blackberries Grapes Raspberries Gooseberries Melons Cranberries Black currants	Pineapples Plantains Durians Pomegranates Litchis Kiwi fruits Papayas Passion fruit Rambutans Cowberries
Stone fruits	Cherries Peaches Plums Nectarines	Apricots Avocados Dates Mangoes Olives

— **Winter apples,** which are harvested from the beginning of October and are available until the end of May (Golden Reinette, Lombarts Calville, Jonathan, Jonagold, Golden Delicious).

In the industry, apples are processed into sauce, juice, wine, vinegar or syrup, and pectin is obtained from apples rich in pectin.

Pears

The consumption of pears is considerably lower than that of apples. Pears cannot be stored so long and so easily; they become mealy quickly or start to rot. The presence of so-called stone cells in the fruit's flesh is a (less favourable) characteristic of many pear varieties. These hard corns are less pleasant when eaten.

Fig. 21.1 — Persimmons.

Fig. 21.2 — Kumquats.

Fig. 21.3 — Litchis.

Kinds of fruit

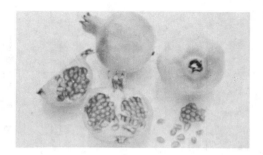

Fig. 21.4 — Pomegranates.

Fig. 21.5 — Passion fruit.

Fig. 21.6 — Dates.

Table 21.3 — Availability of apple varieties

Apple variety	1	2	3	4	5	6	7	8	9	10	11	12	Appearance	Taste and characteristics	Use
Yellow Transparent							☐	☐					Light green to yellow. Medium large. Somewhat irregular in form. Tender skin	Flesh a little crisp. Fairly sour	Eating and cooking apple
Stark's Earliest								☐					Medium large. Flat-round. A little irregular in shape. Bright red on light yellow background	Fairly firm, juicy light acid flesh with good aroma	Good eating apple
James Grieve									☐	☐			Medium large. Yellow with red stripes and sometimes a red blush	Soft yellow-white flesh. Very juicy and with good aroma	Very good eating apple
Ellison's Orange	☐								☐	☐	☐		Medium large. Taller in shape than Cox's Orange. Yellow with red stripes and often a blush	Pleasantly acid and aromatic	Very good eating apple
Cox's Orange Pippin	☐	☐								☐	☐	☐	Fairly small, flat-round. Light green to yellow with orange-red blush and stripes	Pleasantly acid and aromatic. Very juicy, fairly firm flesh	Especially fine eating apple
Glorie van Holland									☐	☐			Medium large. Beautiful yellow with red blush and stripes. Regular in shape	Light acid, juicy, firm flesh	Good eating apple
Laxton's Superb	☐									☐	☐	☐	Medium large and regular in shape. Green with brown-red blush	Juicy, pleasantly acid and aromatic. Cox's-taste	Fine cooker
Golden Reinette	☐	☐	☐	☐	☐					☐	☐	☐	Large, regular shaped fruit. Green to green-yellow in colour, russetted, sometimes rough, often a blush	Firm, fresh sour flesh	Very good eating apple
Lombarts Calville	☐	☐	☐	☐						☐	☐	☐	Fairly large fruit. More wide then high. Colour yellow-green, later yellow. Greasy peel	Excellent in taste. Mild acid. Fairly soft creamy white flesh	Good eating apple
Jonathan	☐	☐	☐	☐	☐						☐	☐	Fairly small but regular in shape. Yellow with beautiful, even, red blush	Fairly firm, light acid flesh	Good eating apple, reasonable cooker
Golden Delicious	☐	☐	☐	☐	☐					☐	☐	☐	Medium large. Regular shape. Long stalk. Green-yellow in colour, later yellow	Fairly firm, soft acid flesh. Sometimes apple nearly sweet	Very good eating apple
Winston	☐	☐	☐	☐	☐								Fairly small, regular shaped fruit. Dark green, later yellow with blush and stripes	Firm flesh with thick skin. Good taste	Good eating apple

Key:
1=January; 2=February; 3=March; 4=April; 5=May; 6=June; 7=July; 8=August; 9=September; 10=October; 11=November; 12=December.
Source: Centraal Bureau van de Tuinbouwveilingen.

Pear varieties vary in form (round to long and thin and somewhat bent), colour (grey-green to yellow with sometimes a red blush) and consistency (soft and juicy to hard and gritty).

The consumer distinguishes, on the basis of intended use, between:

— **Eating pears,** which are soft and aromatic and are suitable for eating raw (Conference, Doyenné du Comice, Clapp's Favourite);
— **Cooking pears**, which are very hard and only slightly juicy. During cooking or stewing these (old) pear varieties become aromatic and soft, and also they sometimes colour red (Gieser Wildeman, Brederode, Saint Rémy).

The fruit grower makes a division into three groups on the basis of the harvest time and keeping qualities:

— **Early pears,** harvested from the end of June and available until middle October (Clapp's Favourite, Triomph de Vienne);
— **Autumn pears,** harvested from the end of September and available to the middle of March (Beurré Hardy, Conference, Doyenné du Comice);
— **Winter pears,** harvested from the end of October and available to the end of April. These are mainly the cooking pear varieties.

Some pear varieties are suitable for industrial processing into syrup, juice, wine and liqueur.

Citrus fruits

Citrus fruits belong to the rue family and grow especially in subtropical areas, but also in the tropics. They have a leathery peel, which contains volatile oils. The fruits are very mould-sensitive after harvesting and are often treated with mould-preventing compounds (see section 21.1). This has to be declared on the trade packaging. For the consumer who buys loose, this is not clear.

There are many kinds of citrus fruits which originate from crosses, as a result of which accurate naming is often difficult. The more usual kinds are oranges, mandarins, lemons and grapefruits.

Oranges can be divided into groups on the basis of sweetness (sour to very sweet) and/or on the basis of intended use in squeezing or eating varieties. Squeezing oranges have a tough membrane around the fruit segments.

Blood organges contain besides the yellow-orange pigment the violet-red anthocyanin also. This influences the colour and also the taste.

Navels have under the peel at one end a growth which looks like a small fruit.

Seville or bitter orange kinds are especially suitable for the production of marmalade or liqueur.

Some oranges are sometimes for sale under certain trade forms; for instance, Salustianas from Spain, Shamoutis from Israel, or Outspan from South Africa.

Mandarins are very sweet and aromatic citrus fruits. They are smaller than oranges and have a more flat-ended shape. They peel easier than most citrus kinds. There are race varieties with and without pips. A distinction is made between 'real'

Table 21.4 — Availability of pear varieties

Pear variety	1	2	3	4	5	6	7	8	9	10	11	12	Appearance	Taste and characteristics	Use
Précoce de Trévoux								☐					Fairly large. Green to green-yellow with light blush	Fresh	Good eating pear
Clapp's Favourite								☐	☐				Large. Green to green-yellow with nice blush	Juicy and sweet	Good eating pear
Triomphe de Vienne									☐	☐			Large. yellow-brown with much russetting	Soft, juicy and aromatic	Very good eating pear
Beurré Hardy									☐	☐	☐		Fairly large. Broad. Rust-brown	Soft, juicy, sweet and aromatic. Firm peel	Very good eating pear
Bonne Louise d'Avranche										☐	☐		Fairly small, narrow. Green often with nice dark brown-red blush	Sourish, juicy and very aromatic	Very good eating pear
Conference	☐									☐	☐	☐	Medium large. Slim. Green to green-yellow with russetting	Juicy and sweet. Light orange flesh. Thick skin	Good eating pear
Légipont										☐	☐	☐	Medium large. Fat. Light green to yellow with little russetting	Soft, melting and sweet flesh	Eating pear
Zwijndrechtse wine pear										☐	☐		Fairly small. Green with faint blush	Gritty flesh around core. Fresh taste	Eating and cooking pear
Doyenné du Comice											☐	☐	Large. Wide. Yellow-brown with sometimes faint blush	Very juicy, melting, sweet and aromatic. Tender skin	Very fine eating pear
Beurré Alexandre Lucas	☐									☐	☐	☐	Large. Wide. Green to nice yellow	Coarse flesh. Juicy. Fairly sweet	Eating pear

Key:
1=January; 2=February; 3=March; 4=April; 5=May; 6=June; 7=July; 8=August; 9=September; 10=October; 11=November; 12=December.
Source: Centraal Bureau van de Tuinbouwveilingen.

mandarins, satsumas of Japanese origin, and tangerines. They differ especially in size and sweetness.

Lemons are round, oblong-shaped fruits with very sour juice and a fairly thick, yellow skin. There are several race varieites which are offered under the name of lemon. Besides these there are related kinds, for instance lime and citron.

The **lime** grows in the tropics. It is rounder and smaller than the yellow, subtropical lemon and green to green-yellow in colour. The fruit is tinted light green and very aromatic and mildly sour in taste.

Citrons are not eaten fresh. The peel of the large green 'lemon' is very thick and meaty and is processed into candied peel (see section 21.6).

Grapefruit: this fruit is probably a mutant of the pomelo. The grapefruit grows especially in the United States (95 per cent of the world production). The round, somewhat flat-ended fruit has yellow-white or light pink to red flesh. The redder the flesh colour, the sweeter is the juice. These citrus fruits are especially eaten fresh or processed into juice.

The exotic kinds of citrus are also less available here.

Kumquats are brightly orange-coloured, small oval fruits with a diameter of 2–4 cm and have many pips. The peel is thin and edible. The taste is sweet and spicy.

Limequats or dwarf lemons look like kumquats and can be used in the same manner.

An **ugli fruit** is a cross between a mandarin variety, a grapefruit variety and a bitter orange.

Pomelos developed from a cross between the pompelmuss and the grapefruit; and a **mineola** or **tangelo** is a cross between a grapefruit and a mandarin variety.

Melons

Melons are berry fruits which grow on climbing or trailing plants. A distinction is

Fig. 21.7 — Melons.

made between water melons and other melons. Melons are cultivated a lot in tropical and subtropical areas and outside these also in greenhouses (for example, in the Netherlands). The trade uses the following division of families and varieties.

- **Cantaloupe melons** have a round or slightly flat-ended shape, sometimes with more or less clear segments; sometimes the peel is somewhat rough. The fruit flesh is yellowish to green. Some examples are Galia and Charentais melons. Cantaloupe melons have to be harvested ripe and can only be kept for 1–2 weeks after the harvest.
- **Netted melons** have a cork-like network over the peel. They are smooth, round or segmented. The flesh is mostly apricot coloured; an example is the Dutch netted melon.
- **Winter melons** have a hard, mostly yellow peel. The fruits are more or less oblong in shape. They are for sale as sugar or honey melons. The fruit flesh varies from yellow to green. Some examples are the white sugar melon, Honey Dew and yellow honey melon. The hard peel makes it possible to keep these kinds for a month or longer.
- **Ogen melons** are small. They have a smooth peel with light green stripes. The flesh is light-green and very aromatic.

Melons are ripe when the fruit on the underside (the end opposite the stalk end) gives under light pressure and the stalk detaches easily. Most kinds ripen further after picking at a temperature of 15–20°C.

Water melons consist of about 95 per cent water. They have a dark green, sometimes marbled peel. The flesh is red; the pips are spread in the flesh. The taste is faintly sweet. It is difficult to determine ripeness: a muffled sound on knocking is a sure indication of ripeness. The whole fruit has good keeping qualities; cut, it should be packaged in aluminium foil and stored in the refrigerator. The largest supply comes from the area of the Mediterranean.

Cherries
There are two sorts of cherries: the sweet cherry and the sour cherry (morello). The most important producing countries are Greece, Italy, France, southern Germany, Belgium and the Netherlands. It is increasingly difficult for locally grown cherries to compete against the cherries from more southern countries.
    Sweet cherries vary in colour from black to purple, bright red or yellow with a red blush. They are used fresh or are processed into jam, jellies, juice, syrup or wine, and distilled (about 65 per cent of the harvested cherries go to the preservation industry).
    Some varieties are:

- The **May cherry** (a cross between the sweet cherry and the morello). It is harvested in June, has a bright to dark red colour.
- The **varikse zwarte** (black cherry) with dark red flesh and a dark stone. The juice colours mouth and lips blue. The supply is in July. They are grown in the Netherlands and Belgium.
- The **wine cherry** is an old Dutch variety. This smaller cherry, red-brown in colour, tastes very sweet and has its own typical aroma.
- **Yellow cherries**, known in the Netherlands as 'Udense Spaanse' or 'Japanese',

are yellow with a red blush. The flesh is firm and hard. This makes the cherries suitable for industrial processing.

The **sour cherries** are known as morellos.

— The **morello** is a red to dark red cherry, so sour that they can only be eaten fresh when completely ripe. They are particularly processed into juice or jam or preserved with juice or alcohol.

Plums

The origin of plums is not clear. The European kinds originate nearly for certain from a wild variety from the Caspian Sea area. They grow in countries with a temperate or subtropical climate. Plums are round or oblong stone fruits with a smooth skin. The colour varies from green yellow to deep blue or dark violet. The flesh is, as a rule, very aromatic and soft.

Some varieties are:

— The Reine Claude Verte, a small round green fruit which is harvested in August and September. They are often preserved with juice.
— The Reine Claude d'Althan, a fairly large round fruit, red-violet in colour, very tasty and with a thick bloom.
— Ontario, a round green-yellow plum with very sweet and juicy flesh.
— Queen Victoria, a very common round oval plum. The taste is often mediocre. This variety often suffers from gum formation.
— Mirabelles, which are yellow, cherry-sized fruits. They are sweet and juicy and are sometimes coloured and canned in juice, candied or processed into jam.
— Damsons are often seen as a different variety. They are small and oval, but very aromatic. They are industrially processed into jam or into distilled spirits (slivovitz).

Peaches

The peach originates from Asia. It has a downy skin, which makes the surface look mat. The colour of the skin is light yellow to yellow with a red blush, depending on the variety. Some producing countries are Belgium, France, Italy, Greece, Spain, Israel and the United States. There are varieties with orange-yellow flesh, including Junegold and Springcrest, and with white flesh, for example Amsden. This last variety is grown in greenhouses in the Netherlands.

The **nectarine** is a mutant (a spontaneously occurring variant) of the peach with a smooth skin. It is smaller in size than the peach. The flesh is yellow-coloured.

Apricots

Apricots grow especially in the subtropical climate areas. The round egg-shaped or somewhat flat-ended fruits have to be harvested ripe, otherwise they are not juicy. Apricot varieties vary in colour from light yellow to orange. The skin is lightly hairy. The smooth stone releases easily. Some producing countries are France, Italy, Greece, New Zealand and the United States. They are eaten fresh or processed into jam, juice or sauce, or dried or canned.

Berries

A number of family varieties are: black currants, bilberries and cranberries, red and white currants and gooseberries.

The **black currant** has a somewhat spicy taste which is not appreciated by everyone. They grow in trusses like red currants, but arrive loose on the market. The largest part of the harvest is processed into juice and fruit drink (cassis).

The **black bilberry** has deep purple flesh and juice. The taste is sweet. The berries grow singly, not on trusses. They are especially processed into jam and medicinal juice.

Fig. 21.8 — Bilberries.

The **cranberry** is red in colour. The taste is pleasantly acid but can become somewhat bitter after processing. It is solely processed into cranberry sauce or jelly. The uncooked fruit is imported deep-frozen and sold as 'cranberry'.

The **cowberry** has a red colour and can be eaten raw as well as cooked. Originating particularly from Germany, it is sold there under the name 'Preiselbeeren'.

The **red currant** has bright red flesh with an acid to sweet-acid taste and contains yellow-brown pips. Red currants are especially processed into jam, jelly, juice or syrup.

The **white currant** differs in little but colour (cream colour) from the red currant.

The **gooseberry** is a green or red, sometimes hairy, berry. The taste varies from acid-sweet to sweet. They are not much eaten fresh, but are processed into jam or into a filling for pastry.

Grapes

The number of grape varieties is very large. Of the grapes produced, about 80 per cent are processed into wine, about 10 per cent are used as table grapes, and the rest are dried (raisins) or processed in the preserving industry. The varieties suitable as table grapes are grown mainly in Greece, Italy and in the United States.

Fig. 21.9 — Gooseberries.

In the Netherlands, Belgium and England grapes grown in greenhouses are an important export product. Greenhouse grapes have the advantage that the individual grapes have an undamaged wax or bloom layer and that the trusses are beautifully formed.

There are black and white varieties of grapes. Black varieties vary in colour from dark blue through blue-red to brown-red. In the Dutch greenhouses the black Black Alicante, Frankenthaler, Gros Maroc and West Frisia grapes are mainly grown. White Dutch greenhouse grapes are Golden Champion and Muskaat van Alexandrie. Some names of imported kinds from the more southerly situated countries are the black Cardinal and Olivette Noire and the white Almeria, Italia (muscat of Roma), Rosaki and Sultana (seedless).

Strawberries, blackberries and raspberries
Strawberries, blackberries and raspberries belong to the rose family. These fruits were originally native to Europe and still grow in the wild. Only cultivated (greenhouse) kinds are supplied to the auctions.

Strawberries are the swollen flower bases of the strawberry plant (false fruit).

There are two distinct kinds of strawberry plant, one which bears only once per season and a second kind which bears continually. A continuous bearer keeps on flowering and forming fruits for a longer period.

The **forest strawberry** or wild strawberry is also cultivated in France. The fruits are very small, but very aromatic.

Blackberries and **raspberries** grow in the wild, but are also cultivated. The fruits are similar in appearance, are respectively glossy and mat hairy; blackberries are coloured blue-black, and raspberries pink-red.

The three kinds of fruit are eaten fresh in the summer months. For use at other times, they are deep-frozen or processed into jam, pie filling and juice preserved with juice or alcohol.

Bananas
There are many kinds of bananas. In practice, a distinction is made between **dessert bananas**, which can be eaten raw and **plantains,** which are inedible raw. Bananas

Fig. 21.10 — Raspberries.

grow in tropical areas especially in South and Central America and in Africa. Some special kinds (rice and apple bananas) come from tropical Asia.

Dessert bananas are yellow or green-yellow and nearly always somewhat bent. They vary in length from 15 to 30 cm. The flesh is somewhat mealy and is faintly sweet in taste. During ripening, the peel colours brown and the flesh becomes softer. They are imported the whole year through and are especially eaten raw or processed into salads. The small (8–10 cm) **apple banana** is sometimes canned. The **rice banana** is also suitable for baking.

Plantains are large (30–40 cm) bananas and less bent than the dessert bananas. After peeling the flesh is cut into slices or strips and cooked, fried or deep-fried. They are often used in the Surinam kitchen. They are also dried and candied.

21.4 QUALITY DETERIORATION, SPOILAGE AND STORAGE

Quality deterioration and spoilage

Ripe, harvested fruit, in so far as colour, smell and taste are concerned, are in optimal condition. After the ripening, fruit quickly loses its quality characteristics. The most noticeable causes of quality loss are:

— **Damage**, which occurs from rough handling during harvesting, sorting, packaging and transport. The damage can provide openings for moulds, yeasts and other spoilage-causing bacteria.
— **Moisture loss**: evaporation of the moisture takes place through stomata. A moisture loss of 5–6 per cent is already clearly obvious. The fruit becomes limp, the fresh appearance goes.
— **Post-harvest metabolism**: respiration and metabolism continue after harvesting. Carbohydrates are changed into energy, carbon dioxide and water with the help of oxygen from the air. The fruit uses its reserves and becomes crinkly and shrivelled.
— **Low-temperature spoilage** occurs especially in fruits of tropical origin which are stored too cold (lower then 8°C). The metabolism is disturbed and watery spots

develop in the peel, which can be quickly attacked by microorganisms (rotting, mould).
— **Enzymatic processes**: for instance pectolytic enzymes cause structural and consistency losses and tyrosinase causes brown colouring.
— **Colour changes**: the green chlorophyll bleaches under the influence of light and oxygen (photo-oxidation). The yellow or red carotenoids get a deeper colour, the unstable violet-red anthocyanins become brown. All these discolourations happen under the influence of enzymes.
— **Microbial spoilage**: yeasts, moulds and other spoilage-causing microorganisms attack damaged fruit especially and cause spoilage.
— **Rotting** of fruit can be seen as the final phase of all the above-mentioned processes through which the fruit becomes unusable.

Storage
Fresh, ripe, harvested fruit can spoil. This can be slowed down by:

— Careful treatment of the fruit and adequate packaging. This prevents damage and drying out.
— Storage of fruit at temperatures of about 4°C at which the reactions proceed more slowly. In tropical fruit the metabolism becomes disturbed by these low temperatures, and the fruit acquire an off-taste (see also: low-temperature spoilage).
— Storage at the right humidity and oxygen or carbon dioxide concentration (CA-storage), whereby the assimilation processes are slowed down. The oxidation will be incomplete if the oxygen content in the storage room is too low and ethyl alcohol will develop in the fruit tissue. The result is an off-taste.
— Harvesting fruit unripe. The fruit is then firmer, easier to transport and able to be stored longer. The determination of the harvest time of unripe fruit is difficult because the ripening process has had to have started in the fruit. In fruit harvested too ripe the keeping quality is diminished, while in fruit harvested too unripe insufficient taste develops. Unripe fruit can be ripened more quickly by storing it with ripe fruit. Ripe fruit produces ethelene, which speeds ripening. In the home, for preference not fully ripened stocks of fruit should be stored in cool, dark, not too dry surroundings. Ripe fruit is best stored under refrigeration and for as short a period as possible.

21.5 NUTS AND SEEDS

Nuts are dry, one-seeded, non-opening fruits, of which the fruit wall is hard and brittle and also sometimes woody. In daily life there are all kinds of edible fruits which are called nuts because they look nut-like. For instance, the peanut is a pulse, the almond is the seed of a stone fruit. In temperate climates, hazelnuts, walnuts and chestnuts grow.

Consumption
In the Netherlands about 4 kg nuts and seeds per person per year are used (mainly peanuts). The figure in the United Kingdom is about the same.

Nuts and seeds are mostly eaten as a snack, but are also used in main meals. Almonds, pistachio nuts and pine kernels were already in use in ancient Egyptian dishes.

Composition

Nuts are rich in fats, proteins, vitamins and minerals. On average they contain per 100 g edible product (source: *Nederlandse Voedingsmiddelentabel* (34), 1983):

> water about 6%
> protein about 14%
> fat 40–70%
> carbohydrates 7–32%
> minerals: calcium, iron and potassium
> vitamin B

The oils in the nuts and seeds mostly contain many poly-unsaturated fatty acids, but coconut fat contains a fair amount of saturated fatty acids.

Kinds of nuts and seeds

Almonds are seeds which grow on the almond tree especially in the Mediterranean area. There are two varieties: the bitter and the sweet.

The *bitter* kind contains a bitter compound. This can be broken down by enzyme systems into, among other substances, the very poisonous hydrocyanic acid. Because of this, bitter almonds must not be eaten raw. On heating, the hydrocyanic acid is decomposed and becomes safe. This kind is especially cultivated for almond production.

The *sweet* almond is suitable for consumption and is sold unshelled or shelled and sometimes roasted and salted. They are also for sale as flaked almonds and in the ground form, mixed with sugar as almond paste or marzipan.

Cashew nuts: the kidney-shaped cashew nut grows as an appendage on the cashew apple, a stone fruit. They originate in Brazil. At the moment they are also cultivated in Africa (Tanzania and Mozambique) and India. They are mostly shelled for sale, sometimes salted.

Hazelnuts are the seeds of hazelnut tree varieties. The shrubs grow throughout the whole of Europe, but Italy, Turkey and Spain in particular have a large production. They are sold shelled and unshelled. Hazelnuts are used a lot by bakers and by the chocolate industry.

Chestnuts: the edible or sweet chestnut varieties are also collected in the wild by devotees. Commercial production takes place especially in France, the Mediterranean area and North Africa. Chestnuts can be eaten raw, but are mostly cooked in the shell or roasted. They then become more aromatic, softer and mildly sweet in taste. After peeling they can be processed into chestnut puree. They are also candied. Chestnuts contain only about one per cent fat.

The **coconut** comes from the coconut palm. Producing countries include the Phillipines, Sri Lanka and Indonesia. An adult tree gives 30–50 nuts per year and this for about sixty years. In the hollow of the nut is coconut milk, a liquid rich in minerals. Coconut flesh (about 40 per cent) is sold grated and dried, and also processed into coconut bread, a sweet sandwich spread. Santen is the milky liquid which is pressed out of the coconut flesh. It contains water, oily compounds and minerals and is used in the preparation of Indonesian dishes. It is for sale in concentrate form in the Netherlands.

Brazil nuts are the seeds of the brazil nut tree from Brazil. The nuts lie like segments of oranges next to each other in a fruit. They are sold shelled and unshelled.

Pecan nut: these nuts show similarities to the walnut, but are longer in shape and have reddish-coloured shells. The seeds grow on the pecan nut or hickory tree, particularly in the United States and Canada. In America they are very much processed in pastry, ice-cream and stews. Pecan nuts are also candied.

Other names for **peanuts** are: earth nuts, ground nuts, arachide nuts, monkey nuts and katjangs. They are not nuts, but pulses. The beans grow undergound. An annual plant, it is grown mainly in the United States and China. They are roasted for the taste and sometines shelled and salted. Peanuts are also used for the production of peanut butter: roasted and unroasted nuts are ground, after which the mixture is homogenized, perhaps by adding extra oil, honey and malt. Peanut butter is a very smooth product, especially used as a sandwich spread and in the preparation of peanut sauces.

Pistachio nuts come from the Middle East. The seeds have a green colour. They are brined in the shell and then dried in ovens; the shell then splits open. They are used as snacks, but also for the production of ice-cream, chocolate and baker's wares.

Walnuts are fruits from the walnut tree. They are harvested in southern Europe and California, among other places. They are sold shelled and unshelled.

Sunflower seeds are the seeds of the sunflower. The shelled seeds are available fresh or in roasted form.

Pine kernels are the seeds of a kind of pine, originating in the Mediterranean area, America and Mexico. They are shelled and roasted and are particularly often used in the countries around the Mediterranean.

Quality deterioration, spoilage and storage

Unshelled nuts are partially dried after harvesting. Moist nuts go mouldy quickly. There is then the chance of a mycotoxin forming. Nuts should be stored in a cool, dry place. Sometimes the inside is attacked by insects without any indication on the outside.

Do not store nuts in the shell too long: they dry out further and so become less tasty.

Shelled nuts lack the protection of the shell, so they will be more sensitive to drying out, attack by insects and rancidity of the fat. This quality deterioration can be slowed down by vacuum-packaging of the shelled nuts, or storage in a gas-flushed packaging.

Shelled nuts bought loose should be stored dry, dark and airy. This last serves especially to avoid attack by mites. The best storage temperature is around 10°C.

Roasted or baked nuts gradually become soft or tough again. They also lose the lovely smell and taste. They should be bought freshly roasted or baked and should be stored dry after cooling.

21.6 FRUIT PRESERVES

In the preservation of fruit not only is the keeping quality lengthened, but products are also obtained with a different usage; for instance jam, raisins and crystallized fruit.

Preservation methods used on fruit are: drying, pasteurizing, addition of sugar, addition of alcohol and deep-freezing.

Drying

Drying is probably the oldest method of keeping fruit longer and is still a good and cheap method. Dried fruit is especially imported from the Mediterranean area, the United States (California) and Australia. Dried fruits contain more sugars than fresh fruit and supply more energy through this. Also, the contents of vitamins, minerals and food fibre per 100 g product is higher; only vitamin C has disappeared in the drying. When dried fruit takes up moisture again during soaking or cooking, the content of foodstuffs is practically equal again to that of fresh fruit, with the exception of vitamin C.

The drying of fruit is mostly done in the outside air, preferably in the sun (sun-dried), but also in drying tunnels. The moisture content in the fruit decreases during drying to 20–25 per cent and the weight decreases. As a result of this the fruit shrinks. The colour becomes darker. Smell and taste alter somewhat and aromatic compounds can be formed which do not occur in the fresh fruits.

Before drying, the fruit is sometimes treated with sulphurous acid. This is especially applied on light-coloured fruit such as white grapes, apples, pears and apricots. The presence of sulphur prevents the brown colouring which would normally occur during the drying process and slows down the growth of moulds and other microorganisms.

On some types of fruit (such as grapes) a waxy layer is present. This hinders quick drying. To remove the wax layer the fruit is submerged in a warm soda solution (dipping) as a result of which the wax layer disappears and the skin bursts. Not only will the fruit dry faster after this treatment, but the colour is also more stable.

Types of dried fruit

Raisins are dried grapes. The largest producers of raisins are America (California), Greece and South Africa. There are many types of raisins, because all sorts of grape varieties are used.

Raisins vary in colour because the grape varieties used have a white, green or blue colour; but especially also because the grapes are dried without a treatment beforehand or are treated by being dipped and/or sulphured. Black grapes which are dried give a blue to very dark (brown) coloured raisin. If one begins with white grapes, and dries them in the sun, then dark (blue to brownish) raisins are obtained. The white grapes discolour less if they are first dipped and then dried, in which case they become yellow-brown. The white raisins stay yellow if they are sulphured after dipping and then mechanically dried.

Sultanas are raisins without pips, sold blue as well as yellow.

Currants are dried small black grapes, called after the Greek port of Corinth. The seedless fruit is dried in the sun or in the shade. For drying in the shade, the cut trusses are hung on strings back among the currant vines. Currants dried in this way have an especially nice, deep blue colour.

Dates are totally, or partly, dried stone fruit from the date palm, which come especially from Iran and Tunisia. The fruit has a very high sugar content of 50 per cent or more. A light drying is sufficient because it preserves itself in sugar. These half-dried dates are sometimes wrongly regarded as crystallized fruit (see later in this section). The half-dried dates are sometimes glazed (see also later) to prevent drying out during transport and storage.

Figs are the dried fruits of the fig tree and come mainly from Turkey. The figs are gathered after they have fallen off the tree and have shrivelled a bit. They dry for some days in the sun, are pressed flat, and packaged.

Prunes are especially imported from the United States (California), West Germany and Yugoslavia. They are dried in tunnels. They must not contain more than 18–20 per cent moisture for extended storage, but they are then very hard and dry; which is why the moisture content is sometimes brought up to about 24 per cent, before they are sold.

'Tenderized' or pre-soaked prunes contain about 35 per cent moisture. Because these prunes are extra-sensitive to mould spoilage, a preservative is added.

Apricots are especially imported from Turkey and the United States. They are sulphured after picking. Then they are dried for a few days in the sun, after which they are halved or de-stoned, washed and further dried in tunnels. Very rarely there are very dark coloured, unsulphured apricots for sale.

Peaches are imported from Argentina, South Africa and Italy. The drying process and the other treatments follow the same lines as described for apricots.

Apples are imported from Italy and the United States. They are peeled mechanically and divided into pieces. They are then sulphured and dried in tunnels. The sweeter apples are especially suitable for drying.

Pears are imported from South Africa and the United States. They are mostly halved, then sulphured, blanched with steam and dried in tunnels.

Tutti frutti is a mixture of dried plums, peaches, apricots, apples and pears. Sometimes, crystallized coloured fruits are added to enhance the appearance.

Quality deterioration, spoilage and storage of dried fruit
Dried fruit has to be stored in a dry, cool and dark place; preferably no longer than one year.

The discolouration of dried fruit which sometimes occurs during storage is a result of the influence of light.

Sometimes during storage a whitish film appears on the fruit; this is the result of sugar crystallizing out.

Dried fruit has to be packaged in such a way that they cannot be reached by rodents and insects.

Dried fruit can start to ferment or go mouldy when it has become moist.

Pasteurizing
This preserving method is used on whole fruit or pieces, fruit puree and sauce, compotes, fruit desserts and sauces. Pasteurization of these products in cans or bottles is sufficient. Pathogenic bacteria have no chance to multiply at the low pH (about 4.5) of fruit. Heating up to about 80°C also ensures that the fruit does not over-cook (become too soft).

Fruit in cans or bottles are described as fruit (whole, halves or pieces) in:

— water, sweetened or unsweetened;
— light, heavy or extra-heavy syrup;
— lightly, heavily or extra-heavily sweetened fruit juice.

More than one fruit in a packaging may be described as mixed fruit, or fruit salad. Fruit cocktail is composed of at least four different kinds of fruit cut into small pieces.

Fruit puree and sauce is obtained by macerating and sieving of the fruit. Puree is made of one fruit without addition of sugar. In the preparation of, for instance, apple sauce, besides apples a small part of a different fruit and sugar may be added for the taste.

Compote looks like sauce or puree, but a part of the fruit has been coarsely cut.

Fruit dessert is a mixture of sieved and/or pieces of fruit, sugar, water and thickeners and perhaps fruit acids.

Fruit sauce is a mixture, of sauce thickness, which contains fruit, sugar, thickeners and perhaps fruit acids and flavours and dyes.

Fruit or fruit-taste dessert sauce: the first-mentioned sauce contains sugars, water and some fruit; the second has a fruit taste and colour through the use of flavours and dyes.

Addition of sugar

Preserves which are rich in sugar have a sugar content which varies between 55 and 80 per cent. Three assortment groups are described:

— crystallized fruit;
— jams, jellies and syrups;
— syrups (see Chapter 30).

Crystallized fruit

Crystallized fruits are preserved because the fruit — or a part of it such as the peel — is saturated as much as possible with sugars (65–80 per cent). They are eaten as such, or are an ingredient in compotes and fruit salads, puddings, ice-cream and pastry. They often serve as decoration, in meat dishes as well as pastry. They are imported from Italy and France.

Some fruits are, before crystallization, soaked in a solution of calcium bicarbonate and sulphurous acid to decolourize. At the same time calcium pectate forms which gives the fruit more firmness.

Mixed peel, citrus fruit, melon, angelica and ginger are for some time soaked in a salt solution. Through a slight fermentation, the tissue becomes softer, transparent and a perhaps bitter taste disappears. Also, water is extracted from the fruit, which improves the crystallization.

Crystallization takes place in stages, because a too highly concentrated syrup would cause the moisture to be extracted too quickly, so that the fruit would shrivel. At each successive stage the solutions are more concentrated and sometimes contain dyestuff. The fruit reaches in the end a sugar concentration of about 70 per cent.

When finished, sometimes candying or glazing follows. In candying, the product is covered with a thin layer or very fine sugar crystals, for instance dry crystallized ginger. Glacé products have a smooth, glossy sugar layer, for instance crystallized pineapple slices.

Types of crystallized fruit
Some of the types sold are: crystallized pear, pineapple, melon, fig, cherry, citrus fruit and nuts.

Crystallized cherries or *bigarreaux* or French fruit are especially used in pastry or used as decoration.

Sukade is the crystallized peel of the citron, a lemon variety. Orange peel is the crystallized peel of a certain orange variety; lemon peel is also crystallized.

Crystallized nuts, particularly chestnuts (*marrons glacés*), walnuts and almonds, serve as decoration and as sweets.

Cocktail cherries are after crystallizing soaked in an alcoholic aromatic solution and therefore a lower sugar concentration will suffice.

Maraschino cherries are soaked in a syrup flavoured with maraschino essence or soaked in sugar and liqueur.

Not only fruit and fruit parts are crystallized but also parts of plants. On the market are, for example: ginger root (ginger), angel root (angelica stalk), violets, rose petals, mint sprays and mimosa.

Quality deterioration, spoilage and storage
Crystallized products have a nice, clear colour, are firm and supple, smell aromatic and have a good taste.

They have to be stored sealed, because they can attract moisture and can ferment, or they can dry out, so that the sugar crystallizes out.

Jam, jelly and syrup
Jam is made from fruit and sugar which is then usually boiled down, but sometimes is not. The fruit acids and pectin from the fruit form together with the sugar to form a jelly-like thickened mass. The end product is also preserved when it reaches a minimum content of 63 per cent sugar.

Not all fruit contains pectin and acid in sufficient quantity. That is why sometimes pectin obtained from apples or other fruits is added. This also provides a fresh acid taste. Sometimes sugar is partly replaced by glucose syrup, especially in the cheaper types of jam.

The industry does not always start out with fresh fruit. Fruit is often preserved in sulphurous acid or canned. These fruit masses with sulphurous acid (pulp) have to be heated during the jam production, so that the sulphurous acid can disappear. This long, high-temperature heating makes the fruit soft and less aromatic. Also the canned pulp is heated, and so is soft and less aromatic. A much more aromatic jam is obtained by using deep-frozen fruit; however this is more expensive.

After boiling down in open or vacuum kettles, filling and sealing of the glass jars follows. They are mostly subjected to a warm treatment in a pasteurization tunnel. Yeasts and moulds on the jam surface or on the inside of the lid are killed. This pasteurization is always necessary in jams which contain less than 63 per cent sugar;

for instance, halva jam and jam for diabetics. Extra-fruit jam contains more fruit than ordinary jam, while halva jam contains less sugar.

A **jelly** develops on boiling down — with or without pectin — of juice or watery extracts of fruit. Besides jelly, extra-fruit jelly is also for sale, which contains more fruit in the starting product.

In bakeries, coating or baking jelly is used, a product with a jelly-like consistency. This jelly with a fruit taste is composed of sugar and water and such things as thickener and fruit flavours.

Marmalade is produced from one or more citrus fruits. Halva marmalade is also available.

Syrups: apple and/or pear juice is used in the production. This is boiled down, with the addition of sugar and perhaps beet juice (a sugar-containing liquid from the sugar beet), to the thickness of jelly with a sugar content of 70 per cent. Apple syrup, pear syrup or mixtures of these are available. Apple syrup which has been boiled down with beet juice has the description '*rins*' in the Netherlands.

Addition of alcohol

Addition of alcohol to dried and fresh fruit is used on a small scale and in the home. Some examples are:

— dried apricots in brandy ('farmers' girls')
— raisins in liqueur
— various fruit in rum or gin
— sloes in gin (sloe gin).

The product is preserved by a heat treatment if the end product contains less than 15 volume per cent alcohol and has a limited keeping quality after opening.

Deep-freezing

This is used on soft fruit such as strawberries, raspberries, blackberries and currants on a limited scale in industry. After thawing, the fruit is limp and the colour less attractive.

21.7 LEGISLATION

The legislation is usually divided into regulations for fresh fruit and regulations for preserved fruit.

The Ministry of Agriculture and Fisheries gives general regulations concerning technological and quality aspects. The Marketing Board for Vegetables and Fruit may describe the requirements for each product more closely in quality Ordinances. A division is made into quality class extra, I and II. A number of external characteristics are taken note of such as colour, shape, size and soundness. Imported fruit are given a class labelling as well as a name labelling with the mention of the country of origin.

Various other regulations may deal with levels of pesticides and herbicides, preserved products and sauces, their composition and permitted additives.

22

Fungi

22.1 INTRODUCTION

Toadstools are the fruiting bodies of fungi. They form one of the largest groups in the plant world. Only a small number of species are edible, because most toadstools are tough, have a bad taste, or are poisonous. Toadstools are composed of a network of threads or mycelium, which is present in the soil. In the late summer and autumn, during moist warm weather, buds develop on these, which grow out of the soil and into toadstools. In the fruiting bodies there are spores through which the fungi propagate themselves. Most of the fungi disappear when the first heavy night frosts occur.

The green pigment chlorophyll is not present in fungi. That is why they are not able to build up a sugar reserve. Fungi extract these compounds from other organisms or dead material by means of the mycelium. They can do this by living as parasites on other plants or by feeding as saprophytes on rotting plant or animal material.

Most fungi grow in woods, meadows and dunes. The gathering of fungi is not advised, because the edible types are easily confused with the poisonous types. For the consumer, a number of cultivated types are sold, of which the mushroom is the most well known. In the last few years the oyster mushroom has also gained in popularity.

Consumption
In the Netherlands about 1.9 kg mushrooms are eaten per person per year, fresh as well as preserved. The UK figure is about 1.5 kg. For the other edible fungi no consumption figures are known.

Composition
The mushroom contains roughly the following compounds:

water	90%
protein	3%
fat	0.5%
carbohydrates	1%
B-vitamins	traces

minerals, including phosphorus

Possible health hazards
A small group of fungi contain poisonous compounds. Perhaps the most well-known is the death cap, which looks like the meadow mushroom, and of which the poison is deadly. The poison works on the central nervous system, and results in cramps and heart palpitations.

Fungi which grow in the wild can contain heavy metals (lead and cadmium), and other environment and air pollutants.

22.2 KINDS

Fungi are divided according to the way in which they produce their spores:

— **Basidiomycetes**, in which the spores are situated on a stalk on a club-shaped endcell;
— **Sac fungi** or **Ascomycetes**, in which the spores are formed in long stretched cells.

The Basidiomycetes include the mushroom, the chanterelle, the oyster mushroom, the shii-take, the shaggy ink cap, the blewits and the ceps. Depending on the structure of the cap, in which the spores are, the Basidiomycetes are divided into *gill fungi*, which are recognizable by the thin plates or gills on the underside of the cap and *hole* or *tube fungi*, which look sponge-like on the underside of the cap and have many holes. The sac fungi include the truffles and the morel.

The **field mushroom** exists in the wild as well as in the cultivated form, in many varieties. The wild types are found in meadows where horses are kept, or have been kept, and in woods.

In the Netherlands mushrooms used to be cultivated on the ground in compost beds in the marl pits in South Limburg, where there was an average temperature of 10°C and it was very moist. As a soil covering, marl was used. Marl or grotto mushrooms are still sold on a limited scale. They are beige-brown in colour and have a thickening on the end of the stalk. They shrink less than the common mushrooms and have more taste.

Since the 1950s, mushrooms have been cultivated in special cultivation chambers, which consist of well-insulated units, supplied with heating, ventilation and the correct humidity.

In the cultivation chambers, troughs filled with fermented horse dung or compost from compost producers, mixed with mushroom spawn, are stacked above each other. The spawn consists of a pure culture of mushroom mycelium which has been put on sterilized cereal grains. If the mycelium after two weeks has largely grown through the compost (temperature 25°C), the troughs are covered with a layer of cover soil, which functions as a water buffer so the compost is protected from drying out. The temperature of the compost is reduced to 16–18°C. About three weeks after the covering, the first mushrooms can be harvested. The mushrooms however do not appear continuously, but in weekly flushes. After four to five harvest weeks the cultivation is stopped, because the flushes become smaller and smaller. The total harvest from one cultivation, which lasts for about 12 weeks, can on average amount to 22 kg per m^2. The cultivation chamber is sterilized by steaming and emptied at the end of the harvest, after which a new cultivation can be started.

A new development in mushroon cultivation is the use of compost which has already been inoculated with mycelium at the compost producers. This means for the grower a shorter but labour-intensive cultivation. The cultivation lasts, in this case, nine weeks.

The largest part of the harvest (70 per cent) is processed into preserves, which are exported mainly to West Germany. The rest is sold loose or packaged in boxes.

Besides small mushrooms, giant mushrooms with a cap of 3–10 cm diameter and a firm stalk are also cultivated.

Fig. 22.1 — Mushroom cultivation.

The **chanterelle** is an egg-yellow-coloured fungi with a funnel-shaped cap and a short firm stalk. The gills underneath the cap look like fat pleats. Chanterelle cannot be cultivated. They are found in woods in the autumn. They are imported both fresh and preserved from eastern Europe and Canada. The taste of chanterelles is rich and somewhat peppery.

The **oyster mushroom** grows in the wild like a roof tile on the trunks of trees and stumps. The cap of the fungus is 5–20 cm in diameter and has an oyster-shell shape with a pleated edge. The colour of the cap is blue-grey to brown-black, depending on the kind. In most countries only cultivated oyster mushrooms come to the market. The cultivation takes place on wet straw bales, which are inoculated with spawn (wheat grains with mycelium from the oyster mushroom). The straw bales are packed in plastic bags which are not completely closed, to provide aeration. Because oyster mushrooms, in contrast to common mushrooms, need much light, the bags are removed after three weeks and there are already small buds visible along the vertical side of the straw bale. Within 14 days the oyster mushrooms are full grown. Oyster mushrooms appear mostly only in three flushes. They are sold loose or in boxes of 200 g.

The **shii-take** is a fungus originating from Japan. It is cultivated on trunks of deciduous trees. The fungus has a light brown cap of 6–12 cm diameter. This is initially bell-shaped. Later it has the shape of an upside-down saucer. The short stalk and the lamellae are white to brown. It is imported in the dried form and used mainly in Chinese restaurants. Recently, growing these fungi in cultivation chambers has been started in some European countries and they are now available fresh.

The **shaggy ink cap** has an elongated scaly cap (length about 12 cm) with white gills and a hollow white stalk of about 20 cm. The white gills liquify in older specimens into a black dripping mass. This process starts at the edge of the hood. Ink caps grow in the wild, but can also be cultivated in the same way as mushrooms. Ink caps are available preserved in jars. The taste of the fungus is very delicate.

The **blewit** has a cap of 6–12 cm diameter with the shape of a turned-over saucer, and a stalk of about 10 cm. The flesh of the fungus is violet in colour with a sweetish smell and taste. Blewits grow in coniferous and deciduous woods. The cultivation of blewits is still at an early stage. The fungus is sold on a very limited scale in the preserved state.

The **boletus** or **cep** has a glossy brown-coloured cap with a diameter of 5–25 cm and a fat, bulging, light-coloured stalk with surface reticulation. The underside of the cap is light yellow to green (in older specimens) in colour. The flesh is white in colour with a nutty taste. Ceps grow in the wild in deciduous and coniferous woods. Dried ceps are available.

Truffles grow underground on the roots of trees, particularly oak and beeches. They are more-or-less round in shape without a stalk and can be black or white in colour.

Truffles cannot be cultivated: they must be gathered in the wild. From October to the end of January black truffles are found with the help of specially trained dogs or pigs in the French area of Perigord. White truffles are found in the Italian Piemonte area also from October to January. Here, too, they make use of specially trained dogs to find the truffles.

Because of their aroma and lovely taste, truffles are used in luxury dishes. They are also used in pâtés for the decorative element (black colour).

Because the truffle is very expensive an attempt has been made to produce a substitute. This copy or imitation truffle is a firm, black-coloured paste which contains animal fat, broth and salt.

The **morel**, a fairly small fungus (5–10 cm), can be recognized by its pointed cap

Fig. 22.2 — Chanterelle.

Fig. 22.3 — Oyster mushrooms on straw substrate.

which has a spongy structure with a maze-like surface. The colour of the cap is greyish-yellow; the stalk is yellow-white and hollow inside. In the spring, the morel is found on chalk-containing sandy soil (dunes). The morel is regarded as a very fine fungus and is available canned and dried.

Fig. 22.4 — Shii-take.

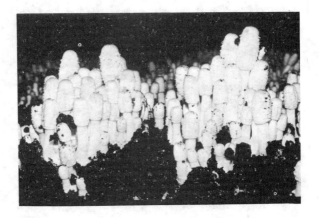

Fig. 22.5 — Shaggy Ink Cap.

22.3 PRESERVED FUNGI

Fungi are available, preserved in cans or jars, for instance whole, sliced or diced mushrooms. Truffles can be sterilized whole, peeled or unpeeled, or cut into pieces. The truffle peel is also available preserved. It is too tough to eat, but very suitable for flavouring dishes with the lovely truffle taste.

Several fungi are available in the dried form, for instance chanterelles, shii-take and mushrooms. Dried fungi or parts of them are also processed into soup and sauce powders.

22.4 QUALITY DETERIORATION, SPOILAGE AND STORAGE

Fresh fungi should have a sound exterior without dark lamellae. There should be no insects or other parasites in them and the stalks should be free of earth. Mushrooms

Fig. 22.6 — Truffles.

which comply with the above requirements may be divided into three classes at the auction. Closed mushrooms are graded Class I; specimens which are on the point of opening are graded Class II, Class III consists of opened mushrooms.

Fungi deteriorate quickly during storage. Under the influence of light, warmth and oxygen, discolouration occurs and the caps open up. The fungal flesh becomes tough and limp as a result of drying out, while moist storage can lead to mould formation.

Fresh fungi can best be stored in a cool and dark room and well packaged. The storage time in the refrigerator is about one week. To improve the keeping quality, mushrooms are sometimes irradiated. The caps stay closed longer as a result of this. Preserves of fungi can be stored for a year or longer. To prevent discolouration of dried fungi and fungi in cans or jars, colour preserving additives are added (e.g. sulphurous acid).

22.5 LEGISLATION

There may be regulations concerning the addition of colour or preservatives to processed fungi.

Quality classes and size description for mushrooms may also exist. On the basis of the quality rules for fresh mushrooms and the diameter of the cap the Extra Quality, Class I or 1st quality, and Class II or 2nd quality are determined. It is also required to mention the drained weight on products which are in a liquid.

23

Herbs and spices

23.1 INTRODUCTION

'Herbs and spices' describes those plant parts which are especially valued because of their taste, their aroma, their medicinal or preservative action and/or colour.

The name **herbs** is more specifically applied to all herbaceous plants of which mostly the **fresh** leaves are used.

Spices are the aromatic or sharp-tasting **dried** parts of plants, including root, fruit, tree bark or leaf. They are of tropical origin and are used in small quantities in food preparation and by the food processing industry.

In prehistoric times, and also in the time of the ancient cultures, the borderlines between plants which were used for food (vegetables), or as herbs in medicine, or as beautifying aids were not sharp. About 4000 years ago the medical application of herbs such as marjoram and mint had already been described. The use of spices such as cinnamon and cassia (Chinese cinnamon) and of creams and ointments in which volatile oils were used also originates from this early period.

From the Far East and southern Arabia, spices and aromatic compounds were transported to Egypt. The Phoenicians supplied the Greeks with spices and aromatic types of resin. The Greek doctor Hippocrates (about 400 BC) had 400 medical remedies based on spices described in books. During the peak of prosperity of the Roman Empire an extensive trade in herbs and spices was started. From Asia, spices like cinnamon, cardamon, cloves and pepper were brought to Constantinople.

From the twelfth century onwards, the views and thoughts on health and sickness changed. Epidemics were no longer the result of bad air, which could be got rid of with the scent of incense and spices. More interest was shown in everyday food. Especially in the monasteries, the cultivation and use of herbs and vegetables was stimulated. In the beginning, the cultivation for the most part was done in monastery gardens. Later, more and more people started cultivation. The demand for herbs became so great around 1800, that they were cultivated on a larger scale to supply the demand from doctors, pharmacists and others. The still-famous gardens such as the Leidse Hortus Botanicus, Jardin de Plantes in Paris and the Botanical Garden in Oxford originate from this period.

Applications
The present-day applications of herbs and spices are manifold.
— Medical science makes use of the **medicinal compounds** from herbs and spices. For instance: eucalyptus oil has a slight expectorant action; valeric oil acts on the central nervous system; and quinine has an anti-microbial action. Medicinal spices, such as were employed in the herbal methods of olden days, are once more employed in homeopathic and anthroposophical healing.
— Some herbs contain compounds with a **preserving** action. For instance: marjoram, nutmeg, pepper and paprika act as anti-oxidants; and the oils in cinnamon, cloves and garlic have germicidal characteristics. Extracts from species of the onion genus contain anti-microbial agents and garlic and onion juice have a retarding effect on the toxin production of *Clostridium botulinum*.
— Herbs, and extracts or compounds from herbs, find a use in the cosmetic and perfume industry because of the **aromatic compounds** which they contain, or because some compounds have a **beneficial action** on the skin, for instance camomile.
— In the food and drink industry and in the kitchen, herbs and spices are mainly used to give a typical smell or taste to products or dishes or to enhance their appearance. The **hedonistic value** of a product will be increased by aroma and smell compounds, and through a varied use of herbs and spices more variation in smell and taste is possible.

Fresh herbs
In northern Europe herb cultivation takes place in open soil or under glass. The assortment of fresh garden herbs, distributed by way of the vegetable auctions, is very limited. The most commonly traded leaf herbs are parsley, celery, chives, cress and chervil.

The consumer will have to grow his own for a more varied use of herbs.

Use in the kitchen
For preference, herbs are used fresh. The leaf herbs have to be fresh green or firm; root and bulb plants not shrivelled and unsound. In the case of a number of herbs, the smell and taste are at their best if they are added, as fresh as possible, at the last moment to the dish; for instance, parsley, chives, dill, chervil and watercress. Other types have to cook with the dish to liberate the aroma; for instance, lovage, rosemary, marjoram, celery and thyme.

If, in everyday food, a generous portion of fresh herbs is used (for instance, of parsley, chives or cress) a contribution can be made to the requirement for vitamin C, β-carotene and the minerals iron, calcium, sodium and phosphorus.

Preserved herbs and spices
Herbs can only be kept fresh for a short time after harvesting, and most of them are not available during the winter months. A number of preserving techniques are used for fresh herbs, among others deep-freezing, canning and drying.

Spices are always dried in the country of origin before they are exported.

Deep-freezing is especially suitable for the leaf herbs such as parsley, chives, celery and chervil. The fresh herbs are selected, washed, finely chopped, packaged in portions and frozen. The storage times and the use of deep-frozen herbs are the same as for fresh deep-frozen vegetables.

The **canning** of fresh herbs or spices by means of heat treatment (sterilization) is used for chervil and green peppercorns. The keeping quality is the same as for canned vegetables.

Drying is the method most used for leaf herbs, seeds and spices. This is done in drying chambers or in the open air (tropical or subtropical climate) depending on the place of harvest. During the drying process, appearance and taste often change considerably, especially under the influence of enzymatic processes. Sometimes the enzymic changes are stimulated; in such fermented products changes in appearance and taste are most noticeable, as for instance with vanilla pods, cloves and black peppercorns.

Use in the kitchen
The quantities of dried herbs and spices added are usually small. The nutritional value of the dish will be scarcely altered. The dried herbs and spices, however, are important in food production. The smell and taste components which are transferred into the dish activate the secretion of digestive juices. Through this, the appetite is stimulated and digestion aided.

Dried herbs and spices (not crumbled or powdered) are mostly cooked with the dish. Some keep their delicate smell and taste better if they are added in ground form just before serving the dish, for instance nutmeg.

23.2 FLAVOUR COMPOUNDS IN HERBS AND SPICES

The smell, the flavour or the bitter taste in herbs and spices is caused by compounds of very different type. The aroma of the dominant taste can be the result of the following.

— **Volatile oils**: These are volatile compounds (aldehydes, alcohols, esters, acids and terpenes) which give a strong smell and are difficult to dissolve in water. Each type has a characteristic aroma. They can occur in all plant parts. For instance: in cinnamon, nutmeg, dill, peppermint and vanilla.
— **Alkaloids** (among others capsaicin and piperine): These compounds have mainly a sharp taste. They determine the smell in paprika and types of pepper (chillis, white and black peppercorns).
— **Sulphur-containing esters**: These compounds are broken down by the enzyme myrosinase into, among other things, mustard oil, glucose and sulphuric acid. Through this, the herbs obtain a very characteristic smell and a sharp taste. Mustard oil is a volatile oil, but is also sulphur-containing and has, because of this, a characteristic smell. Sulphur-containing esters occur in horseradish, capers and black mustard seed.
— **Sulphur-containing amino acids**: These are divided by the enzyme allinase into, among others, ammonia and allyl sulphuric acid (a lachrymatory). In garlic, allicin (typical garlic smell) can develop from this. The sulphur-containing amino acids occur in onions, shallots, chives, spring onions and garlic.

- **Acid-tasting compounds**: Some plant parts contain organic acids such as iso-citric acids or oxalic acid; for instance: sorrel and wood-sorrel.
- **Bitter-tasting compounds** of very diverse types. For instance: the bitter compounds in chicory, hop, bitter almonds and thistle types.
- **Several other compounds** which cannot be easily classified. For instance: the pleasant, nutty taste in the leaf of salad burnet, or the typical sweetish smell of lady's bedstraw.

23.3 TYPES: DIVISION OF HERBS AND SPICES

The division of herbs and spices into groups is dependent on which criterion is adapted.

It is possible to group herbs and spices according to smell and taste components (see section 23.2). However most of them contain several smell or taste compounds with often one predominating.

In the herb healing science, a botanical division is sometimes made, into families which have the same healing action. For instance, the Umbelliferae (cumin, aniseed and dill) of which the seeds are used, contain compounds which have a beneficial influence on the stomach and digestion.

For use in the kitchen, a division according to plant part used (as is also customary for vegetables) seems more appropriate (see Table 23.1 — herbs, and Table 23.2 — spices).

Herbs
Types of herbs of which the leaves are used

Basil is used fresh as well as in the dried form. It goes with soup, sauces, veal, fish and tomato dishes.

Chives is a member of the onion family. The long, thin, tubular leaves are only used fresh or deep-frozen in salads, potato dishes and cold sauces.

Savory is used fresh as well as dried in the cooking of broad beans and other legumes.

Borage, because of its taste, is also called cucumber herb. Dried borage loses practically all its smell. It is only used fresh, in salads and cabbage dishes.

Lemon balm can only be used fresh, in salads, in marinating fish and in cooking fish.

Dill: the leaves are used fresh in salads and in sauces which are served with fish dishes or white meat. The umbels and seeds are used in the pickling of gherkins and cucumbers.

Tarragon is used fresh in the preparation of, among other things, tarragon vinegar, and is used in salads, marinades and cold sauces.

Hyssop has a fairly bitter taste and is never cooked with a dish. Fresh leaves are added to meat sauces.

Chervil has a very fine soft leaf. Bunches of chervil have to be cut or snipped and are added to soups, cold sauces and salads.

Mint has a light peppermint taste. Mint sauce goes with lamb and mutton. Mint is processed into sauces. In the dried form a tea-like extract is made with it.

Table 23.1 — Types of herbs: division according to plant part used

Leaves	Flower umbels, fruits, seeds	Roots, bulbs, tubers	Flower buds
Basil	Aniseed	Angelica root	Capers
Chives	Dill	Garlic	
Savory	Juniper berry	Horseradish	
Borage	Carraway seed	Onion	
Lemon balm	Cumin		
Dill	Coriander		
Tarragon	Mustard		
Hyssop	Fennel		
Chervil			
Mint			
Lovage			
Marjoram			
Peppermint			
Parsley			
Salad burnet			
Rosemary			
Sage			
Celery			
Cress			
Thyme			
Fennel			

Table 23.2 — Types of spices: division according to plant part used

Flowers, arils, seeds, berries	Bark	Leaves	Fruits	Roots
Mace	Cinnamon	Bay laurel	Paprika	Turmeric
Cardamon			Sweet peppers	Ginger
Cloves			Vanilla	
Nutmeg				
Allspice				
Pepper				
Saffron				
Star anise				

Fig. 23.1 — Several herbs.

Lovage or **maggiplant** has a strong taste and is very suitable for the making of broth. Its strong taste overpowers others quickly.

Marjoram has a strong taste and is used fresh as well as dried, in soups, brown sauces, beans and tomato dishes.

Peppermint is added fresh or dried in the cooking of cabbage and Brussels sprouts.

Parsley: there are plain-leaved and curly-leaved varieties of parsley. They are both suitable to be added fresh to soups, sauces, fish dishes and potato dishes. The curly-leaved parsley is especially used for the garnishing of dishes. Parsley loses a lot of its smell and taste on drying.

Salad burnet is used fresh in salads and cold sauces because of its cucumber-like taste. It is also used in herb vinegar.

Rosemary, of which the dried leaves look like pine needles, is especially used for meat, game and poultry, but also for fish dishes and salads.

Sage, with its silver-grey leaf, is suitable for fish and meat sauces or in the preparation of sage milk (stand leaves for 15 minutes in hot milk).

Celery is related to the parsley plant, but is distinguished from it by its typical celery smell and the darker green and the coarser cut of the leaf. Leaf celery is used for the making of stock and in soups and sauces.

Cress is always available fresh nowadays, sown in small boxes. It is suitable for salads with grilled dishes and in sandwiches. It is also used for garnishing dishes.

Thyme has a pronounced taste and is suitable for adding fresh or dried to meat, game or poultry dishes. Thyme is also nice with tomato dishes and in broths.

Fennel: the leaf of fennel is used as a herb in salads, the root as a vegetable with an aniseed taste and the seed umbel for pickling gherkins and cucumbers.

Types of herbs of which the flower umbel, fruit or seeds are used

Aniseed: from the seeds, the volatile oils are extracted and added to sweets (aniseed balls). The seeds are also surrounded with a sugar coating or in the ground form are a component of 'koek' spices (see section 17.5).

Juniper berries are, among others, used in marinades for game dishes, in the sauerkraut production and also in the smoking of meat.

Carraway seed (or **kummel**) is used in the preparation of goulash and white cabbage salads, and is a component (ground) in 'koek' spices and in some liqueurs.

Cumin is somewhat similar in taste to carraway seed. In the Netherlands it is used in cheese production and in the sausage-producing industry.

Coriander is a very aromatic seed that (mostly in combination with other herbs) is used in meat dishes, sauces and marinades, and also in sweets and liqueurs.

Mustard: There are brown, white or yellow mustard seeds. The seeds contain a high percentage of mustard oil; the seed after grinding becomes paste-like or wet because of this oil. The fat has, for the main part, been removed from dry mustard flour. Mustard flour or powder is very little used in the Dutch kitchen; however, wet mustard is used (see section 23.4). The whole seeds are used in pickling onions.

Types of herbs where the root parts are used

Angelica root is used in the preparation of liqueurs and pastry. The stems of this plant are sold crystallized as angelica stems and serve as a decoration on pastry.

Garlic is related to the onion family and has of them the most pronounced taste (very dominant). It is used fresh or as a powder and added to salads, herb butters, cold sauces and fresh cheese. It has many uses in the southern European, Turkish and Moroccan kitchens and in Indonesian and Chinese (meat) dishes.

Horseradish has a very sharp taste which is similar to that of radishes. It is especially used grated in horseradish sauces and served with meat, fish and cheese dishes.

Onion is used fresh in the preparation of vinaigrette or served with salt herrings. It is also added in the preparation of stocks.

Types of herbs where the flower buds are used

Capers are blue-green flower buds, which are often bottled with (tarragon) vinegar. They originate from the south of France or North Africa. Capers are used in sauces and salads, and are particularly good with fish dishes.

Spices
Flowers, arils, seeds and berries which are used as spices

Mace is the seed covering of the nutmeg. Harvesting takes place when the fruit (peach-size) bursts open. Under the fruit flesh, the mace can be found around the nut. Good quality mace has an orange colour and comes from Indonesia. It is used in broths, meat and game dishes and meat products (see also nutmeg).

Cardamon is the seed of the cardamon plant from India and Sri Lanka. The seeds have a penetrating taste and are available whole or ground. The powder is processed into spice mixtures for meat, sausage and 'koek' (pastry).

Cloves are harvested as flower buds from the clove trees on the Moluccan Islands. The red flower buds are dried on mats in the sun after harvest. They have a high volatile oil content and are sold whole or ground. They are used, for example, in the preparation of red cabbage, meat and game dishes, biscuits and meat products.

Nutmeg is the seed of the nutmeg tree. The seed remains after the mace has been peeled off. The seeds (nuts) are dried and sometimes white-chalked to prevent pests nibbling them. The nuts are also available ground and are used in the preparation of vegetables, meat, game and sausages.

Allspice is a brown- to black-brown-coloured berry which grows on the allspice tree in Jamaica and Mexico. The taste is a mixture of nutmeg, clove, pepper and cinnamon. It is used whole or ground in meat preparation and in sauces.

Pepper is the ripe or unripe harvested berry from the pepper tree, on which the fruits grow in trusses.

White pepper is obtained if the red berries are harvested ripe and then are soaked in (sea) water until the shell can be removed (about seven days). The grey-white corns are then dried and perhaps ground.

The unripe berry is green in colour and has a less strong taste than the ripe pepper berries. *Green pepper* is for sale in cans or in jars or is processed to obtain *black pepper*. The picked green berries are piled in heaps after harvesting; they become wrinkled and black in colour through the action of enzymes. After several days of fermentation the fruits detach from the trusses. They are spread out in the sun to dry. Black pepper has more smell and aroma than white pepper.

Saffron is the dried stamen of a certain kind of crocus from Spain and the south of

France. 70 000 flowers are needed to obtain 1 kg saffron. Saffron is used to give dishes, for instance rice, a yellow colour and a fine smell and taste.

The seeds of the **star anise** appear in so far as taste and looks are concerned like the aniseed of the herb which grows in our climate area. This aniseed gets its name from the star-shaped, tropical fruit in which the seeds are situated. The extracted volatile oils are used in, among others, soy sauces.

Bark which is used as spice

Cinnamon is the inside bark of the branches of the cinnamon tree, which grows in Sri Lanka, Java and China. The branches are peeled, the bark is rolled and dried in the air until it colours brown. The word 'cinnamon' means 'little pipe' (*canna*). Cinnamon is used in the preparation of stewed fruit, red cabbage, fruit juices and, in the ground form, in pastry and 'koek' (component in 'koek' spices).

Leaves which are used as spice

Bay leaves are the dried leaves from the bay tree. They are imported from Italy and Turkey. The dried leaves should be pliable. The leaves are used in the making of stock and the preparation of meat, game and fish dishes.

Fruits which are used as spices

Paprika is the dried, ground fruit of an herbaceous plant from Hungary, Bulgaria, Rumania or Spain. The seed can be removed, partly removed, or not removed at all before grinding, which produces paprika powders with different tastes, for example Edelsüsz (mild taste, no seed), strong (a little stronger in taste, few seeds) and Rosen (very strong taste, a lot of seeds). Paprika is related to the sweet peppers. It is used in the preparation of meat dishes (goulash).

Sweet peppers are related to chilli peppers. The sweet peppers are green to red fruits with a strong taste. There are many varieties. They are used fresh, but also dried and sometimes when ground, processed into kinds of sambal. *Cayenne* or *chilli pepper* is a small pepper with a very hot taste. Ground chilli peppers (or Lombok-peppers) form the basis of sambals and chilli sauce.

Chilli powder is a mixture of ground chilli peppers, cumin, oregano, garlic and other compounds and is used in the Mexican kitchen. In the Surinam kitchen a chilli pepper variety 'Madame Jeanette' is used.

Vanilla is the unripe harvested bean of a climbing orchid. It grows in Java, Mauritius, and Bourbon. After picking, the pods are left to ferment for a few days before drying. The black vanilla pods often get a burnt-in quality stamp before being packaged.

Roots which are used as spices

Turmeric is the ground dried powder of a plant which mainly grows in Indonesia and India. Ground turmeric has a brilliant yellow colour and is the main ingredient in curry powder. It is used as a substitute for saffron and serves as a taste and colouring agent in the Indonesian and Indian kitchen.

Ginger is the dried ginger root ground into powder. The ginger plant grows in China, Japan, Jamaica and Africa. Ginger powder is often used in the Indonesian and Chinese kitchen. Sometimes the root is cut into pieces and crystallized (see section 21.6).

23.4 PRODUCTS

Herb and spice mixtures
Dried herbs and spices may sometimes be sold cut or ground and mixed with salt and/or glutamates. A mixture or blend of several spices and/or herbs often indicates its use through its name. The spices are often sold under such names as the following:
— bami and nasi spices
— frying and grilling spices
— minced meat spices
— pickling spices
— curry spices
— chicken spices
— herb bags
— 'koek' spices
— cooks' spices
— mixed spices
— fish spices
— meat spices
— sausage spices.

Some herb and spice mixtures contain salt. Special products are for sale for a low-sodium diet.

Products
In herb vinegar, herb butter, herb jelly, herb teas and herb liqueurs, the taste is produced through the herbs or spices used. In sambals, mustard and tabasco they are the most important basic compounds.

Sambals consists of ground fresh red chillies to which several kinds of other herbs and spices are added.

Tabasco is a piquante sauce based on chilli peppers.

Mustard is a paste prepared from a mixture of mustard seed, vinegar, herbs and often salt. Mustard used to be made by mixing the mustard seed and must (fermented grape juice) into a piquante sauce. Dijon (France) obtained in 1634 the exclusive right to make Dijon mustard. Dijon mustard is still a protected and defined kind which is made with wine or fruit wine instead of vinegar.

Mustard production is mostly done in small factories, in which the seed is ground in the mustard mill sometimes with the help of wind power.

Labelling for the different kinds of mustard may be connected with the herbs used (French, Dutch) and/or with the area where the factory is situated.

Herb extracts
Extracts can be produced from spices and herbs by such methods as steam distillation, biotechnical processes, pressing or with the aid of extraction liquids such as ethylene-di-chloride.

The **natural extracts** so produced can be chemically synthesized. These aromatic compounds or **essences** have their use in improving the smell and taste of foodstuffs.

In industry there are advantages in the use of herb extracts and/or essences: they are microbe-free, can be added very accurately and can be processed homogeneously in the end product. Disadvantages are that the extracts are relatively expensive and that residues from the extraction liquid can be found in the end product.

23.5 QUALITY DETERIORATION, SPOILAGE AND STORAGE
Fresh herbs
Harvested garden herbs remain a live product. The metabolic processes can be slowed down by correct storage surroundings and temperature. A deterioration in quality of the fresh product, however, cannot be avoided. The most noticeable changes which occur after harvesting are:
— *loss of moisture:* the herb becomes limp and dries out;
— *loss of aroma and smell:* the volatile oils disappear and with them part of the characteristic smell. Enzymatic processes can also slowly start which can affect the smell, taste and appearance of the leaf.

Storage advice
For use in the kitchen, herbs should be bought or harvested fresh each day. Leaves can be kept for a few days if the stalks are put in water. Roots and tubers can be kept for several weeks or longer, dependent on the storage circumstances. A dry (but not too dry) cool and dark place is usually desirable.

Dried herbs and spices
During the drying process microorganisms and insects can develop in large numbers. Regulations may set out rules on the level of infestation by insects and the related contamination with microorganisms. To reduce the number of insects and microorganisms, the following are used.
— *Ethylene oxide* to reduce microorganisms. Ethylene oxide is poisonous and can form, by combining with chlorine in products to which it is added (including sodium chloride) the toxic ethylene chloro hydrin.
— *Methyl bromide* to destroy insects. This is a poisonous compound.
— *Ionizing rays* through which insects, their larvae and eggs are killed. Depending on the dosage, microorganisms can also be killed. Irradiated herbs do not have to be labelled as such. But permission to irradiate is necessary from the appropriate Ministry.

During transport and storage the chance of deterioration in quality of dried herbs and spices is still great. They are sensitive to:
— discolouration
— taste and aroma loss
— moisture uptake
— mould formation
— oxidation processes under the influence of light
— insect attack.

Comminuted herbs and spices, which are labelled as broken, ground, powder or flour, are extra-sensitive to the forms of quality deterioration mentioned.

Storage advice
A good, adequate packaging and storage place will slow down the deterioration in quality. This means: storage in tins, glass or plastic foil in a dark dry place.

23.6 LEGISLATION

Regulations may exist to describe individual spices or mixtures of spices and the composition of mustard, taste enhancers and flavours. Labelling requirements and levels of contaminants will also be outlined.

24
Salt

24.1 INTRODUCTION

Salt is the collective name for a large group of chemical compounds which include chalk, gypsum, magnesium and ammonium salts and also kitchen salt. By salt, in ordinary speech, is meant kitchen salt or sodium chloride (NaCl). Sodium is interesting from a nutritional standpoint (there is 1 mg Na in about 2.5 mg NaCl). Salt is used:

— as a taste maker;
— to improve the structure of foodstuffs (for instance in bread production, see section 17.2);
— to increase the keeping quality (for instance in cheese, see section 11.2);
— as an additive in the technological processing of products (for instance in processed cheese, see section 11.6).

Consumption

According to estimates, salt consumption is eight to nine gram per person per day. The body only needs 1 g per day. Three percent of the quantity of salt which we take in originates from rain water, 17 per cent occurs naturally in the food, 30 per cent we add to the food during preparation, and 50 per cent reaches our body through industrially produced products. Bread, meat products, cheese, soups, sauces, tasty snacks, cakes and pastry especially contribute an important part to the sodium supply. Sodium is also added in industry as, for example, a component of preservatives (sodium benzoate or sodium nitrite) and as a taste enhancer (sodium glutamate).

24.2 PRODUCTION AND DISTRIBUTION

Salt is found in the soil, in sea water and in spring water.

Production from the soil
Salt layers have formed in the Earth's crust from seas which have been enclosed and have then evaporated, through which the salts have crystallized out. Owing to earth movements, the layers have disappeared deep into the earth. In the Netherlands this is the case in the neighbourhood of Hengelo and Boekelo (and in the UK in the Cheshire area). The layers here are at a depth of 400 m. The salt deposits (also called stone salt) can be extracted in two ways. If the layers are very thick, the deposit can be mined. Another method is to free it with water. Water is pumped into the ground and the salt dissolves in it. The brine is pumped up and this is cleaned (refined). After this, the water is evaporated, and the salt crystallizes out. The slower the evaporation the finer the crystals formed.

Production from sea water
Sea water contains dissolved salt. By evaporating the water with the aid of the heat of the sun and wind in large vessels or salt-pans the crystallized salt remains. This way of salt production is mainly seen in tropical areas such as the Antilles.

Production from spring water
From spring water, salt can be obtained through artificial crystallization. This occurs, for example, in salt-works in the area of Salzburg (Austria).

After the recovery of the salt from the soil, sea water or spring water, it is packaged in plastic, cartons and strong paper and distributed to wholesalers or retailers.

24.3 KINDS OF SALT

Kitchen salt is refined and is available with or without iodine. It may contain additives to ensure that it remains free-running.

Sea salt is unrefined salt which is sold in the wholefood shops. It still contains residues of different minerals, for instance potassium, iodine and iron. At the same time there can also be residues of heavy metals such as lead, mercury and cadmium, originating from refuse which is dumped in the sea.

Table salt is produced as very fine crystals and remains more free-running because it contains special additives. Like kitchen salt, table salt may contain iodine.

Bread salt is iodated salt which is mostly used in bakeries.

Pickling salt is used in the production of meat products for the colour formation (see section 5.7). It consists of kitchen salt with a small quantity of nitrite.

Aromatic or **herb salt** contains, besides salt, dried and ground herbs or spices. The salt content is a minimum of 50 per cent. Most of the mixtures also contain anti-caking agents. Sometimes a taste enhancer has been added.

Smoke salt is smoked on top of a fire of nut wood and acquires a typical taste. It is for sale under the name Hickory Salt and is used on grilled meat.

Mineral salt contains 40 per cent less sodium than kitchen salt, but contains other minerals including potassium and magnesium.

Gomasio is a mixture of sea salt with roasted and ground sesame seeds.

Diet salt contains 50 mg sodium per 100 g salt. Diet salt is used as a replacement for kitchen salt for low-sodium diets.

24.4 QUALITY DETERIORATION, SPOILAGE AND STORAGE

Salt is hygroscopic and has to be stored in dry surroundings, for preference in a pot made of Cologne earthenware, glass or plastic. In other earthenware the glazing will be attacked; tin also reacts with salt. To prevent the absorption of moisture somewhat and as a result to prevent the salt from sticking together, an anti-caking agent may be added to kitchen salt and table salt. Salt can be kept indefinitely.

24.5 LEGISLATION

Regulations may set down the requirements for, among other things, the composition of salt, including the iodine content of iodine-containing kitchen or table salt.

25

Sugar, syrup, confectionery and sweeteners

25.1 INTRODUCTION

Sugar is a collective name for a number of compounds which are grouped under the carbohydrates, such as glucose, fructose, maltose. What is called sugar in the home is sucrose, built up from glucose and fructose. Sugar is used for table luxuries, as a sweetener, energy supplier and preserver.

Sugar is mainly obtained from sugar beet and sugar cane. It is obtained on a limited scale from the sap of the Acer (or maple), which grows in North America and Canada, or from the sap of the sugar palm, which grows in Indonesia.

Sugar beet was already known to the Greeks and Romans. The leaf was eaten as a vegetable. It was not until 1747 the the Berlin chemist Marggraf succeeded in isolating sugar from the beet. The quantity which he obtained was very small, 3 per cent. By improvements in the production methods and selective breeding, at the present 13 per cent sugar is obtained. In the Netherlands the cultivation of sugar beet started around 1870. Sugar beet grows on heavy clay, but also on sandy soil.

Sugar cane is also an old plant. In 300 BC it was being grown in India. During the Crusades the Crusaders brought the cane to Europe. Columbus introduced the sugar cane to Central and South America. Sugar cane requires a tropical climate. The main countries where sugar cane is grown now are Cuba, Indonesia and India.

Consumption

In the Netherlands, about 40 kg sugar is consumed per person per year. Of this total 9–10 kg is used directly as sugar in the home. (The figures for the UK are approximately the same.) The rest is consumed in the form of industrially produced products such as chocolate, sweet sandwich spreads, confectionery, pastry, ice-creams, jams and beverages.

Composition

White sugar consists 100 per cent of sucrose. The raw and refined kinds of cane sugar still contain traces of vitamin B and minerals, including iron. The energy value of beet and cane sugar is equal, as is the detrimental effect of both on the teeth.

Some types of syrup contain 75–80 per cent sucrose and about 20 per cent water.

Fig. 25.1 — Sugar cane plantation.

Sugar price

Within the EC a stable price development is aimed at. It contains two elements: namely the annual setting of the intervention or guarantee price for the producers within the EC, and the setting of a production quota. This is a regulated market. Only about 20 per cent of the sugar trade takes place on the free market. Further, the sugar price is increased by the charging of a sugar levy since the Law on Sugar tax (1964) is in force.

25.2 THE PRODUCTION OF BEET SUGAR

The sugar beet is a biennial plant. In the first year, only leaves and roots are formed. In this period much sugar is formed (about 17 per cent) which is stored in the root as a reserve food for the second year. However the beet is harvested in the first year. The harvest time is from September until the end of December (the sugar beet harvest). The sugar beet factories work continuously during the harvest to process the beet as quickly as possible into sugar. This is necessary because during storage the sugar-content decreases as a result of the continuing metabolism. From 1 kg beet, about 130 gram crystal sugar are obtained.

The production process for beet sugar

The beet arrive by lorry, train or ship. Each consignment is weighed and sampled. The sample is analysed for sugar content and its tare (this is the attached soil). The

Fig. 25.2 — Sugar beet.

farmer is paid on the basis of weight and sugar content. The beet are then sprayed clean with water cannons. Stones and leaves are removed. By means of water channels the beet float into the factory, are then washed and cut into slices.

The sliced mass is heated in troughs to 70°C. Through this the cells become transparent so that it is easier to extract the sugar. The warm mass is brought from the heating trough into a diffusion tower and pushed up through it. From the top of the tower, warm water (temperature 70–75°C) is pumped, which flows in the opposite direction to the sliced beet. During this process the water takes the sugar out of the beet. The sugar-rich juice leaves the diffusion tower at the bottom and is called 'raw juice'. The residual sliced beet (the pulp) is processed into cattlefeed.

The blueish-grey and turbid raw juice, which, besides sugar, also contains residues of pigments, minerals and proteins, is treated with lime and carbon dioxide. During this process the proteins from the raw juice attach themselves to the small crystals of calcium carbonate, formed from the lime and carbon dioxide. The raw juice is clarified by filtration. The light yellow filtrate is called 'thin juice'. The filtered-off mass is used as lime fertilizer in agriculture.

The thin juice, which contains about 15 per cent sugar is passed through charcoal filters to remove pigments from the juice. Then the thin juice is evaporated into 'thick juice', in which 65–70 per cent sugar is present. Thick juice is treated with sulphur dioxide (SO_2) after the evaporation process to remove the so-called oxygen-active pigments, which have developed as a result of the evaporation. The thick juice is reduced in heated pans in stages until the solution is super-saturated and the sugar crystallizes out onto icing sugar which is added just before to initiate crystallization. In the cooking pans there is now a crystal mass (crystals, covered with a layer of syrup).

By means of centrifuging, the crystals and syrup are separated from each other. The syrup can be evaporated two or three times again, as a result of which further crystallization occurs. In the end, a dark-coloured syrup, called molasses, remains. This contains about 70 per cent sugar, 20 per cent water and many minerals. Molasses is industrially processed into cattlefeed, yeast and alcohol. The sugar crystals, after drying, cooling and sieving, are packaged.

25.3 THE PRODUCTION OF CANE SUGAR

Sugar cane is a perennial plant, which belongs to the family of the grasses, to which cereals also belong. The plant grows in moist areas with an average temperature of 28°C. The cane can reach a length of 4–6 metres with a diameter of 2–7 cm. The stalks contain a strong sugar-containing juice (7–20 per cent). When the leaves of the sugar cane become yellow and die back, the canes are harvested. This happens 10–24 months after planting.

Fig. 25.3 — Sugar cane.

The production process for cane sugar

In the country of origin, the canes, after cutting, are chopped smaller, crushed between rollers and the juice pressed out.

The juice, which is yellow-brown in colour, is mostly treated with large quantities of sulphur dioxide (instead of lime and carbon dioxide), to remove the proteins. The juice is then evaporated by boiling until crystallization occurs. The raw crystals and syrup are separated by centrifugation and separately packaged and transported.

In the country of consumption, the raw cane sugar crystals are treated with steam (refined) to remove contaminations such as pigments and the surplus sulphur dioxide, so that this sugar complies with the requirements of the Sugar and Syrup Regulations. The sugar which remains is light-brown in colour. Dark cane sugar is obtained by adding cane sugar molasses to the crystals.

The cane-sugar syrup, which is more aromatic and better in taste than the beet-sugar syrup, is processed into products.

25.4 KINDS OF SUGAR

The different kinds of sugar can originate either from sugar beet or from sugar cane. In the latter case this will be mentioned on the packaging. (This is also the case for the different kinds of syrup — see section 25.5.)

Crystal sugar consists of white, gleaming crystals, which, after moistening, can be pressed into cubes for use in coffee or into granulated or decorating sugar, which is used in baking. Dye is also sometimes added to finer or coarser grains of crystal sugar (red or green). Red-coloured crystal sugar, also called pear sugar, gives its red colour to stewing pears. These coloured crystals are also used as decorating sugar. Refined cane sugar crystals have a light brown colour, are somewhat moist and aromatic.

Table sugar has very fine granules and is obtained by sieving crystal sugar. Table sugar dissolves faster than crystal sugar.

Preserving sugar is table sugar to which pectin and citric acid are added. Preserving sugar is used in the production of jam and jelly.

Vanilla sugar is fine crystal sugar, mixed with synthetic vanilla.

Icing sugar is finely ground crystal sugar. To avoid clumping, a drying additive may be used.

Caster sugar is produced by adding an invert sugar syrup to small sugar crystals (see section 25.7), which are coloured to various degrees by caramel formation. Caster sugar is available in three colours; white, beige and dark brown.

Candy sugar is a product which is obtained by growing sugar crystals in a warm super-saturated sugar solution. Sometimes the crystals are grown on a cotton thread. White, yellow, light brown, dark brown and black candy sugar is produced. The candy sugar becomes darker in colour through the addition of caramel. Black candy or Boerhaavese candy is called after doctor Boerhaave who prescribed candy for sore throats and colds.

Goela djawa or **Java sugar** is a product of cane and palm sugar. After boiling down, the juice is left to crystallize in halved coconuts or other shapes so that, after cooling, a brown ball or other shapes develop. This sugar is often used with Indonesian dishes. It is used as sweetener but is also very aromatic.

25.5 KINDS OF SYRUPS

The kinds mentioned below can originate from sugar cane as well as sugar beet. The following kinds are sold.

Sugar syrup is the syrupy liquid which is left over after centrifuging the crystals. The syrup is very viscous. Sugar syrup is also available in pourable form as **pouring syrup**.

Candy syrup is the syrup which remains during the candy sugar production. This glazy syrup is light in colour, not very strongly flavoured and very sweet, and is sold under the name 'Golden Syrup'.

Molasses is a dark coloured sugar syrup, obtained as a result of boiling down several times. This syrup is strongly flavoured because of the concentration of minerals (about 4 per cent). Molasses is called **salt syrup** if the mineral content is still higher (max. 7 per cent).

Household syrup is a mixture of candy syrup, sugar syrup or molasses mixed with glucose syrup (see section 25.7). By the addition of glucose syrup the end product becomes less sweet and less aromatic. Stored cold, this is more viscous in consistency. Rules are laid down concerning the sucrose content, because household syrup has not been completely produced from cane or beet sugar. This has to be a minimum of 30 per cent.

Kitchen syrup, like household syrup, is a mixture of candy syrup, sugar syrup or molasses mixed with glucose syrup. The sucrose content has to be between 15 and 30 per cent.

Syrup can not only be produced from cane and beet sugar but also from other plants. For example, **maple syrup** is produced from the drained and boiled-down juice of the maple tree (also called Acer), which grows in Canada and North America. This syrup is very aromatic and very sweet. Maple syrup is also obtainable mixed with glucose syrup.

25.6 CONFECTIONERY, LIQUORICE AND WINE GUMS

Confectionery

Confectionery can be divided according to ingredients and method of production.

- **Hard** confectionary, also called acid drops, consist of sugar and glucose syrup, which are dissolved in water and boiled down to a final moisture content of 2 per cent. Then flavour and colour are added and the warm mass is poured into moulds and cooled.
- **Soft** confectionery, such as fondant and marzipan are products based on sugar, glucose syrup, cream, milk and/or butter.
- **Peppermint** is produced from a mixture consisting of sugars, a gelling aid, starch and peppermint oil. The mixture can be rolled into a layer from which tablets are cut (hard peppermint). The mixture is also shaped into granules of which tablets are pressed (soft or digestive peppermint).
- **Dragees**, or sugar-coated products, are finished with a layer of sugar, which contributes towards the taste, and beautifies the exterior, adding for instance gloss and colour or protection against the absorption of moisture.

Liquorice

Liquorice is produced from sugar, glucose syrup, gums, water, extract from the liquorice root, ammonium chloride, and colour and flavour compounds. The ingredients are dissolved in water and boiled. Then the mass can be pressed into liquorice (soft kinds of liquorice, such as shoe laces and liquorice allsorts) or poured (hard kinds of liquorice). Hard liquorice contains more gum and less sugar than the soft kind. For soft liquorice a combination of starch and gelatine is chosen as a thickener and more sugar is added than to hard liquorice. Liquorice is often glossed with oil.

Wine gums and soft gums
These products are produced from sugar, glucose syrup, gums, flavours and colours. Instead of gum, gelatine is often used, perhaps in combination with starch.

25.7 SWEETENERS

Among sweeteners are numbered products which because of their sweetening ability are used instead of sugar, syrup or honey. There are several reasons for replacing sugar or sugar-containing products.

— Sugar can be detrimental to certain users, for instance diabetics.
— Some sweeteners supply little or no energy.
— In the foodstuff industry the processing of sweeteners offers technological advantages over the use of sugar.
— Some sweeteners are less harmful to the teeth than sugar.

Sweeteners are processed on a large scale in industry, for sale to individuals for adding to their own drinks (tea and coffee), and for use in chewing gum, drinks and jams.

Sweeteners can be grouped on the basis of common characteristics. A much-used division is the separation into energy-supplying sweeteners and non-nutritive sweeteners, which supply little or no energy. One of the most complex characteristics of sweeteners concerns the sweetness. This is expressed as relative sweetness in comparison to a 10 per cent sucrose solution. In Table 25.1, the relative sweetness, the energy value, the cariogenicity, the laxative effect and the maximum permitted dosage per day (Acceptable Daily Intake — ADI) of several sweeteners are mentioned.

The energy-supplying sweeteners
To these belong the saccharides and the polyalcohols.

Saccharides
The **saccharides**, also called the starch sweeteners, are produced by the hydrolysis of polysaccharides of potatoes, sweetcorn or cereals using acids and/or enzymes. During this process, starch is changed into glucose via dextrin and maltose. If the conversion is total, i.e. all the links of the starch chain are broken, glucose only results (high DE glucose syrup). If there is incomplete hydrolysis, mixtures of the above-mentioned compounds develop (lower DE glucose syrups).

Saccharides are mainly used in the food industry to replace the more expensive crystal sugar or because of technological characteristics. Used are: glucose, fructose, invert sugar, maltose and dextrin-maltose.

Glucose or grape sugar is for sale as dextrose and glucose syrup.

Table 25.1 — Summary of some characteristics of sweeteners

Sweetener	Energy value (kJ/g)	Sweetness compared to sucrose	Cariogenicity	Laxative effect (use per day)	Acceptable Daily Intake (ADI) (mg/kg body weight)
Dextrose	17	0.7	+	—	—
Glucose syrup	17	0.5	+	—	—
Fructose	17	1.1–1.4	+	—	—
Invert sugar	17	0.9	+	—	—
Maltose	17	0.5	+	—	—
Dextrin-maltose	17	0.2–0.5	+	—	—
Aspartame	17	100–400	—	—	0–40
Sorbitol	13–16	0.5	—	+(30–50 g)	—
Xylitol	13–16	1	—	+(50–70 g)	—
Mannitol	8–11	0.7	?	+(10–20 g)	—
Saccharin	—	200–700	—	—	0–2.5
Cyclamate	—	20–30	—	—	0.11
Lactitol	8	0.3–0.4	—	+(40 g)	—

Key to table: +, known; —, not known/not determined; ?, insufficiently investigated.
Sources: Compendium Dieetpreparaten en voedingsmiddelen, 1986–1987; Rapport van het Wetenschappelijk Comité voor de Menselijke Voeding, 1984; Voedingsinformatie, uitgave van het Voorlichtingsbureau voor de Voeding, 1983, nr. 1; Fieliettaz Goethart, R. L. de, Zoetstoffen; eigenschappen en toepassingen, artikel in V.M.T., 2 mei jr. 18, nr. 9, 1985.

Dextrose is pure glucose in crystal form. This is mainly processed into sweets and cubes of grape sugar.

Glucose syrup (white or confectioner's syrup) contains besides glucose also maltose and 20 per cent water. Glucose syrup in solid form is called *massé* (cubes). Some noticeable characteristics of glucose are as follows.

— Strong brown colouring during heating. This effect is desirable in bread and some baker's wares, for instance to obtain a brown crust.
— Slowing down the crystallization of sucrose, because glucose syrup forms a thin protective layer around the seed crystal. This property is used in jams, ice-cream and sweets.
— Greater fermentation ability, which is wanted in the baking and brewing industry. Single sugars ferment completely and leave no residue.
— Greater depression of freezing point, through which the freezing and melting behaviour of ice-cream can be adjusted as required.

Fructose or fruit sugar occurs naturally in small quantities in honey and figs. With the help of enzymes, fructose is produced from glucose. Less fructose can be used to achieve the same sweetness as sucrose, because the sweetening power of fructose is higher than that of sucrose.

Invert sugar is a mixture of glucose and fructose, in which the mono-saccharides can react separately. Each of glucose and fructose in invert sugar has its own melting point and crystallization point. This is in contrast to sucrose, also consisting of glucose and fructose, which behaves as one compound. So sucrose has only one melting point and one crystallization point. Invert sugar is obtained from starch. Invert sugar is also found in honey, which under the influence of enzymes from the bees' stomachs is formed from nectar (sucrose). Invert sugar absorbs moisture quickly. It is used in the production of chocolate with a fondant filling to keep the whole pliable.

Maltose or malt sugar consists of two molecules of glucose. It is mainly used in the production of bread and beer.

Dextrin-maltose is a mixture of dextrin and maltose. It is processed into food for babies and sports people. Some characteristics of the sweetener are a less sweet taste, good solubility in liquids, and reduced fermentation activity in the stomach and intestines.

Polyalcohols

The **polyalcohols** occur naturally in several kinds of fruit and vegetables. However, for use in industry some are specially produced. They are sorbitol, xylitol and mannitol.

Sorbitol is produced from glucose. In contrast to sucrose, is it not completely absorbed in the human body. The largest part is again excreted. If more than 40 g per day is used, diarrhoea can occur, because sorbitol absorbs moisture in the intestines. The part of the sorbitol which is digested is absorbed by the blood slowly. As a result of this, the blood sugar content does not suddenly increase. Sorbitol is processed as a sweetener into products for diabetics (including chocolate, lemonade, biscuits and jam). It is also used in industry to improve the consistency of products.

Xylitol is produced from hemicellulose, from straw and birch wood. Because the absorption of xylitol is slow, with too great a use diarrhoea also occurs. Xylitol is used in sugar-free chewing gum, among other things.

Mannitol is produced from glucose. Mannitol is not used as a sweetener because it causes even faster diarrhoea complaints. However it is used as a thickener and moisture stabilizer.

Non-nutritive sweeteners

This group includes saccharin, cyclamate, aspartame and lactitol.

Saccharin is made artificially from toulene (a coal-tar product). Saccharin is bound to calcium or sodium. Saccharin has, compared with sucrose, a very high sweetening power. It can stand up to cooking and baking. The disadvantages are poor solubility and a bitter after-taste. In very high dosage and prolonged use carcinogenic abnormalities have been noted in animal experiments. For these reasons, a maximum permitted dose per day has been set for human consumption

Saccharin is processed into tablets, powder and liquid and is used in beverages.

Cyclamate or cyclohexylsulphamate are also artificially produced. Cyclamates are seen as ideal sweeteners, because they dissolve well in water, are stable with acids and bases, can stand cooking or baking and have no after-taste.

Cyclamates are prohibited in some countries, because of their carcinogenic characteristics. In the Netherlands the use of cyclamates is limited to diabetic products.

Aspartame is a low-calorific, artificial sweetener, consisting of two peptides (aspartyl and phenylalaline). On heating above 150°C the compound decomposes and the sweet taste disappears. This renders the sweetener unsuitable for the production of pastry in the home. In industry it can be used, if worked at high temperatures for a short time, directly followed by quick cooling. Under these circumstances no decomposition takes place. Aspartame is not stable in an acid environment.

Lactitol is obtained from milk sugar (lactose) by reduction of the glucose part of the disaccharide. Lactitol is only partly absorbed into the blood stream, and is broken down in the intestines by the intestinal flora. A consumption of more than 40 g per day can lead to intestinal interference.

Lactitol can stand acids and high temperatures (about 145°C). The sweetener is used in slimming aids, products for diabetics (in ice-cream and pastry) and in sugar-free sweets, liquorice and chewing gum.

25.8 QUALITY DETERIORATION, SPOILAGE AND STORAGE

Sugar, syrup and confectionery scarcely spoil because of their high sugar content.

Sugar will clump together if stored in a moist place; especially icing sugar and caster sugar.

Syrup can start to ferment in warm, moist surroundings. Syrup which is stored too cool and too long eventually crystallizes.

Confectionery can start to run (become sticky) under the influence of moisture.

In unopened packaging the above-mentioned products can be stored for years. In opened packaging, stored cool and dry they can be kept for a maximum of one year.

Sweeteners have an unlimited storage quality.

25.9 LEGISLATION

The **Sugar Products Regulations** contains information concerning the naming and composition of diverse sugar products. There may also be Confectionery Regulations as well. Some countries lay down regulations about the processing of saccharine, cyclamate and other sweeteners in foodstuffs.

Within the European Community differences exist in legislation for sweeteners and the permitted use of them in certain foodstuffs.

26

Honey

26.1 INTRODUCTION

Honey is the product produced by honey bees from nectar or honey dew, and serving as food for the bees. The use of honey as a sweetener has been known for centuries. Beeswax was also used for making writing tablets, candles and ointment. In the Middle Ages apiculture came to its greatest peak. Guilds of beekeepers were even set up. Besides supplying honey, bees are also responsible for the cross-pollination of fruit trees. This is of vital importance to the agricultural world.

Bee-keeping as a profitable business scarcely exists any more in many countries. Many beekeepers keep bees as a subsidiary business or as a hobby. In blossom time, they very often rent their bee populations out for the pollination of fruit trees.

Consumption

The use of honey in the Netherlands is about 500 g per person per year. The greater part is processed industrially in, for example, 'koek', pastry and sweets. The consumer uses honey mainly as a sweet sandwich spread. Healing properties are also attributed to honey, such as mucus-dissolving properties in colds. Also, value is attached to the minerals and vitamins which are in honey although they are in very small quantities. Especially in wholefood and anthroposophical nutrition, honey as a natural product is often used to replace refined sorts of sugar. In the UK honey consumption is about 300 g per person per year.

Composition

Honey consists of: invert sugar (35–49 per cent fructose and 32–39 per cent glucose) 80 per cent; sucrose 1–5 per cent; water 15–20 per cent; traces of protein; vitamin B; minerals, including potassium. (Source: *Bij en honing*, Mellona/Adelshoeve, Santpoort.)

Honey also contains aromatic compounds and enzymes, such as glucose-oxidase, which has a bacteriocidal action; invertase, which is added to the nectar by the bees;

and diastase, which breaks starch down to sugar. Pollen also occurs in honey. By examining this, it can be determined from which flowers the honey originates.

26.2 PRODUCTION AND DISTRIBUTION

Honey is produced by bees. A bee population lives in a skep or a hive and consists of a queen, 10 000 to 60 000 worker bees and, only in the summer, some hundred drones. The workers produce beeswax with which they build hexagonal cells. To help this process the beekeeper often places wooden frames with artificially made combs in the hive. These man-made combs consist of thin plates of beeswax with machine pressed cell imprints. The workers finish these combs. In the cells of the combs in the bottom of the hive (the brood chamber), the queen, after having been fertilized by a drone, lays eggs in the months of May and June.

The workers fly out and collect pollen and suck nectar, a sugar-containing liquid, from the fowers. In the bee's stomach the nectar is mixed with enzymes. Pollen and nectar is transferred into the combs. The nectar undergoes a fermentation process, whereby sucrose is changed into invert sugar. As a result of the high temperature in the hive (35°C) some of the water evaporates from the nectar. When the moisture

Fig. 26.1 — Worker.

content is reduced to about 18 per cent the cells are sealed, which means closed with a wax lid, and the maturing process takes place. The fermented, evaporated and ripened product is honey. A bee colony can produce about 15 kg honey per season.

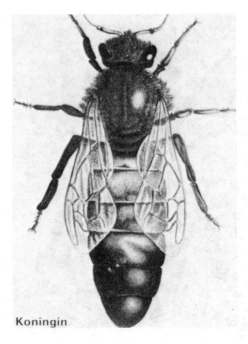

Fig. 26.2 — Queen.

Honey can be produced not only from nectar but also from honey dew. This is a sweet sticky liquid with which leaves and stalks can be covered during the summer and which is secreted by insects, expecially greenfly. This honey is called honey-dew honey.

If during a year, as a result of bad weather conditions, few of the flowers are available from which the bees obtain the nectar, they are supplemented with special honey sugar. The supplementation is also used to obtain a higher honey yield. The honey is collected from the honeycombs and in place of it the bees receive sugar as (reserve) food.

Honey can be harvested in several ways, resulting in comb honey, drain honey, centrifuge honey, press honey and melt honey.

Comb honey is sold still in the sealed cells of the comb, which is totally produced by the bees and is not allowed to contain man-made combs. Comb honey is available wrapped in cellophane and is eaten with the comb. Beeswax is undigestible by humans, but not harmful. Comb honey is also for sale in jars, mixed with honey.

Drain honey is harvested by removing the lids of the comb and draining the honey. This produces a very pure honey. The comb can later be re-used in the hive.

Centrifuge honey is flung out of the unsealed comb by means of a centrifuge or honey extractor. In this method also, the comb can be re-used.

Press honey is, perhaps under slight warming, pressed out of the comb. As a result of the pressing, contaminants such as pollen and wax particles can be present in the

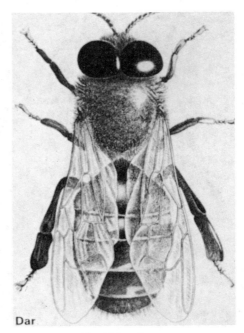

Fig. 26.3 — Drone.

honey, and are sieved out. The comb is thereby lost. Press honey is used for the production of 'breakfast koek' (see section 17.5).

Melt honey is obtained by heating the combs above 70°C, so that the wax melts. By doing this the enzymes in the honey are destroyed. After melting, honey and wax are separated. This honey is available as bakers' or industrial honey.

26.3 KINDS OF HONEY

The colour, taste, yield and quality of the different kinds of honey are influenced by the kind of flower from which the nectar originates. Local honey comes mainly from fruit trees, rape seed, white clover, lime and heather. Most of the available honey is imported from North and Central America, China, the Soviet Union and the Balkan countries. Varieties often available include the following.

Heather honey, a honey with a soft yellow to light brown colour; clear, thick, liquid and very aromatic. This honey rarely crystallizes.

Clover honey, a creamy honey, ointment-like in consistency and slightly sour in taste. Clover honey crystallizes quickly and becomes hard.

Millet honey, a dark brown, ointment-like honey with a strong, spicy taste.

Flower honey, a honey with a golden-yellow to dark brown colour; clear, and prepared from nectar of different kinds of flowers. The honey is liquid, mild in taste and crystallizes fairly slowly.

Acacia honey, a light, liquid honey with a fresh, not-too-sweet taste.
Lime honey, very fine mild honey, light golden-yellow in colour. This honey crystallizes fairly fast into coarse crystals.

Some byproducts of bee-keeping are pollen, mead (an alcoholic drink prepared from honey), honeykoek, candles, beeswax and cosmetics based on honey.

26.4 QUALITY DETERIORATION, SPOILAGE AND STORAGE

Honey has a good keeping quality. Honey can crystallize as a result of fructose crystallizing from the invert sugar; the glucose stays liquid. The more fructose in the inverted sugar the sooner the process of crystallization occurs. Acacia honey, for instance, contains more glucose and will not crystallize as quickly as clover honey, which contains more fructose. The crystallization also occurs sooner at lower temperatures. Although the crystallized honey has not changed its chemical or nutritional value, it is still less valued. To retard the crystallization, honey is heated. The higher the heating temperature, the longer the crystallization is kept at bay. However, the heating influences the taste and the enzyme content. Honey which comes from abroad is nearly always crystallized when it arrives in northern European countries. This honey has to be heated slightly (to a maximum of 40°C) to make it liquid and to enable it to be put in jars.

It is possible to determine if honey has been heated above 70°C for a long time. This can be done according to the diastase-index, which indicates the enzyme content and presence of the compound hydroxymethylfurfural, which develops on heating. A different method of combating crystallization is by seeding honey which crystallizes quickly, with 3–5 per cent fine crystallized honey of the same kind. Through this, a quick even fine crystallization takes place, so that the end product stays easily spreadable. This seeded or cream honey cannot crystallize any more. This procedure is used for clover and lime honey which otherwise will quickly crystallize.

In the kitchen, crystallized honey becomes liquid again by placing the jar in a warm-water bath.

If honey is stored too moist the possibility exists of mould formation and fermentation. Honey is best kept closed, dry and not too cold. The storage time is about one year.

27
Coffee

27.1 INTRODUCTION

Coffee, the name of the seeds of the coffee plant as well as that of the drink produced from cleaned and roasted beans, has established itself thoroughly in European food consumption patterns. The origin of coffee, however, does not lie in Europe. The coffee plant originates in Kaffa, a province of Ethiopia.

Around the thirteenth or fourteenth century, the Arabs took the coffee plant from Ethiopia and became specialists in the cultivation of coffee in the Yemen. Europeans acquainted themselves with the 'wonder-drink' which removed tiredness and sleepiness, during journeys of discovery in the Middle East (sixteenth century). Around 1700, through the initiative of the Dutch traders, coffee plantations were started in Java. In the eighteenth century, coffee plantations were started in South and Central America. Towards the end of the nineteenth century, planting started in Africa. At the moment, Brazil is the largest coffee producer.

Other important coffee countries are Columbia, the Ivory Coast, Angola, Indonesia and Guatemala.

Consumption

In 1985 a total of 7.9 kg coffee per head was used in the Netherlands. Of this total, 116 g was instant coffee (1.46 per cent). On the basis of 7 g ground coffee per cup (40–50 g per litre water) and 1 g instant coffee per cup (10 g per litre water), this means a consumption of four to five cups of coffee per person per day. In the UK, a different picture emerges. About 1.0 kg of coffee was used per person per year, of which 0.77 kg was instant (77 per cent).

Composition

Roasted coffee beans contain the following compounds (source: Franke, W., *Nutzpflanzenkunde*, 1976):

water	2.7%
protein	13.3%
fat (coffee oil)	12.8%
carbohydrates	67%
minerals (mainly potassium)	4.1%
caffeine	1–2.5%
chlorogenic acid	4.1%

During the making of coffee, the protein, fat and carbohydrate do not diffuse into the drink, but stay behind in the sediment. Because of this, black coffee supplies no energy. The coffee oil becomes volatile during the percolation and spreads aromatic compounds (the coffee smell).

Caffeine, a purine-type compound, which has no smell and no taste, has a stimulating effect on the nervous system, through which among other effects, sleep is dispelled.

In the making of coffee, 80–90 per cent of the caffeine passes into the coffee drink. This supplies, on average, 60–100 mg caffeine per cup. Too much can cause nervousness, heart palpitations, and raised gastric and bile secretion.

The stimulating action of caffeine is lessened by the addition of milk to coffee. For those who cannot tolerate caffeine, decaffeinated coffee is available.

Chlorogenic acid causes the bitter taste of coffee. It increases the gastric juice secretion in the stomach and has a stimulating effect on the stomach and intestine wall. The chlorogenic acid can be made inactive by a special procedure.

27.2 PRODUCTION AND DISTRIBUTION

The coffee plant grows mainly in moist tropical areas with an average annual temperature of 18–22°C. The plant is kept at a height of 2.5–3 metres by pruning. It has large, leathery, green leaves and is covered with white blossom when it flowers. After about 10 months the cherry-like fruits, the coffee berries, have matured (see Fig. 27.1).

In the fruit are two seeds or beans, which each have a flat side with a groove. Both beans are surrounded by a hard fruit wall, the parchment or pergamino. Each bean is separately surrounded by a thin seed-skin, the silver skin. In about 10 per cent of coffee berries only one bean develops, which grows into a round bean. In this case they are called pearl beans or pearl coffee.

For the world trade, two coffee bean varieties are of importance: Arabica and Robusta.

Arabica coffee grows mainly in America and supplies about 75 per cent of the world production. The bean is large, flat and oval in shape. The bean gives coffee with a very fine aroma and of high quality. The coffee trade distinguishes between the Centrals or Milds and the Brazils.

The Centrals or Milds are Arabica beans from Central America but also from Kenya, Tanzania and Columbia. The coffee is very fine in smell and taste and is very expensive. The Brazils Arabica beans come from Brazil (Santos) and produce a coffee which is soft and aromatic in taste and belongs, pricewise, in the middle range.

Robusta is especially cultivated in Indonesia and Africa (Angola) and produces 25 per cent of the world production. The name indicates that this tree has more

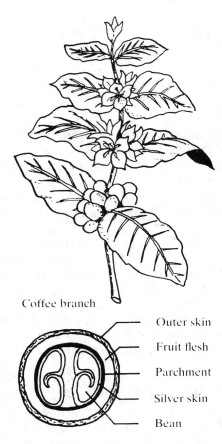

Fig. 27.1 — Transverse section through a coffee berry.

resistance against harmful influences such as leaf diseases. The Robusta are also less dependent on climate and soil.

The yield is higher than Arabica but the quality is not as good.

The Robusta beans are small and round in shape. They contain more caffeine (2–2.5 per cent) than the Arabica beans (0.8–1.3 per cent). The chlorogenic acid content is also higher. The taste of Robusta coffee is full-flavoured and bitter.

A very small part (about one per cent) of the world trade in coffee is taken up by the **Liberica** bean from Africa.

Cultivation and harvest

For cultivation, selected coffee beans are grown in seed beds. The seedlings are planted out in the plantation. After four years the plants start to fruit.

Because the coffee plant flowers three to four times a year, the berries do not all ripen together evenly. As a result, they are harvested several times a year. This is still done by hand. An exception is Brazil where only once a year the berries are ripped off in trusses as soon as they have ripened on average as much as possible. After picking, the beans are freed from the berries.

The processing of the harvested coffee is done in two ways: the wet and the dry manner.

The **wet** treatment is used in Central America, Mexico and Kenya, on Arabica beans. The berries are first washed in channels of running water. The unripe and bad berries then float and are skimmed off. Then the berries are cleaned of the soft flesh around the parchment by pulpers. After this, a fermentation process of one to three days takes place. The aim is to remove the slimy, sugar-containing residues which still stick to the parchment, by setting them to ferment and afterwards to remove them by washing. After the beans have been washed they are dried in the sun.

The bean which is left is still surrounded by parchment and silver skin. This is also called pergamino coffee. Just before the coffee is transported to the consuming country, the parchment and the silver skin are removed. The skins preserve the colour of the fresh beans, varying from blue to green, very well. Faded coffee has a low market value. Beans processed according to the wet method are recognizable after roasting by their having a white line. This is caused by residues of silver skin which stay behind in the groove and become light in colour during roasting.

The **dry** treatment is used in Brazil and West Africa, on Robusta beans. The beans are spread out on large drying floors (patios) and for about two to three weeks laid out to dry in the sun. Sometimes the drying is done artificially and then the process takes only a few days. When they are dry and shrivelled, the berries are taken to a peeling machine, which removes the pulp and parchment.

The bean covered in the silver skin is what remains. This skin is firmly attached to the bean and cannot be removed by peeling. After roasting, this coffee has a dark groove.

The raw green coffee beans are then sorted, packaged in bales of 60–69 kg and transported.

The buying of raw coffee takes place through dealers, agents or directly through the exporter. Raw coffee is not aromatic. Only during roasting is the smell developed. In the coffee-roasting industry, several kinds of bean mixtures are made up. In this way, several differing qualities are characteristically unified. Also, blending has the advantage of levelling out the variations in quality of the raw coffee, so that the resultant drinking quality stays constant.

Roasting takes place in rotating drums using hot air (about 190°C). In the first phase (up to 150°C), mainly evaporation of water takes place. At higher temperatures, brown colouring occurs through the formation of melanoids (reaction products of amino acids, carbohydrates and chlorogenic acid) and aroma formation takes place through volatilization of gaseous compounds. Sometimes some sugar is added during roasting to intensify the colour, or some oil to obtain a nice gloss. During roasting, one per cent carbon dioxide is formed, which acts as a preservative. The roasted beans are quickly cooled, sorted, perhaps ground, and packaged.

27.3 PRODUCTS

Coffee beans consist of Arabica or Robusta beans or mixtures of these.

Ground coffee is available in the standard ground form or the finer, quick-filter ground form. Ground coffee is packaged in plastic or aluminium foil, or in tins, to avoid aroma loss and oxidation processes.

Mocha, also called dessert or dinner coffee consists of a mixture of Arabica beans. The dark colour develops from the addition of extra sugar during roasting. More coffee per litre water is used to make mocha.

Espresso or Italian coffee is deep-roasted and very finely ground coffee. For the preparation, a special espresso apparatus is needed, which pressurizes boiling water through the coffee.

Decaffeinated coffee is treated with an extracting solvent, in which the caffeine dissolves, before roasting. The beans are roasted after the removal of the extraction liquid.

Stomach-friendly coffee (Idee-coffee) has been thoroughly cleaned before roasting and then treated with steam under pressure. The chlorogenic acid is broken down by this or changed into an inactive compound. The caffeine content stays the same.

Coffee extract is a very strong coffee infusion. It is available in liquid form. It can be drunk as coffee after dilution. It is used in hotels, restaurants and industry to give desserts or drinks a coffee taste.

Coffee powder is the concentrated extract in powdered form or granules. The drying process can be done by spray-drying or freeze-drying. As a result of spray-drying, a fine powder develops, which eventually, after being moisturized with water, can be processed into granular coffee powder. An advantage of large granules is that less aroma loss occurs. Porous coffee powder granules develop as result of freeze-drying. Freeze-dried coffee dissolves better and faster.

Coffee substitute or imitation coffee is used as a replacement for coffee and mostly consists of roasted and/or burnt vegetable products such as cereals (malt coffee), chicory, fruit and pulses. Imitation coffee lacks the invigorating effect of caffeine and the typical coffee aroma.

Coffee syrup is burnt, ground sugar from starches (for example, potatoes). It is used to reduce the quantity of coffee per litre. It colours the coffee brown and it gives a little taste.

27.4 QUALITY DETERIORATION, SPOILAGE AND STORAGE

Coffee deteriorates in quality through loss in smell. This occurs sooner in ground coffee. On long storage, the coffee oil can become rancid. Ground and powdered coffee can form lumps and moulds under the influence of moisture.

The carbon dioxide which is present in the bean gives packaged beans a keeping quality of about 10 weeks. Vacuum-packed ground coffee and powdered coffee can be stored for about half a year in closed packages. After opening, the quality decreases noticeably after one to two weeks. The best storage for coffee is packaging which is light, and both air and moisture proof.

27.5 LEGISLATION

The **Coffee and Coffee Products Regulations** lay down rules concerning labelling and composition of coffee, coffee extract and coffee substitutes (e.g. chicory).

28
Tea

28.1 INTRODUCTION

Tea is the name for the leaves of the young parts of the stalks of the tea shrub, as well as of the drink which is prepared from them. In the Netherlands tea is, after coffee, the most popular drink whereas in the UK tea is the most popular.

Originally the tea plant came from Assam (India). The beginning of the culture of the tea shrub and the drinking of tea lies in China. It is mentioned in old legends.

Dutch merchant seamen came into contact with tea during voyages of discovery to the Far East in the seventeenth century. They started tea plantations on Java and Sumatra. Also the English, who had started to appreciate tea, developed their own plantations in India and Ceylon (now Sri Lanka). For tea production on the world scale, these two last-mentioned countries are the most important. Tea is also grown on a large scale in Japan, Russia, Pakistan, Turkey, Kenya, Argentina and Brazil.

Consumption

In 1985 about 650 g of tea were used in the Netherlands per head of the population. This meant a consumption of nearly two cups per person per day on the basis of 1 g tea per cup (8–9 g per litre water). However, in the UK, tea consumption amounted to about 2.50 kg per person per year, nearly four times greater than the Netherlands.

Composition

Black tea contains the following compounds (source: Franke, W., *Nutzpflanzenkunde*, 1976):

water	8%
protein	26%
fat (tea oil)	5.1%
carbohydrates	55.4%
minerals (including fluoride)	5.6%
caffeine	3.3%

The proteins and carbohydrates do not go into solution, but stay behind in the tea leaves during the brewing of tea. The tea oil becomes volatile if hot water is poured on the leaves. Through this, the spreading of the tea aroma is obtained. The stimulative action of tea is caused by caffeine. Black tea contains more caffeine than the same weight of roasted coffee. But because the measure per litre of water differs, the stimulatory action of a cup of tea is less than that of a cup of coffee. If left to brew too long, tea acquires a bitter taste because tannic acid is freed from the leaves. Tannic acid may have a constipating action.

28.2 PRODUCTION AND DISTRIBUTION

The tea shrub needs a tropical climate with an average annual temperature of 18–28°C, reasonable amounts of moisture, and good soil. Tea is grown in the main on estates which lie on the slopes of hills. The higher the field, the better is the quality of the tea; for instance Darjeeling tea from the Himalaya mountains.

The tea shrubs are kept to a height of one and a half metres; this eases the picking of the young leaves, which is done by hand. For the world trade there are two tea varieties of importance:

Camellia assamica, a plant which originates from Assam (India). At the present time, this is grown in the tropical areas of Sri Lanka and India.

Camellia sinensis, which has been developed from *Camellia assamica*. Brought from Assam, the plant was cultivated in China, where such changes occurred that it seemed as if two kinds of tea plants existed. *C. sinensis* is now grown in temperate zones in China and Japan.

The tea estates are divided into gardens. Each day a different garden is picked. Every eight to nine days the tea pickers return to the same garden. In this time, new young top leaves have been formed. During picking, only the leaf bud and the four following leaves are gathered. The youngest leaves produce the most aromatic tea.

The top leaf, which sometimes has not yet unfolded, is called flowery orange pekoe. The first full-grown leaf is called orange pekoe. Pekoe is the second, slightly hairy leaf. The third and fourth leaf, pekoe souchong and souchong, are the somewhat tougher, curled leaves.

The rest of the leaves on the branch are not used for the production of tea. Depending on the leaf order on the tea shrub, three kinds of picking are known:

— the **fine picking**, consisting of the top leaf and the next two leaves down;
— the **medium picking**, consisting of the top leaf with the next three leaves;
— the **coarse picking**, consisting of the top leaf with the next four leaves.

Processing into black tea

After picking, several treatments of the tea leaves follow in the factory, by which **black tea** is produced. These treatments are as described below.

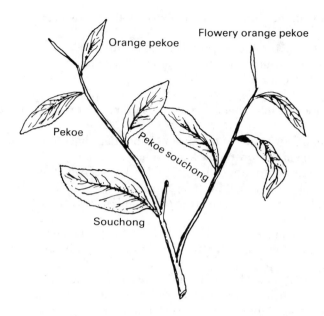

Fig. 28.1 — Tea branch with leaves.

Wilting: The leaves are spread out on racks and warm air is blown through. The wilting, which takes ten to twenty hours is done to reduce the moisture content in the leaves, to make the leaves supple, and activate enzymes which colour the cell sap in the leaf red, among other effects.

Rolling: The leaves are run between two horizontal turning surfaces for 30 minutes. By doing this, the cell walls are bruised, so that the red cell sap is liberated.

Fermenting: The rolled leaves stay for two to three hours in a fermentation chamber at about 27°C. The colour of the leaves is changed from green to red-brown by the enzymes of the red cell juice which is smeared over the upper surface of the leaves. The characteristic tea aroma is also formed during fermentation.

Drying: The still-moist product is dried for 20 to 30 minutes in driers at a temperature of 80 to 90°C. Well-dried tea is black in colour and breaks when bent.

Sorting: The tea is sorted into leaf and broken tea with shaking sieves and packaged in chests, lined with aluminium. In China, the tea is also pressed into blocks or tiles. The tea is then, after sale at auctions, sent to all parts of the world.

In the country of consumption the various tea blends are put together to avoid quality variations. A blend can consist of 80 to 100 different kinds. The tea is then packaged in bags, packs or tin tea caddies.

28.3 KINDS OF TEA

The kinds of tea available can be classified and may be described according to: the country of origin, the size of leaf or order of leaf used, and the method of processing.

Depending on the country or area of origin, the following distinctions are made:

— Assam tea from the Indian province of Assam;
— Darjeeling tea from the Himalaya mountains (India);
— Dimbula tea from Sri Lanka.

Also, various blends from a particular country or area are known:

— China blend, a mixture of teas from China, light in colour, with a fine, not too strong aroma and a low tannic acid content;
— Ceylon blend, a mixture of teas from Sri Lanka, which is drunk a lot in some countries and is very mellow in taste;
— Java blend, a mixture of teas from the island of Java and mellow in taste.

According to leaf size or leaf order, kinds of tea are labelled with the names of the picked leaves (see Fig. 28.1) (for instance, orange pekoe or souchong), or as leaf and broken tea.

Broken tea is obtained by rolling the leaves in the green state several times before processing them further, or by sieving during sorting. These short pieces are called 'fannings and dust'. Fannings and dust is used in tea bags. They infuse quickly.

According to the method of processing there are available: black tea, green tea and perfumed kinds of tea.

Black tea is obtained as a result of the processing in the tea factory in which the fermentation process plays an essential role.

Green tea is not fermented. The leaves are treated with hot steam before rolling and drying. The enzymes are inactivated by this and the leaves stay green. The taste of green tea is sharper, and the brew is lighter in colour. Green tea is mostly drunk in Japan and China.

Oolong tea is produced as black tea, but is only fermented for a short time (20–60 minutes).

Perfumed tea can be produced in several ways. During the processing of the tea, aromatic flowers, leaves or fruit can be used, for instance jasmine flowers in jasmine tea and bergamot (a herb related to the lemon plant) in Earl Grey tea. Or, the tea aroma compounds and/or extracts of flowers or herbs can be added after processing. This aromatization is often done just before packaging by the tea packer. Examples of aromatized kinds are: lemon tea, apple tea, anise tea, vanilla tea and peppermint tea.

In China, tea is sometimes dried over charcoal and, in the course of this, also smoked. This smoked tea with a tar-like aroma is called Lapsang Souchong.

Some less well known kinds of tea or products are:

— **Three-year tea** consists of leaves and low-growing branches which are only picked after three years. There are many more minerals in the leaf and there is more balance in the composition than in the young leaves. The caffeine content is also less (about 0.5 per cent). This tea is used in macrobiotic nutrition among others.
— **Caffeine-free tea** is treated with a solvent by which the caffeine is removed.

- **Low-tannic-acid tea** originates from soil (for example, in China) which contains few compounds from which tannic acid can be formed in the leaf.
- **Powdered or instant tea**: The production of dried tea extract consists of extracting black tea and then drying the extract.
- **Maté tea** is made of the light roasted and broken leaves of the maté tree, which grows in South America. The Indians have drunk this tea for centuries. The maté leaves contain 0.3–1.7 per cent caffeine.
- **Tea substitute** or imitation tea are names for mixtures of herbs or dried plant parts. Some examples are camomile tea, rosehip tea, stinging nettle tea.

28.4 QUALITY DETERIORATION, SPOILAGE AND STORAGE

Tea should be aromatic, dry and not mouldy. No contamination such as small sticks and grit should occur in tea. Tea can be stored for about half a year or longer if well packaged in cans, glass pots or tins.

28.5 LEGISLATION

Tea Regulations, where they exist, set down the rules with which tea, tea substitutes and tea products have to comply.

29

Cocoa and chocolate

29.1 INTRODUCTION

Cocoa is the powdered product from roasted ground seeds (cocoa beans) from the cocoa tree. It is used as the raw material for chocolate among other things.

The trade of cocoa culture lies in the Amazon area of Brazil. Columbus on his journeys of discovery came across the cocoa plant along with the Indians who ate the seeds of the tree and also used them as a means of payment. Through the conquest of Mexico in 1520 the Spaniards got to know about the cocoa bean from the Aztecs. They already had several preparative methods for cocoa. They prepared a drink from it; and the squares which they formed from the thick, solidified brew from ground cocoa beans, mixed with corn flour, honey and spices looked even then a bit like our chocolate.

The Spaniard Cortez took the beans to his country. At first the use and preparation of them was limited to Spain. It did not become known in other European countries until the seventeenth century. The consumption of cocoa escalated so enormously that, not only in Central and South America but also in Ceylon, in Indonesia and, in the nineteenth century, Central and West Africa, large plantations were established. Africa produces at the moment more than half of the world production of cocoa, particularly the countries of Ghana and Nigeria. Other cocoa-producing countries are the Ivory Coast, the Cameroons and Brazil.

Consumption

Because cocoa powder or cocoa is not only used as an end product but on a much larger scale as a basic ingredient for numerous products such as chocolate, bonbons, chocolate cake, pudding powder and ice-cream coating, a consumption figure is hard to determine. In the Netherlands the consumption of chocolate products, including chocolate sandwich spread, in 1985 was 4.5 kg per head of the population, whereas the production of chocolate confectionery was 492,000 tonnes in the UK in the same year, equivalent to about 10 kg per head.

Composition

In Table 29.1 an analysis is given of the composition of cocoa beans, cocoa powder and pure chocolate, the components being expressed as a percentage of the whole.

Table 29.1 — Composition of cocoa beans, cocoa powder and chocolate in per cent

Component	Cocoa beans	Cocoa powder	Chocolate (bitter)
Water	5	5.6	—
Protein	11.5	19.8	5
Fat	54	24.5	32
Carbohydrates	17.5	43.6	51
Minerals	2.6	6.5	0.6
Theobromin	1.2	2.3	?

Sources: Franke, W., *Nutzpflanzenkunde*, 1976; NEVO-table, 1986/1987.

Cocoa fat is also called cocoa butter. In the liquid state it has a bright yellow colour; in the solid form it is a yellowish white fat.

Outside the chocolate industry, cocoa butter is used in the pharmaceutical industry as a basis for creams.

Cocoa contains among others the minerals potassium and copper.

Theobromin is a compound related to caffeine. The stimulating effect of cocoa-containing products such as chocolate milk and chocolate is low because of the addition of other compounds such as sugar and milk.

29.2 PRODUCTION AND DISTRIBUTION

Cocoa bean trees need a moist, tropical climate (20°–35°C). They are very sensitive to wind and are, when young, in need of shade. Mostly banana trees are planted to provide the shade. The cocoa tree can reach 15 metres, but is kept to a height of 6 metres by pruning so that the fruit can be harvested with a long stick, without a ladder. The tree flowers during the whole year with small pink-coloured blossoms. From the blossoms, fruits develop in about six months (see Fig. 29.1) which are oblong in form with a length of 20 cm, containing 30–40 seeds. The peel of the fruit is yellow to red-brown in colour. On opening the fruit, one finds white- to purple-coloured seeds or beans, enveloped in bitter pulp.

The seeds are collected in chests or piled in a heap and for four days covered with banana leaves. In this period a fermentation by lactic acid bacteria takes place, followed by an acetic acid fermentation through which alcohol and then acetic acid is formed from the enveloping pulp.

The fermentation has the aim of inhibiting germination. The bean also changes its colour (from white to red-brown), the bitter taste decreases, and flavour precursors

Fig. 29.1 — Cocoa fruit with seeds.

are formed, compounds which later during the roasting are changed into aromatic compounds. From unfermented beans a good cocoa aroma cannot be obtained.

After the fermentation the beans are dried in the sun, or in drying tunnels at 70°C. The colour of the beans becomes browner during the drying process. The final moisture content is about six per cent.

Cocoa and chocolate production
The dried beans go in bales of 60–70 kg to the harbour to be transported. The beans are sometimes treated with pesticides, because they can be attacked by insects such as cocoa moths, during drying and transportation. On arrival at the factory, in the country of consumption, the beans are sprayed with an insecticide and so protected against possible attack by the cocoa moth. Then different kinds of beans from several countries are mixed to obtain the right blend. The beans are crushed after they have been freed from impurities such as woody debris and stones. A mixture of broken shells (peel), germs and pieces of cocoa cotyledons, also called 'nibs', is obtained. The shells and germs are separated from the nibs by a high velocity air stream. The nibs are roasted in rotating drums (135°C). During this process, the aroma, colour and taste develop. In the end the nibs are ground. During this, as a result of the increase in temperature through friction, a brown half-liquid mass is produced, the cocoa mass (about 54 per cent fat). This mass is the basis for the preparation of cocoa powder, cocoa butter and chocolate.

For processing into **cocoa powder**, about half of the fat (cocoa butter) is pressed out of, or extracted from, the cocoa mass. The rest of the cocoa press cake is finely ground into cocoa powder or cocoa. The quality of the cocoa powder can be improved by treatment with a dilute alkali solution. The more concentrated the alkaline solution, the softer the taste and the darker the colour of the cocoa. It also prevents cocoa particles coming out of solution. The alkali process is a discovery of the Dutchman van Houten. This process is also called the **Dutch process** and **Dutching**.

The **cocoa butter** which is pressed or extracted out of the cocoa mass is filtered many times. With steam and under vacuum the cocoa butter is deodorized to remove aromatic compounds. The cocoa butter is stored in liquid or solidified form.

40–60 per cent (icing) sugar and extra cocoa butter are added to the cocoa mass for processing into **chocolate**. Depending on the kind of chocolate other ingredients are added, for instance milk powder for milk chocolate. The mixture is then kneaded until it has the right fineness and structure. The mass is then warmed to 50°C in large basins, called conches, and beaten by rotating paddles. The conching can take two to three days. Finally, the chocolate is poured in the desired shape and cooled. After it has solidified, the contents are tapped out of the moulds.

29.3 COCOA AND COCOA PRODUCTS

Cocoa powder or cocoa is the cocoa press cake processed into powder. It can serve as the semi-manufactured product for the production of, for instance, chocolate. As an end product, mixed with milk and sugar it makes a nutritious drink. Skimmed cocoa powder contains less fat than ordinary cocoa.

Cocoa fantasy looks like cocoa but contains less of the cocoa bean. Other, mostly cheaper, kinds of fat may be used and also flavour and colour compounds and sugar are added. Cocoa fantasy is processed into a powder to obtain a chocolate drink, sandwich granules, flakes and imitation chocolate.

Chocolate is a mixture of cocoa powder, cocoa butter and sugar. To distinguish it from milk chocolate, the labelling 'pure', 'bitter' or 'plain' is sometimes used. No dyestuffs or synthetic flavour compounds may be added to chocolate, with the exception of ethylvanillin. Natural flavour compounds **may** be added; for instance to produce chocolate with an orange flavour.

Milk chocolate contains (besides cocoa powder, cocoa butter and sugar) milk powder.

Praline is milk chocolate with whole or fine ground nuts, mostly hazelnuts.

White chocolate contains cocoa butter, milk powder, sugar and vanilla.

Chocolate couverture is chocolate to which extra cocoa butter and sugar have been added. Couverture melts more quickly than ordinary chocolate. It is used in factories and bakeries for the production of chocolate articles, gateau decorations and to cover bonbons. Besides pure couverture, milk couverture also exists, and mocha couverture with coffee extract.

Filled chocolate has to contain one quarter of the total weight of chocolate. Candy bars can also be classified as filled chocolate if they comply with the above-mentioned regulation.

Chocolate bonbons consist of a minimum of one quarter part chocolate. The filling of bonbons can consist of such things as almond paste, nougat, marzipan, soft cream or liqueur.

Ice chocolate contains, besides cocoa butter, some coconut fat, which melts in the mouth quicker than the cocoa butter. It gives a cooling impression through this. Ice chocolate is sold in bar shape and in cups. For preference, ice chocolate should be kept in the refrigerator because of its soft consistency, resulting from the coconut fat.

Chocolate butter consists of spreadable fats with cocoa powder and sugar and is used as a sandwich spread.

Chocolate spread is a spreadable product based on chocolate. It contains a minimum of 15 per cent cocoa powder. It is called household chocolate spread if less cocoa powder is used.

Diet products

For diabetics, bonbons, chocolate bars, and spread are produced with sorbitol instead of sugar.

Substituted products

Carob is the powder of the roasted seeds from the pods of the Carob or locust bean tree. It is used as a substitute for cocoa in alternative nutrition. It contains, in contrast to cocoa, no fat and no theobromine. Carob can be processed into a drink or into bars.

29.4 QUALITY DETERIORATION, SPOILAGE AND STORAGE

Cocoa powder can become lumpy as a result of moist storage. Rancidity as a result of fat oxidation can occur because of the high fat content.

The storage time is about one year if the cocoa is stored in a cool dry and dark place.

Chocolate can become white externally. There are two reasons for this.

— If chocolate is stored above 30°C, the cocoa butter melts. This becomes partly liquid and separates out. On lowering the temperature the fat solidifies and is visible as a white deposit (fat-bloom).
— The sugars in chocolate can sometimes crystallize out to the surface. These crystals form a hazy white layer (sugar-bloom).

White-surfaced chocolate is not a health hazard but is less attractive.

Chocolate can spoil through fat oxidation. This occurs quicker in bars or bonbons filled with nuts in which fat-splitting enzymes (lipases) are present. There is more chance of mould formation in the case of filled chocolate.

Chocolate can be attacked by rodents and insects (cocoa moth).

Well-packed, and stored at around 18°C, chocolate can be kept for several months. The storage time of filled chocolate and bonbons is considerably shorter, depending on the type of filling.

29.5 LEGISLATION

The **Cocoa and Chocolate Products Regulations** cover legislation on cocoa and chocolate products, in terms of composition, additives and labelling requirements. Regulations may also exist for chocolate spreads.

30

Mineral water and beverages

30.1 INTRODUCTION

The trade in refreshing, non-alcoholic drinks has increased greatly in the last twenty years. Included in such beverages are:

— natural mineral and spring water;
— fruit juices and fruit drinks;
— lemonades and syrups.

Consumption

At the present time beverages have a firm place in the northern European consumption pattern. Table 30.1 shows the Dutch consumption figures split into groups.

Table 30.1 — Average consumption figures for beverages in 1984 (Netherlands)

Assortment group	Approximate consumption per person (litres)
Natural mineral waters	5.5
Fruit juices and drinks	20.0
Lemonades and syrups	60.0

Source: Marketing Board for Vegetables and Fruit.

The beverages vary in colour, taste, sugar percentage and price. The Dutch consumer drinks at the moment about 20 litres of fruit juice per year, with a preference for orange juice and apple juice.

The consumption of fruit juice is only slightly seasonally or weather dependent.

The sales figures of lemonades are much more seasonally sensitive. However, the total consumption of lemonades has hovered for the last few years around 60 litres per person. Within this group the fruit lemonades are appreciated the most, then follow the caffeine-containing kinds and the clear colourless drinks, known as up-market drinks.

The last few years have shown a noticeable trend within the consumption pattern in the direction of unsweetened fruit juices and mineral and soda waters. A similar pattern is found in the UK. In 1984 consumption of mineral water was about 1 litre per person, whereas the consumption of fruit juice was about 7.5 litres per person.

30.2 NATURAL MINERAL WATER

Spring or natural mineral water, also called table water, is bottled for consumption from a natural spring, or a spa mineral well, respectively.

Much of spring and mineral waters available originate from France, Belgium and Scotland.

Composition
The composition and the associated taste and quality differ greatly between mineral waters, and are dependent on the source and the point of collection. The water which reaches the source has filtered through different soil layers. It is purified by this, and also absorbs mineral substances and gases (including carbon dioxide), to which sometimes a medically healing effect is attributed. The absorbed compounds vary greatly in quantity and kind. Mostly sodium, magnesium, iron and/or iodine ions together with carbon dioxide and hydrogen sulphide are found.

Production
Spring or natural mineral waters are mostly directly bottled from the source without treatment or addition of compounds. The bottling is done in small quantities up to 2 litres. Sometimes a sediment forms during the collection, which can be removed by filtration or decanting. If mineral water by nature contains little carbon dioxide or if more carbonated water is wished for, this can be added under high pressure.

Kinds of mineral water
One distinguishes between:

— natural mineral water or spring water;
— naturally carbonated mineral water;
— mineral water strengthened with natural carbon dioxide;
— natural mineral water with added carbon dioxide.

The place where the water has been obtained has to be clearly mentioned on the label for all four groups.

The mineral content varies greatly, from about 5 mg to 2000 mg per litre or more.

Because sodium is always strongly represented, some mineral waters are less suitable or not recommended for a sodium-restricted diet.

30.3 FRUIT JUICES

Fruit juice is the liquid product from fruit, neither diluted nor fermented, complete with the colour, the aroma and the taste of the fruit.

Composition
By pressing fruit, the juice is separated from the pulp. The juice contains particularly the water-soluble vitamins, sugars and some minerals. Depending on the fruits used, many juices may be rich in vitamin C. The sugars which are present by nature supply a certain amount of energy. Sometimes extra sugar is added, which increases the calorific value of the product.

Production
The term fruit juice is applicable to directly obtained fruit juice as well as to the reconstituted juice, obtained from a concentrate by adding water and flavours.

Both in the production of fresh juice and concentrate, fruit is sorted, washed, destalked and crushed. The pulp obtained is sometimes heated and/or treated with pectolytic enzymes to obtain a higer juice yield.

After these preparations come the following processes:

— **Pressing** the pulp: To prevent oxidation the pressing has to be done quickly. Oxidation has an adverse effect on the colour and the taste.
— **Filtering** of the pressed and perhaps centrifuged juice: The aim is to remove finely dispersed impurities and/or to reduce the count of bacterial or yeast cells.
— **Concentration** of the juice in a vacuum evaporator: The concentrate obtained makes storage and transport easier.
— **Distillation**: Volatile aroma compounds may be separated from the juice. The aroma concentrate obtained (from 100 litres juice about 1 litre is obtained) is used again in the beverage industry.
— **Bottling** and perhaps acquiring a longer keeping quality for the consumer: The improvement of keeping quality for clear juices is done by ultrafiltration; the cloudy juices are pasteurized or deep-frozen.

Kinds of fruit juice
Fruit juice has to be 100 per cent from fruit or be produced from fruit juice concentrate by the addition of water and flavours, which have been obtained during the concentration of the juice.

Sweetened fruit juice: A quantity of sugar specified by law is added to fruit juice.

Fruit drink or nectar is a group name for products obtained from fruit juice or pulp, possibly concentrated. Water and sugar is added. The content of juice or pulp has to be mentioned on the packaging.

30.4 SOFT DRINKS

Composition
A soft drink consists mostly of water, sugar, flavour compounds or fruit juice and other additives.

Production
The taste of the end product is important in soft drinks production. It is determined by the balance between acidity and sweetness, bitterness, fruitiness, aroma, clarity, cloudiness (turbidity) and freshness. The final result is dependent on:

— *The water used.* The presence of minerals also influences the taste.
— *The quality of sugar.* A drink usually contains 9–11 per cent sugar, added in the form of sucrose, invert sugar, glucose and fructose. In diet beverages the sugar is replaced by other sweeteners (see sugar-free beverages in this chapter).
— *The addition of fruit acids* (citric and tartaric acid) *and essences or aromas.* They improve the taste.
— *The content of carbon dioxide.* Most lemonades contain 4 to 8 g CO_2 per litre. It gives a refreshing effect to drinks and improves the storage quality (it is bacteriostatic above 4 g per litre).
— *The use of additives*, for instance clouding agents (the thickeners Carob bean flour and modified starch) and alkaloids (quinine and caffeine). Clouding agents give the impression that fruit particles have been added to the drink. The alkaloids provide a certain bitter taste (quinine) and a stimulating effect (caffeine).

Kinds of soft drinks
Lemonade is a drink which consists mainly of water, sugar, organic acids and/or flavour compounds and perhaps colouring agents. A lemonade with carbon dioxide is referred to as **sparkling** or **carbonated lemonade**.

Lemonade with fruit or plant extract contains, besides water, sugar and carbon dioxide, also fruit or plant extracts and perhaps a small quantity of fruit juice. In this category belong:

— Cola-drinks with a certain amount of caffeine and coloured with caramel;
— types of tonic with a high content of quinine (quinine is a plant extract);
— up-market drinks with fruit extracts of, for instance, lemon or lime for the taste.

Fruit lemonade contains a certain percentage of fruit juice, water, sugar and carbon dioxide. Sometimes a small quantity of quinine is added. The lemonade then also carries the labelling 'bitter' for instance bitter-lemon.

Sugar-free lemonades: In lemonades, the sugar can be replaced by artificial sweeteners (see section 25.7). Permission for this may be necessary from the appropriate legislative body. With such permission the extra labelling for these lemonades suitable for diabetics can be used.

Lemonade syrup is a syrupy product, which consists of 55 per cent sugar and water-soluble aromas and flavour compounds and fruit acids. They are mostly coloured and are cheap.

Fruit lemonade syrup or fruit sorbet is produced from a certain quantity of fruit juice and contains 55 per cent sugar. It is dearer and more aromatic than lemonade syrup. If vitamin C-rich fruit juices are used, then the fruit lemonade syrup is sold with the labelling 'rich in vitamin C' and is especially meant for toddlers and infants.

Alcohol-free liqueur looks like a syrup liquid and contains less than 55 per cent sugar. It is suitable for drinking undiluted.

Sugar water is ordinary water or spring water to which minerals and carbon dioxide are added. Sometimes also called artificial mineral water.

30.5 QUALITY DETERIORATION, SPOILAGE AND STORAGE

The quality of fruit juices depends on the way in which the juice is obtained and the further processing. The quality will change during storage.

— Under the influence of oxygen from the air, the colour and taste will change as a result of oxidation processes.
— The presence of metals has a catalysing effect on oxidations, through which sometimes a cloudiness or turbidity becomes apparent.
— The increase of temperature (pasteurizing) results in a brown colouration (Maillard reaction).
— Microorganisms in the end product can cause spoilage (moulds and yeasts).

The quality of carbonated beverages is determined by the individual raw material. The keeping quality of these drinks is improved by:

— pasteurization of the individual raw materials or of the total drink;
— addition of, for instance, benzoic acid as preservative and of L-ascorbic acid as anti-oxidant;
— addition of 4 g carbon dioxide per litre of beverage: the carbon dioxide has then a bacteriostatic effect.

Storage

Fruit juices can be kept in unopened packaging for three to six months or longer. In opened packaging, they have to be stored cool and to be used within a week.

Carbonated drinks in unopened packaging can be kept for a year or longer, preferably in the dark (light results in quality deterioration) and at low temperature. The pressure within the bottle can become so high at high temperatures, that the bottle can break or explode. Opened packaging has to be stored cool and well closed, preferably no longer than about one week.

30.6 LEGISLATION

The **Natural Mineral and Spring Water Regulations**, where they exist, set rules for the bacteriological status of the bottled water as well as labelling requirements, permitted additives and levels of contaminants. These Regulations were made under the European Communities Act.

Fruit Juices and Fruit Nectar Regulations are also based on EC guidelines and contains rules concerning the labelling and composition.

Soft Drink Regulations contain rules concerning the composition, labelling and use of additives for a wide variety of these drinks.

31

Alcoholic drinks

31.1 INTRODUCTION

Alcoholic drinks originate through the action of yeast cells on sugar-containing liquids.

From the sugars, alcohol and carbon dioxide are formed. Single sugars such as glucose and fructose serve as the nutrient for the yeast cells. These sugars occur especially in fruit, the so-called grape sugar. In cane and beet sugar, the sugars occur as disaccharides, in cereals and potatoes as polysaccharides. These kinds can also be used as the raw material for alcohol production, but then the transformation of the polysaccharides to single sugars is necessary (see malts: section 15.3).

The production of alcoholic drinks has a long history. The forerunners of beer and wine were very likely discovered by accident. As early as 3000 BC wine groves were known in the Mediterranean area.

Alcoholic drinks can be divided into weak alcoholic types and distilled liquors, depending on the volume percentage of alcohol per litre (from about 3 to 60 per cent). The border between the two kinds of drink is at about 20 per cent. This division is connected with the licensing laws of the individual country. The sale of alcoholic drinks in shops and in public buildings is only possible with a special licence.

Consumption

Alcohol has a not unimportant place in the daily life of most people. The consumption figures in the Netherlands, calculated on an alcohol basis was, in 1985, 8.5 litres of pure alcohol per person. In fact, this quantity was consumed by about 85 per cent of the Dutch of 15 years and older. The intake per drinker came to 12 litres of alcohol, or more than 1000 glasses of alcoholic drinks per year. The Marketing Boards give a consumption break-down according to kinds. Table 31.1 shows the consumption figures of beer, wine and distilled liquors per head of the population for the Netherlands and United Kingdom.

Table 31.1 — Consumption figures of beer, wine and distilled liquors

Year	Consumption (litres, Dutch; kg, UK)					
	Beer		Wine		Distilled	
	Dutch	UK	Dutch	UK	Dutch[a]	UK[b]
1976	83.8	118.9	11.4	6.7	2.5	4.1
1978	85.2	121.4	12.2	7.5	3.0	4.3
1980	86.4	116.4	12.9	8.1	2.7	4.4
1982	81.9	109.0	14.2	8.7	2.5	4.0
1985	84.4	108.7	15.0	10.9	2.2	4.3

a: Expressed in litres of 100 per cent alcohol strength.
b: Expressed in litres of 40 per cent alcohol strength.

Composition

Alcoholic drinks are aromatic liquids with a specified alcohol content. Some kinds contain carbon dioxide, others a quantity of sugar. The quantity of alcohol per consumption varies and depends on the type of drink (beer, wine, gin), the quantity per glass (35–250 ml) and the mixing, or not, of alcoholic with non-alcoholic drinks (gin and tonic, rum and coke, cocktails).

31.2 BEER

The basic materials for beer production

Beer is produced from water, malt, hops, yeast and perhaps other adjuncts. These basic materials for beer production each have their own influence on the colour, smell, taste and alcohol content of the beer.

Water has a great influence on the taste. It has to be absolutely clear, pure and reliable from a microbiological viewpoint.

Malt is mostly made of barley. Good brewers' barley is rich in starches, which can be changed into dextrins and single sugars during the malting process. The dissolved proteins in beer, originating from the cereal, are important for the taste and froth formation.

The **hop**, a climbing plant, is grown in southern Germany and southern England, amongst other places. The dried female flowers, the hop cones, contain resins and ethereal oils (for the beer aroma), tannic acid or tannin (for clarity and keeping quality) and humulon and lupulon (for the bitter taste).

Yeast: For the yeast, pure cultures of *Saccharomyces* species are used. The Dutch beer industry uses bottom-fermenting yeast varieties especially, which are active at optimal 4–12°C. German and English brewers often work with top-fermenting yeast cultures with an optimal activity between 15 and 20°C.

Adjuncts: Sometimes crystal sugar is added, so that less barley malt is needed but still the desired amount of alcohol is produced. Also, malt from rice, corn or wheat is sometimes used instead of barley malt. These substitutions are responsible for a different taste.

Beer production
Beer production is shared by the malt house, the brewery and the cellar.

In the **malt house** the barley is sorted, washed, and soaked for about three days. The moist barley will sprout within 7–10 days, during which starch is broken down by enzymes. The enzymatic processes are stopped again in the oast house (drying with hot air). The temperature at which the barley is dried varies. It is possible to obtain:
— light malt or Pilsener malt, oasted at 75–80°C;
— dark malt or Münich malt, oasted at about 105°C.

The dry malt is cooled and stored in silos.

In the **brewery** the barley malt is ground in grist mills. The grist is mixed with water at 50°C, so that the cereal proteins can dissolve. The mixture is heated to 65°C, whereby the enzymes are activated. The change from starches into dextrins and sugars begins. After 15 to 30 minutes the temperature is increased further to 70°C; the conversion to single sugars is then optimal.

Meanwhile the very murky liquid is filtered or 'cleared'. The clear liquid obtained ('wort') then contains especially sugars, dextrins and proteins; the filtered residue serves as cattlefeed. In the wort kettle, the wort is boiled with the hops. The liquid absorbs the hop aroma and the bitter taste and microorganisms are killed. The cooled liquid goes to the cellar.

The fermentation takes place in the **cellar department**. Depending on the type of yeast, the wort is heated to between 4 and 10°C (bottom fermentation) or between 10 and 20°C (top fermentation). The first fermentation takes about 10 days in closed tanks.

The second fermentation takes, including lagering, 4–12 weeks at 0–1°C in closed tanks. Superfluous carbon dioxide is released. Besides the formation of alcohol and carbon dioxide the real beer aroma develops during storage or 'lagering'. Sometimes foam stabilizers, sweeteners, dyes and anti-oxidants are used, with an eye to better quality and storability. After this the beer is filtered, and then bottled or canned or drawn off into barrels or tankers. Beer in bottles or cans is mostly pasteurized at 65°C. The keeping quality of unpasteurized beer is less, but the taste is preferred by some.

Kinds of beer
The **light kinds of beer**, made with light or Pilsener malt, are available under different brand names:
— **Pilsener (pils)**: This kind of beer has a light yellow colour, a clear yeast taste and hop aroma. It contains 5 per cent alcohol.
— **Super pils**: This is light yellow in colour, has a full malt taste and contains 6.5 per cent alcohol.
— **Table beer**: This is light yellow in colour, has a light aroma and contains 3 per cent alcohol.

Fig. 31.1 — Traditional beer.

The **dark kinds of beer** are produced with darker or München malt. They may perhaps be coloured with caramel. Some examples:

- **Old brown** is a dark-coloured beer, with a sweet taste and an alcohol content of 3.5 per cent.
- **Dortmund beer** has a fairly light colour, and a fuller taste than that of Pilsener. It contains 5 per cent alcohol.
- **München beer** is clear brown in colour, has a light malt taste and contains 5 per cent alcohol.
- **Bo(c)k beer** is a dark-coloured beer with a bitter taste. The alcohol percentage is 6.5.
- **Stout** is a dark brown-coloured beer with a bitter taste. The alcohol percentage is 6.5.
- **Trappist beer** is a dark brown-coloured monastery beer with a sweetish taste. The alcohol percentage varies between 3 and 6.5.

Some other (foreign) kinds of beer are: **Gueuze**, **gueuze-lambic** or **Lambic**, names for Belgian beer, which develops by spontaneous fermentation with a minimum of 30 per cent wheat malt. For **Kriek-Lambic**, cherry juice or cherry extract is added.

Low-alcohol beer or alcohol-free beer contains a maximum of one per cent alcohol.

Quality deterioration, spoilage and storage

The quality of beer in bottles or cans is dependent on the production and the further processing. After pouring, beer should be clear, should have a firm foam head or froth collar, should be good in colour and taste. On long storage the quality can be impaired.

The best storage circumstances for pasteurized beers are:

- a cool place (maximum 12°C);
- dark surroundings, because light increases oxidation processes and through this the taste is impaired.

Unpasteurized kinds such as Trappist beer can only be kept for a limited time. The fermentation can continue and proteolysis can occur, especially at high storage temperatures. In unpasteurized beers sometimes an after-fermentation in the packaging is intended. In this case a sediment will form on the bottom. These beers have to be stored quietly for a few days before being poured, so that the solids will fall to the bottom. On consumption, the final portion of beer, which contains the sediment, is left in the bottle.

31.3 WINE

The description 'wine' may only be used for drinks which have been obtained by alcoholic fermentation of the juice of grapes. The wine grape, the fruit of *Vitis vinifera*, grows on wine vines in a temperate climate which is distinguished by an alternation of rain and sun, with above all many hours of sunshine.

The wine vine has tough, strong roots which penetrate deeply into the mostly poor, rocky soil to obtain water. The composition of the water from the deeper layers of the soil influences the composition of the grape and, through that, also the quality of the wine. Besides the climate and weather conditions and the groundwater, the micro-climate also has an influence on the ripening and composition of the grape. ('Micro-climate' describes the situation of the vineyard on a certain hill, the influence of the wind and the angle at which the sun reaches the plants.)

Several kinds of grapes are suitable for the production of wine. Each kind produces a wine with its own characteristics. The grape harvest in the Northern Hemisphere is between the end of September and the end of October. At the moment of harvesting, grapes contain:

— water (about 82 per cent);
— carbohydrates (16–20 per cent), partly in the form of glucose and fructose;
— minerals, especially Ca, P, Fe, Mg and K, which have a large influence on the taste;
— acids, for instance malic, tartaric and tannic acids.

During the ripening the sugar content increases steadily and the grapes will taste sweeter. The total content of acids determines in the end the pH of the wine, which is important for the keeping qualities.

Wine production

The ripe grapes are mostly picked in trusses. The trusses are taken to the presshouses, where the wine production begins with the crushing or pressing of the grapes, which have been mostly destalked. The yeast cells, which by nature occur on the skin or are also added as pure cultures of *Saccharomyces cerevisiae*, come into contact with the contents of the grape. The sweet juice (must) is obtained in two ways. For a **vin de goutte**, only the juice which drains from the crushed grape mass (pulp) is used. For most wines, the juice which can still be pressed out of the pulp is also used: **vin de presse**.

The first fermentation, *'fermentation tumultueuse'*, takes place in an open vat and takes usually 10 to 15 days. Depending on the kind of wine desired, the skin and pips are left in, or not, during the fermentation. The yeasts produce alcohol and carbon dioxide from the glucose and fructose. The carbon dioxide generation causes a considerable movement in the must. The first fermentation comes to a halt when all fermentable sugars are changed into alcohol, carbon dioxide and some side-products such as glycerol, which will later determine the taste and smell of the wine.

The yeast cells will die down if the alcohol content increases above about 15.5 volume per cent (spontaneously or through the addition of wine alcohol) or when sulphurous acid is mixed in.

The young wine is transferred into wooden vats and goes through a first development 'ripening'. Among other reactions, malic acid is changed into lactic acid; and oxygen from the air causes oxidation processes which alter colour and taste. To obtain a clear wine the wine is transferred to clean vats several times.

In the Spring, a second fermentation takes place, sometimes spontaneously, sometimes stimulated by the addition of sugars and wine yeast. If this happens in an open vat, then the carbon dioxide escapes and a **still wine** is obtained.

When secondary fermentation takes place in a closed tank, the gas is dissolved in the wine and an **effervescent** or **sparkling wine** develops.

If the secondary fermentation takes place in closed bottles, a strongly effervescent wine is obtained (*méthode champenoise*).

Secondary fermented still wines are mostly kept in full vats, so that as much as possible the oxygen is kept out. During the storage a sediment is formed, called **lees** or **deposit**. By transferring into sulphured, clean vats or tanks this sediment can be lost and the wine becomes clearer. Cheap, simple wine is cleared by filtration or centrifugation. Clearing using fresh egg white is sometimes used in dearer wines.

The wine is only bottled when it is clear and has developed sufficient taste and bouquet. The bottled wine can be drunk straight away or can develop still more in the bottle. The colour of red wine, during storage, gradually darkens to a brown colour.

The **bouquet** of a wine depends on the type of grape, type of soil, compounds which develop during the fermentation and on the aroma which is formed during the storage in the wooden vats. The bouquet comes to fullness during maturation in the bottle.

Kinds of wine
Red wine is obtained from black grapes in which the fermentation process takes place before draining and pressing. The pigments from the skin give the wine its characteristic colour. With the pigments, tannin (tannic acid) from the pips also comes over into the wine. Because of this it keeps better, but also acquires a slightly bitter taste. During storage, the tannic acid is changed into less bitter compounds. This is one of the reasons why red wine is stored for some time before it is drunk.

White wine is produced from black or white grapes, which after picking and crushing are immediately pressed. Only the colourless, sweet juice is allowed to ferment. This wine contains less tannin, and because of this is less bitter and ready to be drunk sooner.

Rosé wine can be made in different ways. The better quality is obtained by letting the crushed black grapes ferment for a few days before draining. The skin and pips give a slight colour and taste. After this, it is pressed and the pink-coloured juice ferments further. Simple rosé wines are obtained by mixing red and white wine to the desired colour.

Sweet wine is obtained by different methods. The fermentation can be stopped even before all the sugars have changed into alcohol and carbon dioxide. To do this, sulphurous acid is added, or the incompletely fermented wine is filtered. A second method is applicable for grapes which contain so much sugar that after the fermentation a sweet-tasting wine (dessert wine) still remains with an alcohol content of about 15.5 per cent. At this level the fermentation stops spontaneously.

Dry wine is completely fermented wine. This means that all the sugars have changed into alcohol and carbon dioxide. The taste is not sweet but dry (*sec*).

Fortified or **apéritif wine** is a sweet or sometimes aromatic-tasting wine of which, at the desired moment of the first fermentation, the alcohol content is increased to about 18 per cent. These wines can possess any desired degree of sweetness and are well known under area names as port, sherry, madeira and marsala.

The production of wines takes place in the areas where the wine grape grows. In Europe these are especially the EC countries France, Spain, Portugal, Italy, Germany and Greece and wines from these countries are freely available throughout Europe. But wines from other European countries (e.g. Bulgaria), from the United States (California), South Africa and Australia are also available. Each wine area offers its own specific kind of wine, which mostly has evolved according to its own views and tradition.

France

The wine country *par excellence* has many well-known wine areas which can supply wines from good to excellent quality, such as the Bordeaux area, the Burgundy area, the areas of the rivers Rhône and Loire, the Champagne area and the Alsace.

Bordeaux wines come from the surroundings of Bordeaux. The main wine districts in this area sell wines under their own name, such as Médoc, Graves, Sauternes, Pomerol, Saint Emillion and Entre deux Mers. Usually the name of the Château (=labelling for farm with vineyard) is also mentioned.

The Bordeaux area is the largest wine district in the world with wines of good quality. The wines are mainly red, but good whites, mostly dry are also available.

Burgundy wines come from the districts Côte de Nuits, Côte de Beaune, Côte Chalonnais et Maconnais, Côte de Beaujolais en Basse Bourgogne. The Burgundy area in southern France produces, especially, red wines. On the label, the district name is mentioned, but no château name.

The name Beaujolais Primeur (Nouveau) is carried by a very young red wine which is marketed each year at the beginning of November. As a result of a special fast way of wine production, it is ready to be drunk after a few weeks. It is, however, only to be kept until the end of April the following year.

Champagne is a sparkling wine from the area of the same name in northern France with Rheims as its trade centre.

Champagnes are always white because the grapes are pressed. At the first pressing only half of the juice is freed. This first juice gives the best champagne (Cuvée).

The first fermentation proceeds as described in wine production. In the Spring, the light wines are bottled. A solution of cane sugar and some yeast cells are added before the bottles are closed with special corks. The second fermentation lasts about six months in the closed bottles. In this period, carbon dioxide dissolves in the wine,

Fig. 31.2 — Bordeaux wines.

Fig. 31.3 — Wine from different areas.

Fig. 31.4 — Beaujolais wines.

but a deposit also develops. By a special treatment, in which the bottles are placed in racks with the neck downwards, the deposit can be removed. Before re-closure, the bottles are topped up with either sweetened or unsweetened champagne. In this way, champagnes are obtained which vary in sweetness. For instance:

— champagne *brut* (very dry)   0.5–1% sugar
— champagne *sec* (dry)   2–4% sugar
— champagne *doux* (very sweet)   8–10% sugar.

Loire wines come from the areas of Anjou, Vouvray, Muscadet, Sancerre and Pouilly. On the label the area name is mentioned. The wines are mostly white, sometimes rosé.

Rhône wines, especially those produced by the areas Hermitage and Châteaneuf du Pape, are excellent, mostly red, wines. On the label the name of the village or vineyard is mentioned.

Alsace wines originate from the vineyards on the west bank of the river Rhine and are white French versions of Rhine wine. On the label is mentioned, among other information, the name of the type of grape. For instance: Riesling, Sylvaner, Gewürztraminer or Tokay. The description '*Zwicker*' or '*Edelzwicker*' is used for simple wines, made from the juice of different kinds of grapes.

Germany

The German vineyards are found along the rivers Rhine and Moselle and their tributaries. The vines grown are especially the white Sylvaner and Riesling. German wine growers are permitted to increase the sugar content of the must by adding sugar, if the harvested grapes have too low a sugar content. This is called '*Verbesserung*' or '*Anreicherung*'.

Rhine wines are on the whole aromatic white wines of good quality. Some well-known names are Nierstein and Oppenheim.

Moselle wines are slightly green-tinted wines and in comparison with the Rhine wines less aromatic. Sometimes place names are mentioned on the label, for example Piesport and Bernkastel.

Sekt or **Schaumwein** is a sparkling wine.

Italy

Within the European Community, Italy is the largest wine producer, with the well-known wine areas Piedmont, Chianti, Alto Adige and Verona. The red wines from the Piedmont district are especially most loved by the Italians: among others are the Barolo and Barbaresco.

From the area surrounding Turin, the universally known **vermouth wine** originates. This aperitif wine is made with red, white or rosé wine, wine alcohol, herbal extracts and sugar (syrup). The somewhat bitter taste comes from the herb wormwood; hyssop, quinine, coriander, juniper berries, cloves, camomile and lemon peel are also used. Vermouth wine is fortified with alcohol to 15.5 or 18 per cent volume. Red vermouth is very spicy and slightly sweetened; white is sweet and mildly spicy; dry vermouth is unsweetened and is used to mix with other drinks.

Chianti, from the hills around Florence, is produced from four different kinds of grapes. The young wine is livened up after fermentation with a sweet must from dried grapes and bottled in the well-known raffia-covered green chianti bottles.

Alto Adige and **Verona** make mostly white but also red wines of excellent quality. On the label are mentioned the names of the wine areas, for instance Valpolicella, Soave and Bardolino.

Spain

The Spaniards have more acres of vines than the Italians; however their production is only one third that of the Italians. The cause lies basically in the way the wine is produced. Districts which produce quality wines are Jerez, Rioja, Montilla and Catalonia.

Rioja wines can be compared with the French Bordeaux and Burgundy wines in so far as production is concerned. They can be drunk young, after two to three years, but the quality wines (*reservas*) can mature for up to 10 years in oak casks.

Sherry is a fortified or aperitif wine from the region of Jerez de la Frontera. White grapes are used with a very high sugar content. When the must has finished fermentation after about three months, the young wine is strengthened with wine alcohol of from 15.5 to 18 per cent and stored in above-ground sherry 'cellars' (*bodegas*).

In January/February a certain mould (*flor del vino*) develops on the wine with the 15.5 per cent alcohol content. The mould layer excludes the wine from the outside air. It prevents the alcohol from being converted into acetic acid and the golden-yellow colour disappears. It also gives a typical nutty taste to this sherry (**fino sherry**).

No mould grows on the young sherry with the 18 per cent alcohol content because of the high alcohol percentage. This sherry develops a different taste and gets a darker colour (**oloroso sherry**).

When the young sherry has developed sufficiently it is stored in a *solera* system: barrels lie in layers of three or four on top of each other. The youngest wines (three years old) are put in the top barrels; in the bottom ones are the oldest kinds. The sherry for sale comes from the bottom barrels, which are topped up with sherry from the barrel above, and those from the one above again. This circulating system guarantees that each *bodega* delivers a consistent quality.

Fino sherry, because of its alcohol percentage of 15.5 per cent, is very unstable; for export, the finos are strengthened to 18 per cent.

Cream sherry is a mixture of oloroso sherry with a very sweet sherry. To obtain such a sweet sherry, the trusses of a special variety of grape are first dried in the sun; as a result of this, the sugar content of the must will be very high. The fermentation, which takes about two months, stops spontaneously when 15.5 alcohol per cent volume has been reached. The wine is still sweet, because unfermented sugars are left.

Manzanilla is a special, unblended, fino sherry with a slight salty taste which comes from the grapes, which grow close to the coast. In the Netherlands this sherry is sold as manzanilla without the labelling of sherry. Manzanilla has a maximum 15.5 per cent alcohol.

Portugal

The best-known Portuguese wine is **port** (porto, port wine). This red or white fortified wine owes its name to the export harbour Porto on the mouth of the river Douro in northern Portugal.

In the production of red port the fermentation is stopped by adding pure wine alcohol to 18 or 20 per cent volume. The remaining, unfermented sugars ensure a more-or-less sweet taste.

White port is exclusively made with white grapes. The must can finish fermentation; only after that is wine alcohol added. So this is drier in taste than red port.

Types of port wines of different years and vineyards are mostly mixed to obtain a consistent quality. This mixing is done by 'port shippers'.

There are several different types of red port:

— **Ruby**, a ruby-coloured port which is bottled fairly young. The taste is fruity and the colour bright red.
— **Tawny**, a red-brown, tan-coloured wine, which is matured for a long time in barrels and has very good qualities. During this long storage, through a slow oxidation process, this port acquires a soft mellow taste and the typical colour.
— **Vintage Port** is a red port from an exceptionally good year. The wine is not mixed with different years, has a specific year labelling and matures in the bottle. After about 15 years this port starts to display quality. It can become very old.

Madeira is a fortified wine, called after the volcanic island of Madeira on which special grapes grow, which yield a somewhat sour wine. To improve the taste, the young wine is stored for four to five months in open vats in sheds with glazed roofs, which causes the temperature to increase to about 50°C. The wine acquires a brown caramel-like colour and taste through the working of the sunlight and the warmth. The wines are mixed according to the *solera* system (see sherry), with or without the addition of sugar solutions and extra wine alcohol. Originally used as an aperitif wine, now more as a dessert wine and in the kitchen in the preparation of soups, sauces and desserts.

Greece

Greece produces white wines particularly. The most important wines for the Greek population are the retsina kind (mostly white, sometimes rosé). These wines taste of resin and have a pine smell. This typical smell and taste develops during fermentation. The wine is transferred into resin-treated pine vats. The Greeks use this technique for half the wines produced. Outside Greece this kind of wine is less appreciated. In the areas around Samos and Patras the sweet (muscat) wines are produced. The muscat grapes give a very specific smell and taste to wines.

Austria
The vineyards are situated especially in eastern Austria around Vienna and along the River Danube. The mainly white wines, made with Tokay and Rhine Riesling grapes, are most similar to the German Rhine wines. The labelling is the same as that of the German wines: name of the village, kind of grape, quality and perhaps the vineyard. Very young wine which is sold in cafés from the vat is known under the name *'heurigen'*. *'Liebfraumilch'* on the label means that the contents contain a mixture of white wines.

Quality labelling
Information on the label of a wine bottle, besides giving the description of the type of wine (dry, sweet, white, rosé, red), also describes the country and area of origin and the quality.

Each wine-producing country has its own set of standards, regulations and labelling. Within the European Community, the harmonization of the legislation for wines is being attempted (in so far as this is possible). Wines which do not satisfy the quality standards of a country or area are sold as table wine and country wine.

Table wine is a simple wine from an EC member state, but the area or country of origin is not mentioned on the label. However, the alcohol content and the code number of the Head Marketing Board for Agriculture (HPA code) is obligatory. By means of this number, the origin can be determined (*tafelwein*—Germany, *vin de table*—France, *vino da tavola*—Italy).

Country wine is comparable with table wine, but the area or country of origin is mentioned. The alcohol content and the HPA code are also obligatory on the label (*vin du pays*—France, Vin de France, Vin de Greece, Vin d'Italie).

On the label of quality brands, after inspection and approval in the area described, the quality class can be mentioned.

In Germany, besides the *Amtliche Prüfungsnummer* (AP), the quality labelling *Qualitätswein mit Prädikat* is known. The *Prädikat* is usually further described, for instance:

— **Kabinett** stands for wine of high quality, which has not been 'improved' (see *'Verbesserung'* of German wines earlier in this section).
— **Spätlese** stands for wine of which the grapes are harvested by the truss, but fully ripe.
— **Auslese** means that the wine is made of fully ripe grapes (so perhaps may be harvested grape by grape). The wine has a high sugar content.
— **Beerenauslese** means that the wine is made only from over-ripe grapes, which have been attacked by *'edelrot'*: a mould which splits and dries the grapes out. As a result of this, the sugar content increases greatly.
— **Trockenbeerenauslese** means that the wine has been made from late-harvested shrivelled grapes attacked by *edelrot*, which almost look like raisins.
— **Eiswein** is a quality labelling for wine of completely ripe grapes which have been

frozen in the vineyard. They are harvested at a maximum temperature of −6°C and pressed frozen. In this way a small quantity of juice is obtained, which is rich in sugars and flavour compounds.

In France, certificates of origin exist for quality wines. On the labels is given the *Appelation Controlée*, in which the area is also mentioned. For instance, *Appelation de Bordeaux Controlée*. A wine of great class is labelled as *Vin Delimité Qualité Supérieur* (V.D.Q.S.).

In Italy, there are also two quality classes, *Denominazione d'Origine Controlata* (D.O.C.) and *Denominazione d'Origine Controlata e Garantita* (D.O.C.G.). Besides this, there are extra, but not obligatory, descriptions which provide the buyer or consumer with more extensive information about the type of wine. For instance, a **Riserva** is a better quality wine, a **Classico** is a quality wine from central Italy (Verona and Alto Adige).

Fruit wine

Fruit wine is a fermented drink, which has been produced with the juice of fruit other than grapes. The alcohol content has to be a minimum of 9 per cent volume. Sugar may perhaps be added to the juice.

In principle, wine can be made from all sweet or sweetened fruit juices. Usually the wine gets its name from the fruit used; for instance: cherry, currant or apple wine. Fruit wines can differ greatly in colour, smell and taste. These characteristics are dependent on the quality of the fruit, whether or not extra sugar has been added, and whether or not the juices have been mixed. Sometimes, to improve the aroma, spices or spice extracts are added, as happens to **May wine** (fruit wine with sweet woodruff) and **Bishops wine** (wine with extracts of cloves and lemons).

Kinds of fruit wine

Cider is a fermented sparkling drink, which has been produced, without the addition of sugar, from the juice of apples (apple cider) or pears (perry). The alcohol content varies from 1.5 to 6 per cent volume.

Alcohol-free fruit wine and alcohol-free wine can be obtained by extracting the alcohol from the wine.

Quality deterioration, spoilage and storage

Wine is bottled when sufficient bouquet is formed or, to put it another way, when it has sufficiently matured. Each wine, each year has its own maturing rate which varies. The riper the grapes are at the moment of harvest, the more flavour and aroma they contain, and the more can still be formed in the wine during storage.

A stock of bottled wine should be stored lying down in a cool, dark place, preferably in a cellar with a constant temperature (about 10°C). Bottled wine demands, in addition, quiet surroundings (no draughts or vibrations and not too much moisture). Storage lying down is necessary for bottles with corks, so that the corks can be kept moist. A dry or dried out cork is porous, so that atmospheric oxygen can reach the wine, and undesired oxidative processes can be started. Bottles with a non-porous closure can, if wished, be stored upright.

Some wines form a sediment during long storage in the bottle. Bottles with **sediment** have to be carefully decanted into a carafe before use.

Wines have to be brought to the right temperature before use. Red wines are taken into the room at least a day before use, to *chambrer* (18–20°C). Some Italian wines need air before consumption. The bottles are opened and chambred one day before use.

White wines are mostly drunk at cellar temperature (8–12°C), some kinds lightly cooled (some hours in the refrigerator).

Champagne and other sparkling wines are drunk cooled (about 6°C). These wines are quickly cooled on ice in a wine cooler.

Spoilage of wine is practically impossible with well-stored and sealed wines.

A higher temperature results in a quicker ripening.

Light and air in the bottle (porous cork) speed oxidation processes, as a result of which the wine deteriorates in bouquet and colour. When the bottle is opened, the bouquet can be found to have altered in such a way that the wine is called 'shrivelled wine'.

Wine has a limited shelf-life after opening (about one week). Through contact with air and oxygen, acetic acid bacteria can quickly acidify the wine (see Chapter 32). The wine will also quickly lose aroma after opening.

Fortified wines with minimum 15.5 per cent alcohol keep well after opening. Acetic acid bacteria have no chance. The bottles have to be closed each time after use, to protect the contents against dust and dirt and loss of aroma.

31.4 DISTILLED LIQUORS

These drinks take their name from the fact that they are obtained by distillation of alcohol-containing drinks. During the distillation, the aqueous part is separated from the alcohol. The distillates obtained are sold under several names (brandy, gin) and have different alcohol percentages (35–60 per cent). Different sorts of alcohol-containing drinks can serve as a base. Often spices are added to these for aroma and flavour.

Because of their high alcohol percentages, these liquors can be kept for an indefinite time after opening. Sealing the bottles is necessary to make sure that the alcohol does not evaporate and to protect the contents of the bottle against dirt and dust.

Each country has its own old, specific kinds, distilled according to its own tradition and methods, sometimes with a protected name (cognac).

Table 31.2 gives a survey of a number of distillates with the basic ingredient for the production of the alcoholic drink indicated and the country of origin. A number of the most common will be more explicitly described.

Kinds of distilled liquors

Cognac is called after the town in the Cognac district of Charente. The wine distillate from this area alone is allowed to carry the name cognac. Other wine distillates are called eau-de-vie in France.

Cognac is distilled from the wine in two stages, in copper equipment (alembics).

Table 31.2 — Survey of distilled drinks

Distillate name	Basic ingredient	Country of origin
Tequila	Cactus	Mexico
Kirsch	Cherry	Austria
Calvados	Apple	France
Arak	Date	Middle East
Aquavit	Cereal	Scandinavia
Bourbon (whisky)	Cereal	United States
Gin	Cereal	England
Genever	Cereal	The Netherlands
Schnapps	Cereal	Germany
Vodka	Cereal	Russia–Poland
Whiskey	Cereal	Ireland
Whisky	Cereal	Scotland
Grappa	Grape skins	Italy
Marc	Grape skins	France
Arak	Rice	Indonesia
Rum	Cane sugar	Central America
Slivovitz	Plum	Yugoslavia
Vodka	Potato	Finland–Sweden
Akwavit	Potato	Denmark
Schnapps	Potato	Germany
Armagnac	Wine	Armagnac (France)
Brandy	Wine	Italy
Brandy	Wine	The Netherlands
Weinbrand	Wine	Germany
Cognac	Wine	Cognac (France)
Ouzo	Wine	Greece
Raki	Wine	Turkey
Eau-de-vie	Wine	France

Ten barrels of wine produce one barrel of distillate, a colourless, clear liquid with 70 per cent alcohol and a rough taste. The young cognac is matured for at least two years in Limousin oak barrels. A part of the alcohol evaporates, the colour becomes darker under the influence of wood and air (oxidation). Before bottling, the cognac is brought to the appropriate alcohol percentage (40–45 per cent) with distilled water. Colour and sweetness are adjusted with caramel and sugar solutions. After unclassified cognac, which means from one year old, the age is then mentioned. This can vary from 5 to 20 years or more. To obtain a standard quality ('S' brand) kinds of cognac are mostly blended. Cognac only matures in barrels; once bottled, no noticeable changes occur.

Fig. 31.5 — Various distilled liquors.

Armagnac has a great similarity to cognac, including also a protected name. The most noticeable differences are: the different distillation apparatus, with which it is only once distilled to 53 alcohol per cent volume, and the kind of wood for the barrels (black oak). Also, sugar is rarely added, so armagnac is drier in taste. It can be drunk younger.

Whisky is a cereal distillate with an alcohol content between 40 and 50 per cent. Formerly, malted barley was mainly used as the basic raw material; now oats, corn and rye and mixtures of cereals are also used. This is one of the reasons for the large differences in taste. Some kinds of whisky are the following.

— **Scotch whisky** used to be produced from malted barley (malt). The special aroma originated from the smell of smoke and peat, which was used to dry the germinated barley. The distillation took place in two stages. After that, the young whisky matured in oak vats (preferably old sherry casks), from which the distillate coloured brown. Such a whisky is mild in taste. It is still made, but rarely sold pure because of the high price.
— **Irish whiskey** is produced with a high percentage malt from barley and a little rye. This kind of whiskey is very aromatic and slightly sweet.
— **Bourbon** (American whisky) is produced from corn, rye and some barley malt. It is milder in taste than the Irish or Scotch kinds of whisky.

Genever is a Dutch distillate made from cereals (corn, wheat, rye and/or barley). The typical taste of the clear distillate is obtained by the use of juniper berries. Young genever contains about 35 per cent volume alcohol, and old genever about 38.

Sometimes fruit extracts are added, lemon or currants (red and black). The genevers are then sold as currant or lemon genever.

Gin is like Dutch genever, but has, besides the smell of juniper berries, that of coriander as well. No sugar is added for the taste. The alcohol content is about 40 per cent volume.

Rum is produced from the residues of cane sugar production, including molasses. There is light or white rum, and dark or brown rum.

Light rum matures, after distilling, for only six months in oak casks, is not sweetened, is slightly aromatic and clear. This kind is more suitable for mixing with other (alcohol-containing) drinks.

Brown rum matures at least five years in oak casks, through which the colour changes. It is also sweetened, and caramel is added for a consistent brown colour. The alcohol percentage varies from 42 to 70 per cent.

Vodka, in so far as taste is concerned, is rather like genever when cereals are the raw materials. Potatoes can be also used, but give the vodka a more bitter taste. The alcohol content of the kinds of vodka are about 40 per cent volume.

Vieux is produced from neutral alcohol, to which colour and taste compounds are added. It contains about 35 per cent alcohol. This drink is seen as an imitation of cognac. Sometimes real cognac or eau-de-vie is added to the better qualities to improve the taste.

Bitters are distilled drinks based on neutral alcohol and extracts of spices and/or fruit, which both serve the taste and have a beneficial effect. This is one reason why bitters are valued as a drink before a meal: they aid the digestion.

Some kinds are: Berenburg, Underberg, Elsbitter, Oranjebitter and Campari.

Calvados is a distillate from apple cider. This distillate is then matured in barrels for one to five years. The name Calvados is only permitted for the apple brandy from a certain area in France. The alcohol content is 40–50 per cent volume.

Brandy used to be produced only from wine. In Germany the distillate was called Weinbrand, in France eau-de-vie, and in England Brandy. Now it is also produced, including the foreign kinds, from unmalted cereals with spices and peach pips.

Brandy (alcohol percentage 35–38%) is used in the Netherlands for, among other things, the bottling of (dried) fruit, such as raisins in brandy (farmers' boys), and apricots in brandy (farmers' girls) and for the production of advocaat.

Advocaat is a thick liquid drink produced with eggs, brandy and sugar. The alcohol percentage is about 15 per cent. The bottle should be kept in the refrigerator once it has been opened and can only be kept for a short time.

Distillates of fruit wines are sold under many names. Mostly the name of the fruit is recognizable in the name of the product. For instance: Kirsch, Eau-de-Vie de Framboise, Eau-de-Vie de Poire Williams.

Liqueur is a collective name for distilled drinks which are produced with sugar, fruit juices, extracts of spices and/or fruit. The alcohol percentages vary from 20 to 50 per cent. In the kinds produced in northern Europe the name relates to the aromatic compound used. For instance anisette (aniseed), coffee liqueur or banana liqueur. The word 'crème' in the name relates to an extra high sugar content (crème de cacao).

Liqueurs are mostly drunk with the dinner coffee (*pousse-café*). They serve further as ingredients in cocktails (mixtures of drinks). They are also often used in the kitchen in the preparation of desserts, gâteaux and fruit salads.

The foreign **monastery liqueurs** are famous. These mostly herbal liqueurs were and are prepared according to secret recipes. The alcohol percentages are mostly about 40–50 per cent. Some examples:

Fig. 31.6 — Coffee liqueur.

— **Benedictine** is a French monastery liqueur, dedicated by the monks to God, Most Good, Most Great (Deo Optimo Maximo). This labelling, D.O.M., is still to be found on this herbal liqueur.
— **Chartreuse** is a liqueur based on alpine herbs from the surroundings of Grenoble. Chartreuse with a yellow label contains 43 per cent alcohol; with a green label, 55 per cent.
— **Cointreau** is a colourless liqueur containing orange juice.
— **Grand Marnier rouge** is a liqueur based on cognac and orange extract. **Grand Marnier jaune** is based on neutral alcohol and orange extract. It is therefore cheaper than the *rouge* label.

31.5 LEGISLATION

Legislation may produce a number of different regulations which apply to alcoholic drinks and their sale.

Beer Ordinances give the regulations concerning composition, labelling, adjuncts, colour and taste.

Wine Decrees may set requirements for the composition of wine and fruit-based wine, especially concerning the alcohol percentages and additives. Also, organizations are mentioned which can supply certificates of origin or authenticity of wines from countries such as France, Germany, Greece, Spain, Portugal and South Africa.

Ordinances from the Marketing Boards for distilled drinks regulate the authentic naming of genever, advocaat, and other well-known distilled liquors and liqueurs.

They may further lay down the business rules of wholesalers and importers of distilled drinks.

Other regulations describe the sale of alcoholic drinks, and lay down rules on the management of off-licences. These rules include a prohibition on giving or selling alcoholic drinks to persons who are under age (usually 16, 18 or 21).

32

Vinegar

32.1 INTRODUCTION

Vinegar is a liquid which contains at least four per cent acetic acid and sometimes also other aroma and flavour compounds, but no nutritious substances.

Vinegar is made mostly from alcohol and has been known for a long time. It was already being used by old civilizations like the Egyptians, Babylonians, Indians, Greeks and Romans. In the Bible it is spoken of as '*edik*'.

Vinegar was made in a very simple way. Wine or beer was left to stand in an open vat. In this way, the liquid came into contact with air and acidified.

In the Middle Ages, vinegar was used in the preparation of meals, but also to improve the keeping quality of foodstuffs and as medicine for pestilence and fever.

32.2 VINEGAR PRODUCTION

Vinegar can be produced in two ways: by a natural and by a chemical method.

In the **natural method** the acetic acid bacteria are responsible for the change from alcohol to acetic acid. This is **natural vinegar**. The process is called acetification or acetic fermentation; however, it is in fact an oxidation process. From the alcohol in light alcoholic liquids, acetic acid develops through the action of vinegar bacteria, for which oxygen is needed. The acidification develops fairly slowly (weeks to months). This process can be speeded up by increasing the surface area. For a long time, wooden barrels filled with beech shavings have been used for this. The vinegar bacteria attach themselves to the shavings and air can penetrate everywhere. The production then takes only a few weeks. The conversion goes faster (about six days) if a mixture of vinegar, alcohol-containing liquid and nutrients are dumped over the beech shavings. Air is also blown through on a counter-current principle.

In 'fast production' the process tasks 1.5 to 2 days. Alcoholic liquid is pumped in a high tower (acetator) and air is sprayed in as a stream of bubbles, to which the acetic acid bacteria attach themselves. The contact with oxygen and nutrients is very intimate. This saves a lot of time, but produces a vinegar which has matured less and is less aromatic.

There are no acetic acid bacteria necessary in the **chemical acetic acid production**. By way of a number of chemical reactions, strongly concentrated acetic acid is produced from calcium carbide.

Pure acetic acid is also formed (as well as methanol and acetone among other compounds) in the dry distillation of wood. This produces vinegar after dilution with (pasteurized or sterilized) water.

32.3 KINDS OF VINEGAR

Vinegar is water with four per cent acetic acid. This vinegar is sold as a colourless or as a brown liquid. In the latter case it is coloured with caramel. The taste is neutral acid. If the vinegar has been obtained from alcohol (natural method) then the labelling **natural vinegar** is used. Vinegar with X per cent lemon juice is a mixture of vinegar with the mentioned percentage of lemon juice. This mixture may contain sulphurous acid, which however may only originate from the added lemon juice.

Double vinegar contains at least eight per cent acetic acid. This is also labelled as pickling vinegar and is used to preserve foodstuffs such as gherkins. It is less suitable for the preparation of meals, but more suitable as cleaning vinegar.

Herb vinegar is an aromatic chemical or natural vinegar to which natural extracts of herbs are added. If the extract of only one herb is used, for instance tarragon, then it is called tarragon vinegar. Sometimes it is sold with a sprig of the herb in the bottle.

Wine vinegar is produced with wine as the basic liquid. The acetic acid percentage is a minimum of 5.5 per cent. The smell, colour and taste are especially determined by the compounds which exist in the basic solution. If sherry is used as the basic liquid then it is sold as sherry vinegar.

Apple vinegar is produced from apple cider and has a fresh apple taste. The actetic acid content is a minimum of 5.5 per cent.

Fruit vinegar is produced from an alcohol-containing liquid based on fruit. So raspberry vinegar, currant vinegar, and raisin vinegar are for sale. The acetic acid percentage is a minimum of 5.5 per cent.

Vinegar essence is a liquid with 80 per cent acetic acid obtained through chemical means.

Acetic acid solution contains 12.5 to 80 per cent acetic acid. Both kinds have to be supplied with a label with the inscription 'dangerous if used undiluted'. Table vinegar can be obtained after dilution.

32.4 QUALITY DETERIORATION, SPOILAGE AND STORAGE

Vinegar has to be clear and to have a good smell, colour and taste, while wine vinegar may contain a little sediment.

Spoilage
Chemical vinegar cannot spoil. There is no nutrient for spoilage-causing microorganisms because the water has been made germ-free by pasteurization or sterilization.

Natural kinds of vinegar, despite pasteurization, are sensitive after opening to:

— *Bacteria:* through the activity and growth of these, little flakes are visible.
— *Vinegar eel*: these exist on the bacteria which develop in the vinegar. They are 1 to 2 mm long. They make the vinegar murky but are not detrimental to humans.
— *Development of skin mould*: this is caused by a fungus. First, it floats on top of the vinegar, but can, if the quantity formed becomes very large, precipitate in small pieces.

Storage

Unopened bottles can be kept for an unlimited period, provided they are pasteurized. Chemical vinegar will not spoil even after opening.

Opened bottles of natural vinegar have to be stored cool and well closed. The above-mentioned forms of spoilage will then occur less quickly. Wine vinegar and fruit vinegars are more sensitive to spoilage and are therefore better kept in the refrigerator.

32.5 LEGISLATION

Vinegar Regulations set out requirements concerning labelling, type and composition of kinds of vinegar.

Appendix: Literature review

As this book was first produced in Dutch, the original Dutch references are included as an Appendix. For futher information on many of the subjects covered, the reader should consult other books in the Ellis Horwood Series in Food Science and Technology (see the list facing the title page of this book).

Alternatieve Landbouwmethoden, Rapport Commissie onderzoek biologische landbouwmethoden, Pudoc, Wageningen 1977.
Alternatieve Land- en Tuinbouw, Infotitel nr.2, Ministerie van Landbouw en Visserij, 's-Gravenhage 1981. Annual Abstract of Statistics, Central Statistical Office, HMSO.
Bakkum, A., Kruidenboek, Het Spectrum, Utrecht/Antwerp 1976.
Becht, G., Levensmiddelenhygiene, De Tijdstroom B.V., Lochem 1974.
Belderok, B., Ontwikkelingen in de Franse Maalindustrie en bakkerij, Voedingsmiddelentechnology nr. 2, jrg. 19 (1986).
Bailey, A., The blessings of bread, Paddington Press Ltd, London 1975.
Bothma, F., Het brood in Nederland TNO-project, Institut voor Graan, Meel en Brood 1(9) 343–347, 1973.
Bouterse, M., Wie doet wat op voedingsgebied, Nederlands Institut voor de Voeding, NIVV, Wageningen 1983.
Bij en honing, Mellona/Adelshoeve, Santpoort.
Brouwer, W., Suikerwerk en drop, Voedingsmiddelentechnologie, nr. 6, jrg. 19 (1986).
Buishand, T., Groenten uit alle windstreken, Het Spectrum 1986.
Champignonteelt in Nederland, Proefstation voor de Champignoncultuur, Horst.
Coenders, A., Spectrum Kruiden en Specerijenatles, Het Spectrum, Utrecht/ Antwerp 1979.
Compendium Dieetpreparaten en Voedingsmiddelen, de Toorts, Haarlem 1986–1987.

Appendix: Literature review

Dairy Handbook, Alfa-Laval, Sweden.
Day, A. and Stuckey, L., Groot Kruidenkookboek, ed. Wina Born, Born, N.V., Assen-Amsterdam.
Deelen, W. van, Enzymatische winning van groente-en vruchtesappen, Voedingsmiddelentechnologie, nr. 22, jrg. 17 (1984).
Diepvries, Unilever (3) 1975.
Dokkum, W. van, Additieven en contaminanten, Bohn, Scheltema en Holkema, Voeding in de practijk, March 1985.
Dool, H. v.d., Aromastoffen, Voedingsinformatie, jrg. 4, nr. 8 (1981).
Dowell, P. and Bailey, A. The Book of Ingredients, Dorling Kindersley Ltd, London 1980.
EEG-zoetstoffenrapport, Committé voor de menselijke voeding, Voedingsmiddelentechnologie, nr. 7, jrg. 18 (1985).
Eekhof-Stork, N., The World Atlas of Cheese, Paddington Press Ltd, New York and London 1976.
Fieliettaz Goethart, R. L. de, Zoetstoffen, eigenschappen en toepassingen, Voedingsmiddelentechnologie, nr. 9, jrg. 18 (1985).
Franke, W., Nutzplantzenkunde, Thieme Verlag, Stuttgart 1976.
Frederiks, A., Verpakt nog aan toe!, Stichting Verpakking en Milieu, 's-Gravenhage.
Gabriel, J., Kruidengids, L. J. Veen, Wageningen 1970.
Glucose informatie nr. 1, jrg. 12, Vereniging van Nederlandse Glucose fabrieken.
Goede, J. de. et al., Onze Levensmiddelen, Nijgh & Van Ditmar Educatief, Den Haag 1978.
Goock, R., Kruiden en specerijen van A tot Z, Strengholt, Naarden 1974.
Hartog, C. den, et al., Nieuwe Voedingsleer, Het Spectrum, Utrecht/Antwerp 1978.
Hawthorn, J., Foundations of Food Science, W. H. Freeman and Company, Oxford 1981.
Hempsphill, R., Kruiden, Publisher Gottmer, Haarlem, z.j.
Heus, J. de, Koffie en thee, Voedingsmiddelentechnologie, nr. 3, jrg. 14 (1981).
Houwing, H. and Moerman, P. C., Het ontdooien van vis en vlees, Voedingsmiddelentechnologie, nr. 6, jrg. 15 (1982).
Huizinga, J. et al., Handboek Vlees en Vleesproducten, Agon Elsevier, Amsterdam 1972.
Jaarverslag 1982, Institut voor Graan, Meel en Brood TNO, Voedingsmiddelentechnologie, nr. 18, jrg. 16 (1983).
Jaarverslag, Productschap voor Pluimvee en eieren, Zeist 1982.
Jackson, M., Spectrum Bieratlas, Het Spectrum, Utrecht/Antwerp 1978.
Jansen, L. A., Boter, Voedingsmiddelentechnologie, nr. 25, jrg. 16 (1983).
Johnson, H. The World Atlas of Wine, Mitchell Beazly, London 1971.
Jukes, D. J., Food legislation of the UK — A concise guide, 2nd edn. Butterworths, Guildford, 1987.
Katan, M. B. et al., Vetzuursamenstelling, Voeding nr. 4, jrg. 15 (1984).
Koning, P. J. de, De definitie van UHT melk, product-en/of procesgebonden?, Voedingsmiddelentechnologie, nr. 9, jrg. 17 (1984).
Koorengeval, J., et al., Handboek voor de keuken, Kluwer, Amsterdam, z.j.

Kwaliteitsvoorschriften verse groenten en vers fruit, Het Productschap voor Groenten en Fruit.
Langerak, D. and Stegeman, H. Het effect van bestralen, verpakking en bewaren op de kwaliteit van spacerijen en kruiden, Voedingsmiddelentechnologie, nr. 20, jrg. 16 (1983).
Lankhuysen, H. J. van, Veranderingen in de gehalten aan fytinezuur tijdens broodbakken met gist en met zuurdesem, Voeding nr. 3, jrg. 46 (1985).
Leclerq, E., Groente als grondstof voor biotechnologische processen, Voedingsmiddelentechnologie, nr. 8, jrg. 17 (1984) 21.
Lord, T., Spectrum Drankatlas, Het Spectrum, Utrecht/Antwerp 1980.
Maaker, J. de, Invloed van kleinverpakking op het kwaliteitsbehoud van groente en fruit tijdens de afzet, Voedingsmiddelentechnologie, nr. 9, jrg. 17 (1984).
Man, J. M., Principles of food chemistry, Publishing Company, Inc., Westport, Connecticut.
Massart, D. L., Deelstra, H. and Hoogewijs, G., Vreemde stoffen in onze voeding, De Nederlands Boekhandel, Antwerp/Amsterdam 1980.
Melkeiwiten: eigenschappen en toepassingen, Nederlands Zuivelbureau, Rijswijk.
Melkveehouderij en zuivel, Infotitel nr. 12, Ministerie van Landbouw en Visserij, Den Haag 1985.
Melkwegwijzer-van boerderij tot koelmeubel, Het Nederlands Zuilvelbureau, Rijswijk.
Meursing, E. H., Cacao, Voedingsmiddelentechnologie, nr. 3, jrg. 14 (1981).
Moore, E., Food preservation, Unilever educational book (3), 1980.
Nickerson, J., et al., Elementary Food Science, Westport, Connecticut, USA 1978.
Nota Voedingsbeleid, Tweede kamer vergaderjaar 1983–1984, 18 156 nr. 1–2.
Ong, T. L., Margarine, vetten en olien, Voedingsmiddelentechnologie, jrg. 15, mei 1982.
Pauli, E., Het complete leerboek voor de keuken, S.V.H., Zoetermeer, 1981.
Peterson, S., Encyclopaedia of Food Science, Publishing Company, Inc., Westport Connecticut, USA 1978.
Pijpers, D. et al., Fruit uit alle windstreken, Het Spectrum, 1985.
Product gegevens groente en fruit, Mededelingen nr. 30, Sprenger Institut, Wageningen.
Reiningh, W. J. C., Vleeskeuringswet en Destructiewet, Edition Schuurmans & Jordens, Tjeenk Willink, Zwolle, 1973.
Rombouts, F. M. and Kampelmacher, E. H., Voedselvergiftiging, from: Natuur en Techniek cat. nr. 280, nr. 10 (1973).
Samenstelling van melk en zuivelproducten, Wetenschappelijke notities op voedingsgebied, nr. 1, jrg. 12 (1985).
Schild, H. van de, Frites en puree, Landbouwkundig Tijdschrift nr. 11 (1985).
Sluimer, P., Deegbereiding met zes verschillende snelkneders, Voedingsmiddelentechnologie, nr. 24, jrg. 18 (1985).
Smak, C., Broodbereiding in Nederland, Voedingsmiddelentechnologie, nr. 8, jrg. 5 (1972).
Staarink, T. and Hakkenbrak, P., Het Additievenboekje, Staatsuitgeverij, Den Haag 1984.

Staarink, T. and Hakkenbrak, P., Het Contaminantenboekje, Staatsuitgeverij, Den Haag 1984.
Steward, G. T., Introduction to Food Science and Technology, Academic Press, London 1973.
Vletter, R. de, De Suikerindustrie, Voedingsmiddelentechnologie, nr. 25, jrg. 15 (1982).
Vliet, C. B. van, Verpakking en distributie van lang houdbare bakkerijproducten, Voedingsmiddelentechnologie, nr. 24, jrg. 18 (1985).
Voedingsmiddelenjaarboek 1986–1987, Uitgave van Postbus 268, 3700 AG Zeist, Nederland.
Voedingsmiddelentechnologie (VMT), jrg. 13 (1980) until jrg. 20 (1987).
Voedingsmiddelentabel, Voorlichtingsbureau voor de Voeding, 's-Gravenhage (34) 1983.
Voedingsinformatie, Voorlichtingsbureau voor de Voeding, 's-Gravenhage, nr. 1, jrg. 3 (1980) until nr. 3, jrg. 10 (1987).
Voedsel, gezondheid en kwaliteit, Infotitel nr. 6, Ministerie van Landbouw en Visserij (3) 1978.
Voedselconservering door straling, Cahiers Bio-Wetenschappen en Maatschappij, Leiden, nr. 2, jrg. 10 (1985).
Waard, H. de, *et al.*, Het metabolisme van D(−)melkzuur, Voedingsmiddelentechnologie, jrg. 16 (1983).
Warenwet, Koninklijke Vermande B.V., Lelystad.
Werkgroep, Voedsel (V.W.M.), Voedsel, van Gennep, Amsterdam 1983.
Winkler Prins Cullinaire Encyclopedie, B.V. Uitgeversmaatschappij, Amsterdam/Brussels 1984.
Zuivel en Voeding, Nederlands zuivelbureau, Rijswijk, jrg. 4 (1982) until jrg. 8 (1986).

Index

acacia honey, 319
acceptable daily intake (ADI), 63
acetic acid, 60, 361
acetic acid bacteria, 37
acid inoculum in cheese, 153
acid rigor, 83
acrolein formation in fats, 33
additives, 17, 58–62
    in bread, 216–17
    legislation, 27, 62–3
aduki beans, 227
adzuki beans, 227
aflatoxin B1, 66
aflatoxin M1, 66
agar-agar, 60, 205, 208
Akwarius, 22
albumen, 170
alcohol
    as preservatives, 52, 53–4
    in foods, 37
alcohol-free fruit wine, 353
alcohol-free liqueur, 339
alcholic drink
    composition, 342
    consumption, 341
    legislation, 358–9
aldehyde rancidity, 33
algae, 22
alginates, 207–8
alginic acid, 207–8
alkaline rigor, 83
Allison's bread, 220, 221
allspice, 297
almonds, 276
Alsace wines, 349
alternative agriculture, 18
Alto Adige, 350
am choi, 248
Amsterdammer, 158
amylopectin, 199, 205, 206
amylose, 205, 206
anabolic steroids, 70
anchovy, 113, 120
angel root, 281

angelica root, 296
angelica stalk, 281
Anisakis marina, 40, 110
aniseed, 296
annatto, 153
Anog-agriculture, 19
anthocyanins, 255
antibiotics, 64, 70
anti-coagulants, 62
anti-foam compounds, 62
anti-oxidants, 59–60
ants, 41
apéritif wine, 347
apple banana, 275
apple sauce, canned, 23
apple snail, 131
apple vinegar, 361
apples, 46, 262–3, 266
apples, dried, 279
apricots, 271
apricots, dried, 279
arabica coffee, 321
Ardennes ham, 89, 90
armagnac, 356
aromatic salt, 303
arrow squid, 132
arrowfoot, 207
arsenic, 64, 65
artichoke, 250
ascorbic acid, 54, 59, 62
asparagus, 24, 241, 252
aspartame, 313–14
Aspergillus flavue, 39
aspic, 95
aubergine, 250, 259
Austrian wine, 352
autolysis, 84

Babassu oil, 180
baby Gouda, 158
baby milk food, 143
Bacillus cereus, 38
bacon, 90
    salted, 87

Index

bacon pigs, 76
bacteria, 37, 84
bacteriological blowing, 34, 38
bactofugation, 45
baker's yeast, 211, 215
bakers' products, 221–2
baking and frying butter, 149
Balkenbrij, 89
bamboo shoots, 252
bananas, 273–4
Bantu, 202
barley, 186, 188, 189, 196–8
basil, 293
bavarois, 141
bay leaves, 298
beans, 226–8, 241
beef, 73–5, 76
   smoked, 89, 90
beef tomato, 251
beer, 60, 342–5
beer foam, 208
beeswax, 317
beet sugar, production of, 306–8
beetles, 41, 204
beetroot, 245, 246
beetroot red, 60
Benedictine, 358
bent beaks, 226
benzoic acid, 35, 54, 59, 256
benzopyrenes, 53
Bergumer, 157, 159
Berlin liver sausage, 92
beschuit, 221
BHT, 59
Bicheka cheese, 153
bigarreaux, 281
biological-dynamic agriculture, 19, 22
biphenyl, 59
biscuits, 32, 221, 222
bitter oranges, 267
bitters, 357
black beans, 228
black bilberry, 272
black candy, 309
black currant, 272
black grouse, 101
black pepper, 297
black pudding, 87, 92, 93
black rice, 186
black Spanish radish, 245
black tea, 326–7, 328
blackberries, 273
bladder worms, 70
blast-freezing, 47
blewit, 286
blood oranges, 267
blood sausage, 92
blowfly, 83
blue moulds, 157
Blue Stilton, 157
Bo(c)k beer, 344
boar, wild, 99, 100

Boerhaavese candy, 309
boiling chickens, 104
Bold cheese, 158
boletus, 286
borage, 293
Bordeaux whines, 347
borlotti beans, 227
bottling, 38
botulinum, 84
botulinum cook, 38, 56
botulism, 37
Bourbon, 356
Boursin, 156, 164
Brabants ryebread, 220
brains, 80
bran, 193
brandy, 357
brawn, 89
brazil nuts, 277
bread
   composition, 212–13
   consumption, 211–12
   health hazards, 213
   improvers, 216–17
   introduction, 211–13
   iodine in, 63
   kinds of, 218–21
   legislation, 224
   production and distribution, 213–17
   production process based on yeast dough, 217–18
   quality deterioration, spoilage and storage, 222
   retrogradation of, 32
   staleness, 222–3
bread salt, 212, 303
bread wheat, 193
breadcrumbs, 221
breakfast bacon, 90
Bresse poularde, 104
Brie, 156, 157
brill, 113
brining of fish, 119
broad beans, 254
broccoli, 251
broilers, 104
broken rice, 200
broken tea, 328
brown beans, 225, 226
brown bread, 219
brown shrimp, 126
Brunswijker spreading liver sausage, 92, 93
Brussels sprouts, 241, 248
buckwheat, 186, 202–3
buckwheat flour, 203
Bulgarian yoghurt, 141–2
bulgur, 194
Burgundy wines, 347
butter canon, 147
butter
   consumption, 145
   kinds of, 148–50
   legislation, 150

Index

price management, 147
production and distribution, 146–8
quality deterioration, spoilage and storage, 150
testing, 149–50
buttermilk, 60, 140, 142
butylated-hydroxytoluene (BHT), 59
butyric acid, 176

CA method (Conditioned Atmosphere or Controlled Atmosphere), 52
cabbage lettuce, 248
cabbage seed oil, 177, 180
cadmium, 64, 65
caffeine, 321
caffeine-free tea, 328
calamari, 132
calcium chloride in cheese, 154
calvados, 357
Camellia assamica, 327
Camellia sinensis, 326
Camembert, 157
camping margarine, 184
Campylobacter, 84, 105
Campylobacter jejuni, 37
candy sugar, 309
candy syrup, 310
can sugar production of 308–9
canned pommes parisiennes, 236
Cantaloupe melons, 270
capers, 297
caramel, 60, 61, 62
caramel custard, 60
caramelization, 33
carbohydrates
  in fish, 110
  in meat, 69
  products based on, 205–9
carbon dioxide, 62
carbonates, 62
cardamon, 297
cardoon, 253
carob, 334
carob seed flour, 208
B-carotene, 153
carotenoids, 60, 255
carp, 111, 116
carrageen, 205, 208
carraway seed, 296
carrots, 239, 241, 244
casein, 209
cash-and-carry stores, 23
cashew nuts, 276
casino bread, 220
Casseler rib, 89, 90
caster sugar, 309
cauliflower, 239, 241, 251
caviar, 120
Cayenne pepper, 298
celeriac, 245
celery, 252, 291, 292, 296
cellulose compounds, 60
centrifuge honey, 317

cep, 286
cereals
  composition, 187
  consumption, 186
  fruit, structure of, 188–9
  health hazards, 187
  kinds of, 191–203
  legislation, 204
  production and distribution, 189–91
  quality deterioration, spoilage and storage, 203–4
cervelet sausage, 93, 94
Champagne, 347–9
chanterelle, 285, 287
chard, 248
Chartreuse, 358
Cheddar cheese, 156, 157
cheese
  carotenoids in, 60
  cheese-making, 45
  cheese mark, Government, 166–7
  classification of, 155–8
  composition, 152
  consumption, 151–2
  Dutch, production of, 153–5
  fresh, 162–4
  kinds of, 158–60, 160–2
  legislation, 166
  maturing and maturation times, 155
  processed, 164–6
  quality deterioration, spoilage and storage, 160–2
  soft, food poisoning, 39
cheese powder, 160
cheese spread, 165
chemical blowing, 34
chemical preservatives, 52, 54, 256
chemotherapeutics, 64
cherries, 270–1
cherry tomatoes, 252
chervil, 291, 292, 293
chestnuts, 275, 276
chevre, 155
Chianti, 350
chick peas, 228
chicken croquettes, 106
chicken eggs, 172
chicken roularde, 107
chicken schnitzels, 106
chickens, 47, 57, 101, 103–4
chicory, 239, 241, 248, 250
chilli pepper, 298
chilli powder, 298
chilling, 45–6
Chinese cabbage, 248
Chinese restaurant syndrome, 35
Chinese shrimp, 126
Chinese vermicelli, 201
chip sauce, 181
chips, 236
chives, 245, 291, 293
chlaza, 170

Index

chlorogenic acid, 321
chocolate bonbons, 333
chocolate butter, 334
chocolate couverture, 333
chocolate milk, 140
chocolate spread, 334
chocolate, 32, 333
   see also cocoa
Christmas bread, 220
Christmas butter, 148
cider, 353
cinnamon, 298
citric acid, 60
citrons, 269
clarification of milk, 137
Clostridium botulinum, 27, 37–8, 56, 59, 110
Clostridium perfringens, 38, 84, 110
clover honey, 318
cloves, 297
coarse sliced sausage, 93
coastal fishing, 111
Coburg ham, 90
cochineal, 60
cockles, 124, 131
cockroaches, 41
cocktail cherries, 281
cocoa, 50
cocoa and chocolate, 50
   composition, 331
   consumption, 330
   legislation, 334
   production and distribution, 331–3
   products, 333–4
   quality deterioration, spoilage and storage, 334
cocoa butter, 333
cocoa fantasy, 333
cocoa fat, 331
cocoa powder, 332, 333
coconut, 276–6
coconut oil, 180
cod, 114
cod liver oil, 114
Codex Alimentarius, 28
coeliac disease, 187
coffee
   composition, 320–1
   consumption, 320
   legislation, 324
   production and distribution, 321–3
   products 323–4
   quality deterioration, spoilage and storage, 324
coffee extract, 324
coffee milk, 139
coffee powder, 324
coffee substitute, 324
coffee syrup, 324
coffee whiteners, 140
cognac, 354–5
Cointreau, 358
cold-air drying, 49
cold shortening, 82
cold shrinkage, 82

cold-store butter, 149
coleslaw, spoilage in, 39
coley, 114
colouring, 17
comb honey, 317
commercial sterility, 55
composite meat products, 86
condensed milk with sugar, 139–40
confectionery, 310
confectionery shops, 22
conger eel, 116
consommé, 132
Consumer's Association, 30
consumerism, 30
contact-freezing, 47
contaminants, 27, 63–6
cooked meat products, 94–5
cookies, 221, 222
cooking/frying fat, 181
cooking/frying oil, 180
cooking sausages, 92
coquettes, pre-fried, 24
Coquilles Saint Jacques, 130
coriander, 296
corn, 186
corn flakes, 195
corn oil, 180
corn salad, 248
corn weevil, 204
corned beef, 89, 90
cornflour, 195, 205, 206
cornseed oil, 180
Corynebacteria, 157
cottage cheese, 162, 164
country minced sausage, 93
courgette, 250
couscous, 194
cowberry, 272
crabs, 124, 126–7
cracked rice, 200
crackers, 221
cranberry, 272
crawfish, 128
crayfish, 128
cream, 139
cream cheese, 158
cream sherry, 351
creatine, 69
creatinine, 69
crème fraîche, 139
cress, 296
Creuses de Zelande, 130
crevette grise, 126
crisp bread, 220
crisps, 52, 236, 238
croissants, 220
croquettes, 95
crust margarine, 184
crustaceans
   consumption, 124
   composition, 124–5
   definition, 124

Index

health hazards, 125
    kinds of, 125–8
    products of, 133
    quality deterioration, spoilage and storage, 133
crystal sugar, 309
crystallized fruit, 280–1
crystallization, 32
cucumber herb, 293
cucumbers, 239, 241, 250, 259
cumin, 296
curdling of cheese, 153
curds, 153
currants, 278
custard, 141
custard powder, 206
cut koek, 221, 222
cuttle-fish, 132
cyclamate, 313

dab, 111, 114
dairy ice-cream, 142
Danish Blue, 157
dates, 265, 279
decaffeinated coffee, 17, 324
deep-freezing, 45, 46–8
    of fish, 119
    of meat, 84
deep-litter system, 170, 171
deep-sea fishing, 111
dehydration, 32
delicatessens, 22
Demeter bread, 220, 221
department stores, 23
depot fat, 69
dessert quark, 164
deterioration of meat products, 94
dextrin-maltose, 313
dextrose, 312
diet cheeses, 160
diet margarine, 185
diet products
    fish, 122
    meat, 94
    milk, 143
diet salt, 304
diethylstilboestrol, 70
dill, 291, 293
dimentylpolysiloxane, 62
diphenyl, 260
distilled liquors, 354–8
dogfish, 113
Dortmund beer, 344
Double Gloucester, 158
double vinegar, 361
dough products, 193–4
dragees, 310
drain honey, 317
Dremts, 220
dried fruit, 278–9
drum-drying, 49
dry cooking rice, 199
dry-salting of fish, 119

dry wine, 347
drying
    of fish, 120
    of meat products, 89
    preservation by, 48–50
    with hot air, 49
Dublin Bay Prawn, 128
duck, 98, 101, 103, 104
duck eggs, 168, 172
durum wheat, 193
Dutch cheese, production of 153–5
Dutch process, 333
Dutch rusks, 221
Dutch shrimp, 126
Dutching, 333
dyestuffs, 60

E-number, 62
Earl Grey tea, 328
ecological agriculture, 18, 22
ecological bread, 220–1
economy meat products, 86
Edam cheese, 154
EDTA, 256
eel, 111, 117
eel sausage, 92
effervescent wine, 346
egg
    composition, 169
    consumption, 169
    introduction, 168–9
    legislation, 174–5
    production and distribution, 170–2
    products, 172–3
    quality deterioration, spoilage and storage, 173–4
    structure of, 169–70
egg yolk, 209
eggshell, 170
einkorn, 186, 191
Emmental cheese, 156, 157
emmer, 186, 191, 193
emulsifiers, 60
endive, 57, 239, 247
enzymatic browning, 33
ergot, 203
ergotine, 187
ergotism, 66
erucic acid, 177
escargot, 131
espresso, 324
esterification, 179
extracellular proteins, 68
extramuscular fat, 69

faggots, 23
farmers' girls, 282
farmers' minced sausage, 93
farmhouse butter, 147, 149
farmhouse cheese, 155
Farmhouse Leiden, 158
Farmhouse-Gouda, 158

Index

fascine, 226, 240
fats and oil
  acrolein formation in, 33
  auto-oxidation of, 32
  in cheese, 156
  in fish, 110
  hydrolysis of, 33
  in meat, 69
  oxidation, 33
  polymerization of, 33
  products based on, 209
  quality deterioration, spoilage and storage, 181, 185
  types, 179–80
fatty acids, 33
fennel, 296
fertilizers, 18
feta cheese, 159
field mushroom, 284
figs, 279
filet Americain, 86
filled chocolate, 333
fire-blast mould, 203
fish, 108–23
  composition, 109–10
  consumption, 109
  families, 109
  fat content, 109
  gutting, 112
  heading, 112
  health hazards, 110
  kinds of, 112–17
  legislation, 123
  preservation by chilling, 46
  production and distribution, 111–12
  products, 117–22
  quality deterioration, spoilage and storage, 122–3
  salt-water, 113–16
  stripping, washing and cooling, 111–12
fishing, 111
fishmeal, 120
flageolets, 227
flakes, 193, 195
flash-heating of milk, 137
flatfish, 109
flavonoids, 255
flavour precursors, 50
flavours, 61, 95–6
flies, 41
floorbread, 218
Florence fennel, 244–5
flounder, 111, 113
flour, 159, 190, 191, 213–14
flour improvers, 62, 217
flour mite, 203
flour moth, 204
flower honey, 318
fluidized bed-drying, 49
Food and Agriculture Organization (FAO), 28
food colours, 62
food contamination, 35

food infection, 35
food legislation, 26–30
food poisoning, 27, 35, 39
food production, 18–19
foot and mouth bacteria, 136
fortified wine, 347
fowl, *see* poultry
fractionation, 179
free-range pigs, 77
freeze-drying, 49
freeze-dried coffee, 17
freezing and spoilage, 27, 32
French beans, 253
French oyster, 130
French quark, 164
French wines, 347–9
fresh cheese, *see* quark
fresh cream cheese, 162, 164
fresh-water fish, 116–17, 118
fresh-water fishing, 111
fricandeau, 89
fried herring, 121
fried liver sausage, 92, 93
fried mince meat, 89
fried roastbeef, 89
Friesian, 157
Friesian clove cheese, 156, 158–9
frogs' legs, 131, 132
frozen meals, 23
fructose, 312
fruit bread, 220
fruit
  canned, 55
  composition, 259–60
  consumption, 259
  enzymatic browning, 33
  health hazards, 260
  kinds, 261–75
  legislation, 282
  pectinase in, 33
  price, 24
  production and distribution, 260–1
  quality deterioration, spoilage and storage, 274–5
fruit juices, 50, 337–8
  filtered, 45
  legislation, 340
  quality deterioration, spoilage and storage, 339–40
fruit lemonade, 338
fruit lemonade syrup, 339
fruit preserves, 278–82
fruit quark, 164
fruit vinegar, 361
fruit wine, 353
frying blood sausage, 92, 93
fungi
  composition, 283–4
  consumption, 283
  health hazards, 284
  kinds, 284–8
  legislation, 289

Index

preserved, 288
quality deterioration, spoilage and storage, 288–9
fungicides, 18, 64

Gamba, 126
game
  composition, 98
  consumption, 97
  feathered, 100–1
  furred, 98
  hanging and gaminess, 98, 99
  high, 98
  kinds of, 98–101
  legislation, 107
  paunching, 98
  products, 105–7
game galantine, 107
game pâté, 107
gammon, cooked, 90
garden cress, 248
garlic, 245, 247, 297
gas-packaging, 51, 52, 85
gases, storage, 62
gateaux, 47
geese, 101, 103, 104
gelatine, 60, 95, 205, 209
Gelderland smoked sausage, 55
gelling aids, 60
Genever, 356
German millet, 201
German quark, 164
German wines, 349
gherkins, 250, 256
giant crab, 127
gin, 356
ginger, 281, 298–9
ginger root, 281
gliadin, 213
glucose, 311
glucose syrup, 312
glutamates, 61
gluten, 193, 213
gluten-free bread, 221
gluten protein, 187
glutenin, 213
glycerol, 33
goat milk, 134
goats' milk, 80
goats' milk cheese, 151, 155
Goela djawa, 309
goitrogens, 226, 240
Golden Syrup, 310
gomasio, 303
goose eggs, 168, 172
goose liver pâté, 107
gooseberry, 272, 273
Gorgonzola, 157
Gouda, 152, 153, 154, 156, 158, 164
Gouwetaler, 149
Graminaea, 201
Grand Marnier, 358

grape sugar, 311
grapefruit, 269
grapes, 272–3
grated cheese, 159–60
gravy, 96
gravy cubes, 96
gravy granules, 96
gravy tablets, 96
Greek wine, 351–2
green chicory, 248
green consumers, 30
green movement, 27
green peas, 228
green pepper, 297
green tea, 328
grits, 197, 200, 203
groats, 197
ground cheese, 159–60
ground coffee, 323–4
groundnut oil, 179, 180
grouse, black, 101
growth regulators, 261
Gruyère, 156, 157
guar seed flour, 208
Gueuze, 344
gueuze-lambic beer, 344
guinea fowl, 101, 103, 104–5
gum arabic, 208
gum xanthan, 208
gums, 60

Haagse liver sausage, 92, 93
haddock, 114
haddock liver, 121
hake, 111, 117
half cream, 139
halibut, 111, 113
halvarine, 135
ham, 86
  canned, 89
  raw, 87, 89, 90
  shoulder, 90
hamburgers, 23
hare, 98, 99–100
haricots verts, 253
Hausmacher sausage, 92, 93
hazelnuts, 275, 276
heart, 80
heather honey, 318
heavy metal contaminants, 64, 65, 240, 260
heavy salting of herring, 119
hepatitis-A, 39
herb and spice mixtures, 299–300
herb butter, 149
herb extracts, 299
herb salt, 303
herb vinegar, 361
herbicides, 64, 70, 240, 260
herbs
  flavour compounds in, 292–3
  introduction, 290–2
  legislation, 301

Index

quality deterioration, spoilage and storage, 300–1
types 293–7
herring, 109, 113, 121
herring fillets, 121
herringworm, 40, 110
Hickory Salt, 303
homogenized vegetables, 256
homogenizing of milk, 138
honey
    composition, 315–16
    consumption, 315
    kinds of, 318–19
    production and distribution, 316–18
    quality deterioration, spoilage and storage, 319
Hoofdkaas, 89
hormones, 64
horse meat, 69, 80, 90
horseradish, 297
household syrup, 310
hunt sausage, 92
Huttenkase, 164
hydrogenation, 179
hygiene and health, legislation, 27
hyssop, 293

ice chocolate, 334
ice-cream, 60, 142–3, 208
    storage, 48
    viruses in, 39
Iceberg lettuce, 248
icing sugar, 62, 309
imitation marzipan, 229
imitation whipping cream, 139
infant milk, 143
inkfish, 132
inland fishing, 111
inosinates, 61
insecticides, 64
insects, spoilage by, 40–1, 204–5
instant coffee, 62
instant pudding powder, 210
instant tea, 329
Intervention butter, 149
intramuscular fat, 69
inulin, 205
invert sugar, 312–13
iodine, 63
Irish whiskey, 356
iron, reactions of foodstuffs with, 34
irradiation, 54, 56–7, 122
Italian wines, 350

jams, 60, 281–2
Japanese radish, 245
Java sugar, 309
jellied eel, 117, 121
jellies, 95, 281–2
jelly powder, 210
Jerusalem artichoke, 242–3
jugged hare, 99
juniper berries, 296

kaffir corn, 201
kale, 248
Kantercheese, 159
Kapucijners, 228, 254
kasha, 203
Katenspeck, 90
Katjang idjoe, 228
keeping sausages, 90, 93–4
kefir, 140–5
Kernhemmer cheese, 156, 159
ketan, 199
ketone rancidity, 33
kidney loaf, 93
kidneys, 81
king crab, 127
kitchen salt, 303
kitchen syrup, 310
Knakworst, 92
koek, 221, 222
kohlrabi, 244
Kollumer, 159
kosher cheese, 153
Kotta, 164
Kriek-Lambic beer, 344
Kriel potatoes, 234
kummel, 355
kumquats, 264, 269

Laaglander, 159
labelling, legislation, 27
Lachsschinken, 90
lactic acid, 60
lactic acid bacteria, 37, 44, 157, 203
lactitol, 314
Lactobacillus acidophilus, 142
Lactobacillus bifidus, 142
Lactobacillus bulgaricus, 141
lactose, 135–6
lamb, 78–80
Lambic beer, 344
laminates as packaging, 26
land wine, 352
langouste, 128
langoustine, 128
Lapsang Souchong, 328
large Asiatic shrimp, 126
large fishing, 111
laying battery, 170, 171
lead, 64, 65
leaf beet, 248
leavening agents, 62, 215
lecithin, 60, 209
lectins, 226
leek, 245, 249
Leerdammer, 157, 159
legislation
    additives, 62–3
    contaminants, 66
    food, 26–30
Leiden cheese, 158
Leiden cheese with cumin, 157–8
Lemaire-Boucher agriculture, 19

Index

Lemaire-Boucher flour, 221
lemon balm, 293
lemonade syrup, 339
lemonades, 60, 62, 338
lemons, 269
lentils, 228
lettuce, 45, 239, 241, 248
leveret, 99
Leyden farmhouse cheese, 154
Liberica coffee bean, 322
lima beans, 228
Limburger cheese, 157, 159
Limburgs, 220
lime, 269
lime honey, 319
limequats, 269
linoleic acid, 17, 110, 146, 176
linoleic-acid-enriched diet, 185
liquer, 357
liquid yoghurt, 141
liquorice, 310–11
liquors, distilled, 354–8
Listeria monocytogenes, 39
litchis, 264
liver, 80–1, 89
liver cheese, 92–3
liver of haddock, 114, 121
liver sausage, 86, 92, 93
loaded bran, 191
loaf cheese, 159
lobsters, 124, 127–8
loempia, 95
logos, agricultural, 20, 21
Loire wines, 349
Lombok-peppers, 298
lovage, 291, 296
low-alcohol beer, 344
low-fat diet, 94
low far margarine, 185
low fat mayonnaise, 181
low-salt bread, 221
low-sodium diet, 94
low-sodium/low fat diet, 94
low-tannic-acid tea, 329
luncheon sausage, 92
lungs, 81
Lysteria, 27

Maaslander, 159, 160
macaroni, 193
mace, 297
mackerel, 109, 114, 121
macrobiotic agriculture, 19
macrobiotic products, 22
madeira, 351
maggiplant, 296
Maillard Reaction, 33, 136
maize, 188, 194–5
maize oil, 180
maizena, 206
malt bread, 219
malt sugar, 313

maltose, 313
mandarins, 267–9
mangetout, 253
manna, 198
manna bread, 220, 221
mannitol, 313
manzanilla, 351
maple peas, 228
maple syrup, 310
Maraschino cherries, 281
margarine, 17, 60
  carotenoids in, 60
  composition, 183–4
  consumption, 183
  kinds of, 184–5
  low fat, 60
  manufacture, 184
  vitamins in, 63
marinating
  of fish, 121
  meat products, 89
marjoram, 291, 296
markets, street, 23
marmalade, 282
Marnier jaune, 358
marrons glacés, 281
marzipan, 230
mashed potato powder, 236, 238
maté tea, 329
matjes herring, 113, 120, 121
matzos, 221
mayonnaise, 59, 60, 181
meat and meat products
  colour formation and colour preservation, 85–6
  composition, 689
  consumption, 67–8, 69
  legislation, 96
  drying out, 83
  freezing, 72
  hanging and maturation, 81–3
  health hazards, 70
  inspection, 71
  kinds, 73–80
  minced or chopped, 86
  organs, 80–1
  preservation by chilling, 46
  quality, 83–5
  quality spoilage and storage, 83–5
  refrigeration, 84
  rigor mortis in, 82
  slaughter process, 70–3
meat extracts, 95–6
meat flavours, 95–6
meat products, 86–94
  prepared and preserved, 87–90
  weight of fat per 100 g, 88
  see also meat and meat products
meat sausages, 90, 91
meat stock cubes, 60
melba toast, 221
melons, 269–70
melt honey, 318

Index

mercury, 64, 65
mice, 204
microorganisms, killing, 54–7
migration of plastic packaging, 26
migration residues, 64
mihoen, 201
milk, 142
    composition, 135
    consumption, 135
    control, 144
    health hazards, 136
    introduction, 134–5
    legislation, 144
    pasteurized, 55, 138
    production and processing, 136–8
    proteins, 209
    quality deterioration, spoilage and storage, 143–4
    spoilage of, 35
    sterilized, 138
milk chocolate, 333
milk fat, 164
milk food for infants, 143
milk ice-cream, 142
milk powder, 143, 144
milk sugar, 135
milkfat, 135
millet, 186, 188, 201–2
millet honey, 318
mimosa, 281
minced meat, 86
mineola, 269
mineral salt, 303
mineral water and beverages
    consumption, 335–6
    legislation, 340
    natural, 336–7
mini tomatoes, 252
mint, 293
mint sprays, 281
miso, 229
mites, 41
mobile shops, 23–4
mocha, 324
modified starches, 60
moisture, absorption of, 32
molasses, 310
Mon Chou, 164
monastery liqueurs, 357
morel, 286–7
morellos, 271
Moselle wines, 349
moths, 41
moulds, 39, 203, 223
mousse, 141
mousse powder, 210
mozzarella cheese, 151
Mung beans, 227, 228
Münich beer, 344
mushrooms, 251, 283
mussels, 124, 125, 128–9
mustard, 59, 296, 299

mutton, 78–80
mycotoxins, 27, 59, 65, 66
myoglobin, 85

nasibal, 95
natamycine, 154
natural vinegar, 361
nectarine, 271
neeted melons, 270
nigger corn, 201
nitrates, 64, 65, 240
nitrites, 64, 65, 85
nitrogen gas, 62
nitrogen monoxide, 85, 86
nitrosamines, 64
nitrosomyochromogen, 85
NIZO method, 147
non-enzymic browning, 33
noodles, 194
norbixin, 153
North Atlantic shrimp, 126
North Sea crab, 127
North Sea shrimp, 126
Norwegian shrimp, 126
nutmeg, 297
nuts and seeds, 259
    consumption, 275–6
    composition, 276
    kinds of, 276–7
    quality deterioration, spoilage and storage, 277

oakleaf lettuce, 248
oats, 186, 188, 189, 195–6
oblong cheese, 159
octopus, 132
oedang, 126
ogen melons, 270
oils and fats
    composition, 176–7
    consumption, 176
    health hazards, 177
    legislation, 181–2, 185
    production of, 177–9
    products containing, 180–1
okra, 251
old brown beer, 344
oleic acid, 164, 176
olive oil, 180
onions, 57, 239, 241, 245, 247, 256, 297
Oolong tea, 328
oranges, 267
organic agriculture, 18
organic–biological agriculture, 19
Oryza, 199
oxalic acid, 240
oyster mushroom, 283, 286, 287
oysters, 124, 125, 129–30

packaging, 24–6
    reactions of foodstuffs with, 33
    residues, 64
packaging materials, 24–6

Index

pak choi, 248
palm oil, 180
palm-kernel fats, 180
palmitic acid, 176
pan-ready fresh vegetables, 254
pancakes, 193
paprika, 298
par-boiled rice, 200
paratyphoid bacteria, 136
Parma ham, 90
parsley, 291, 296
parsnip, 244
partridge, 101
passion fruit, 265
pasta, 193
pasteurization, 27, 54–5
   of meat products, 89–90
   of milk, 137
pasty, 95
pâté, 39, 86, 92
patent flour, 193
pathogens, 35
patna rice, 199
Paturain, 164
peaches, 271
peaches, dried, 279
peanuts, 39, 52, 66, 179, 275, 277
pears, 263–7, 268
pears, dried, 279
peas, 228, 253–4
peas, dried, 225
pecan nut, 277
pectin, 33, 60, 209
pectinase, 33, 256
Penicillin, 157
pentachlorophenol, 53
pepper, 297
peppermint, 296, 310
perch, 116
perfumed tea, 328
perishable sausages, 90–3
peroxidase test, 136
persimmons, 264
pesticides, 27, 64, 70, 240, 260
pH, changing, 48
pheasant, 100
phosphatase test, 136
phytic acid, 187, 213
pickling salt, 303
pigeon, 101, 103, 104
pilchards, 114
pimaricine, 93, 154
pine kernels, 276, 277
pintail, 100
Pilsner (pils), 343
pistachio nuts, 276, 277
plaice, 111, 114
plant extracts, 96
plantains, 273
plate-freezing, 47
platebread, 218
plum tomatoes, 252

plums, 271
pointed cabbage, 248
poliomyelitis virus, 39
polony sausage, 93
polyalcohols, 313
polychlorobiphenyls (PCBs), 64, 65–6, 260
polycyclic aromatic hydrocarbons (PAHs), 65, 66
polycyclic compounds, 240
polyunsaturated margarine, 184
pomegranates, 265
pomelos, 269
Pompadour, 159
popcorn, 195
popped rice, 201
pork, 76–8, 79
porridge, 141
port, 351
Port Salut, 157
Portuguese oyster, 130
Portuguese wines, 351
positive list, 62
potassium carbonate, 62
potassium nitrate in cheese, 154
potato croquettes, 236
potato starch, 205, 206
potatoes
   composition, 232
   consumption, 232
   cultivation, 232–4
   introduction, 231–2
   irradiation of, 57
   legislation, 238
   pre-peeled, 23
   preservation, 46
   preservatives in, 59
   products, 234–6
   quality deterioration, spoilage and storage, 236–8
   varieties, 233–4
poularde, 104
Poularde de Stiermarken, 104
Poularde Den Dungen, 104
poultry, 101–7
   composition, 101–2
   consumption, 101
   kinds of, 103–5
   meat, inspection of, 102
   quality deterioration, spoilage and storage, 105
   slaughtering, 102
pouring syrup, 309
powdered tea, 329
praline, 333
pre-baked bread, 220
preservation
   by stimulation of enzymes and microorganisms, 44
   importance of, 42–3
   methods of, 43–4
   removal of pathogenic/spoilage microorganisms, 44–5
   retarding enzymes and microorganisms, 45–54
preservatives, 17, 59

Index

preserving sugar, 309
Preskop, 89
press honey, 317
Procureurspek, 90
propionic acid bacteria, 157
Proserpina, 22
proteins, productts based on, 209
proteolytic bacteria, 37
proving compounds, 62
provitamin A, 60
prunes, 279
pudding powders, 210
puddings, 47, 95, 141
puffed wheat, 193
pulpo, 132
pulses
   composition, 226
   consumption, 225
   health hazards, 226
   kinds of, 226–8
   legislation, 230
   production and distribution, 226
   products, 228–30
   quality deterioration, spoilage and storage, 230
pumpkins, 250
purchasing, 22–4
purslane, 248

quail, 101, 103, 105
quails' eggs, 168, 173
quality, concept of, 31
quality of product, 24
quark, 153, 156, 162, 163–4
quick flour, 193
quick-cooking rice, 200

rabbit, 97, 98, 100
radioactive particles, 27, 66
radishes, 245, 246
raisin bread, 220
raisins, 278
rape seed oil, 177, 180
raspberries, 273
rats, 204
rattle, 172
raw ham, 89
raw milk dairy cheese, 154
red bacteria, 157
red cabbage, 248
red chicory, 248
red currant, 272
red deer, 98–9
red gurnard, 114
red kidney beans, 228
red sturgeon shrimp, 126
red wine, 346
Reform Movement, 21
reform products, 19–22
refrigeration of meat, 84
reindeer cheese, 151
rennet, 153
retail businesses, 22

retrogradation, 32
Rhine wines, 349
Rhône wines, 349
rhubarb, 252
rice, 186, 188, 189, 198–201
rice banana, 275
rice crackers, 201
Rioja wines, 350
robusta coffee, 321–2
rodenticides, 64
rodents, gnawing by, 40
roe deer, 99
roe, salted, 120
rollmop herrings, 121
rolls, 220
rolpens, 89, 90
ropoe in bread, 223
Roquefort, 155, 156, 157
rose petals, 281
rosé wine, 346
rosemary, 291, 296
rosti, 236
rotting of meat, 84
roulade, 89
roundfish, 109
ruby port, 351
rum, 357
runner beans, 253
rusks, 221
rye, 186, 188, 196
rye bread, 213–214, 220
rye meal, 213–14

saccharides, 311
saccharin, 313
Saccharomyces cerevisiae, 215, 345
safflower oil, 180
saffron, 297–8
sage, 296
sage Derby, 158
sago, 207
St. Anthony's fire, 66
Saint Paulin, 157
saithe, 114
Saksische spreading liver sausage, 92, 93
salad burnet, 296
salad cream, 60, 181
salad oil, 181
salad onions, 247
salami, 94
salmon, 111, 117, 119, 121
salmon trout, 117
Salmonella, 27, 37, 84, 105, 110, 136, 172
salt beef meat, 90
salt herring, 120
salt
   consumption, 302
   kinds of, 303–4
   legislation, 304
   preservation by, 51
   production and distribution, 302–3
   quality deterioration, spoilage and storage, 304

Index 379

salt syrup, 310
salt-restricted diet, 185
salted butter, 149
salted roe, 120
sambals, 299
samphire, 249
sandwich sausage, 89, 92
sardine, 109, 114, 121
sauces, 95–6
sauerkraut, 44, 255
sausage, kinds of, 90–4
sausage meat, spiced minced, 86
sausage roll, 94, 95
saveloys, 89
savory, 293
savoy cabbage, 248
sawah rice, 198
scad, 113
scallops, 124, 130
scampi, 128
Schaumwein wines, 349
scorzonera, 245, 247
Scotch whisky, 356
sea fishing, 111
sea kale, 249
sea lobster, 127–8
sea salt, 303
sea spinach, 249
sea vegetables, 249
seacat, 132
seaweed, 249
seeds, 259
seitan, 194
Sekt wines, 349
selenium, 65
self-raising flour, 193
semolina, 191, 193, 195
sepia, 132
serotine, 260
sesame oil, 180
Seville oranges, 267
shaggy ink cap, 286, 288
shallots, 247
sheatfish, 116–17
sheep milk, 134
sheep's milk cheese, 151, 155
shellfish
   composition, 124–5
   consumption, 124
   definition, 124
   health hazards, 125
   kinds of, 128–33
   products of, 133
   quality deterioration, spoilage and storage, 133
sherry, 350
Shigella, 37
shii-take, 286, 288
shoulder ham, 90
shoveller duck, 100
shrimps, 57, 59, 111, 124, 125–6
silverskin onion, 247
slaughter, 70–3

slaughter pigs, 77
sliced bread, 220
sliced meat products, storage times, 95
slicing sausage, 93
sloe gin, 282
small Asiatic fresh-water shrimp, 126
small Asiatic shrimp, 126
small fishing, 111
smoke salt, 303
smoked eel, 121
smoked meat products, 53
smoked sausage, 92
smoking, 52–3, 89, 120–2
snails, 124, 131
snipe, 100
sodium bicarbonate, 62
sodium carboxymethylcellulose, 209
sodium nitrite, 59
soft drinks, 338–9, 340
soft gums, 311
soft ice-cream, 142–3
soft scoop ice-cream, 143
sole, 111, 116
sorbic acid, 35, 54, 59
sorbitol, 313
sorghum, 201
sorrel, 248
soups, 61, 95–6
sour cream, 60, 139
sour half-cream, 139
sour whipping cream, 139
sourdough, 211, 215–16, 221
soy bread, 220
soy flour, 229
soy milk, 229
soy oil, 180, 229
soy sauce, 96, 229
soy wholemeal bread, 221
soya, 22
soya beans, 228
spaghetti, 194
Spanish wines, 350–1
sparkling wine, 346
specialist shops, 22
spelt, 191, 194
spiced minced sausage meat, 86
spices, 58, 297–9
   flavour compounds in, 292–3
   introduction, 290–2
   legislation, 301
   quality deterioration, spoilage and storage, 300–1
   types, 294
spicy sauces, 96
spinach, 248
spine lobster, 128
spleen, 81
split peas, 225
split tin, 220
spoilage, 31
   by parasites, 39–40
   by vermin, 40–1

Index

chemical, biochemical or enzymatic, 32–4
    forms of, 32
    microbial, 34–9
    physical, 32
sprat, 114
spray-drying, 49
spring chickens, 103, 104
spring onions, 247
square cheese, 159
squid, 132–3
stabilizers, 60
stand yoghurt, 141
standardizing of milk, 138
Staphylococci, 84
Staphylococcus aureus, 37, 110
starch, 195, 205, 206
starwort, 249
steric acid, 176
sterilization, 27, 54, 55–6
    of fish, 121
    of meat products, 89–90
    of milk, 143
steroids, anabolic, 70
stick bread, 220
sticky rice, 199
still wine, 346
stinging nettle, 248
stir yoghurt, 141
stock cubes, 61
stock protection, 64
stockfish, 49
stocks, 95–6
stomach-friendly coffee, 324
storage and storage time, 50
storage onions, 247
stout, 344
strawberries, 24, 57, 273
street markets, 23
Streptococcus thermophilus, 141
strong flour, 193
sturgeon, 117
Subenhara, 159
sublimation, 49
suckling lamb, 79
suckling pigs, 76
sugar beet, 305
sugar cane, 305
sugar
    composition, 305–6
    consumption, 305
    kinds of, 309
    legislation, 314
    preservation by, 50
    price, 306
    quality deterioration, spoilage and storage, 314
sugar syrup, 309
sugar water, 339
sugar-free lemonade, 339
sukade, 281
sulphite, 54, 59
sulphur dioxide, 50
sulphurous acid, 59

sultanas, 278
summer onions, 247
sunflower oil, 180
sunflower seeds, 277
Super pils, 343
supermarkets, 22–3
sweating of foods, 32
swede, 244
sweet butter, 147
sweet corn, 251
sweet peppers, 251, 259, 298
sweet spidercrab, 127
sweet wine, 347
sweetbread, 81
sweeteners, 61, 63, 311–14
syneresis, 154
synergism, 60
syrup, 281–2
    kinds of, 309–10

tabasco, 96, 299
table beer, 343
table salt, 303
table sugar, 309
table wine, 352
tangelo, 269
tapeworm infection, 39–40, 70
tapioca, 207
tarragon, 293
tartrazine (E3102), 63
tawny port, 351
TB bacillus, 136
tea
    composition, 325–6
    consumption, 325
    kinds of, 327–9
    legislation, 329
    production and distribution, 326–7
    quality deterioration, spoilage and storage, 329
tea crackers, 221
tea sausage, 92
tea substitute, 329
teal, 100
tempeh, 229
Terpcheese, 159
Texelaar, 159
textured soy protein (TSP), 229
theobromin, 331
thermization of milk, 137
thick yoghurt, 141
thickened yoghurt, 141–2
thickeners, 60, 210
three-year tea, 328
thyme, 291, 296
tiger bread, 219
tin, reactions of foodstuffs with, 34
toadstools, 283
toatjo, 229
tocopherol, 54, 59
tofu, 229
tomato ketchup, 60, 96
tomato paste, chemical blowing of, 34

Index

tomatoes, 239, 241, 251, 259
tongue, 81
  pressed, 89
  smoked, 86
tongue pâté, 93
toungue sausage, 92, 93
tortue clair, 132
toxins, 35
toxoplasmoses, 40
tragacanth, 208
tranquillizers, 64
Trappist beer, 344
trichiniasis (muscle infection), 40
triglycerides, 33, 176
Triticum aestivum, 193
Triticum dicoccum, 186
Triticum monococcum, 186
tropical shrimp, 126
trout, 111, 116, 119
truffles, 286, 289
tuna, 116, 121
tunny fish, 116
turbot, 116
turkey ham, 107
turkey in aspic, 107
turkey roulade, 107
turkeys, 101, 103, 104
turmeric, 298
turnip, 244, 248
turtle soup, 132, 207
turtles, 131, 132
tutti frutti, 279
twocorn, 186

ugli fruit, 269
UHT milk, 144
UHT sterilization, 55, 138
Uner, 142
unsalted butter, 149
utensils, residues of, 64

vacuum-packaging, 51
vanilla, 298
vanilla sugar, 309
vanillin, 61
veal, 75–6, 77
vegan agriculture, 19
vegetable oils and fats, 179–80
vegetable preserves, 254–5
vegetables
  consumption, 239
  composition, 239–40
  enzymatic browning in, 33
  health hazards, 240
  kinds, 242–54
  legislation, 258
  pectinase in, 33
  price, 24
  production and distribution, 240–2
  products, 254–6
  quality deterioration, spoilage and storage, 256–8

  wastage factor in, 24
vermicelli, 194
Verona wines, 350
veterinary medicines, residues of, 64
vieux, 357
viili, 142
vin de goutte, 345
vin de presse, 345
vine snail, 131
vinegar, 44, 60
  kinds of, 361
  legislation, 362
  production, 360–1
  quality deterioration, spoilage and storage, 361–2
vinegar essence, 361
vinnig, 39
Vintage Port, 351
violets, 281
viruses, 39
vitamin A, 63, 110
  in cheese, 152
  in meat, 69
vitamin B-complex, 110
  in cereals, 188
  in cheese, 153, 164
  in meat, 69
  in milk, 143
  in rice, 200
vitamin C, 59, 110
  in meat, 69
vitamin D, 63, 110
  in cheese, 152
  in oils and fats, 177
vitamin E, 59
  in cereals, 187, 188
  in oils and fats, 177
Vitamin K in oils and fats, 177
vitamins, 63
  in fish, 110
  in meat, 69
  in oils and fats, 177
vodka, 357

Walcherse beans, 226
walnuts, 275, 277
water ice, 142
watercress, 248, 291
watermelons, 269, 270
waxy rice, 199
weevils, 41
weights and measures, legislation, 28
Weinbrand, 357
Westberg, 157, 159
wet cooking rice, 199
wet rice, 198
wet-salting of fish, 119
wheat, 186, 188, 191–4
wheat flour, 213–14
wheat grains, 193
wheat starch, 206
wheatgerm, 193

Index

wheatmeal bread, 219
whelk, 131
whey, 153
whey butter, 149
whipped cream, 17, 139, 208
whipping cream, 139
whisky, 356
white beans, 226
white bread, 219
white cabbage, 248
white cheese, 153
white chocolate, 333
white currant, 272
White Maycheese, 159
white moulds, 157
white pepper, 297
white rice, 200
white wine, 346
whiting, 116
wholemeal bread, 218
wholemeal flour, 190, 193
widgeon, 100
wild boar, 99
wine, 44, 345–54
wine quality labelling, 352–3

wine gums, 311
wine vinegar, 361
winkle, 131
winter asparagus, 245
winter melons, 270
woodcock, 100
Worcestershire sauce, 96
World Health Organization, 28

xylitol, 313

yard-long beans, 253
yeasts, 39
yellow jaundice, 39
yellow peas, 228
yoghurt, 44, 141–2
  spoilage of, 35, 39
yoghurt and buttermilk ice-cream, 142
yolk, 170
Yssel cheese, 159

Zeeland oyster, 130
Zeeuwse Imperialen, 130
Zinzania aquatica, 200